ACOUSTICAL DESIGNING IN ARCHITECTURE

ACOUSTICAL DESIGNING IN

JOHN WILEY & SONS, INC.
NEW YORK · LONDON · SYDNEY

ARCHITECTURE

VERN O. KNUDSEN, Ph.D.
Professor of Physics and
Dean of the Graduate Division
University of California
at Los Angeles

CYRIL M. HARRIS, Ph.D.
Professor of Electrical
Engineering and Architecture
Columbia University

Second Corrected Printing, September, 1968

Printed in the United States of America

PREFACE

This book is intended as a practical guide to good acoustical designing in architecture. It is written primarily for architects, students of architecture, and all others who wish a non-mathematical but comprehensive treatise on this subject. Useful design data have been presented in such a manner that the text can serve as a convenient handbook in the solution of most problems encountered in architectural acoustics.

The book is divided into two parts. The general principles and procedures on which all acoustical designing should be based are considered in Chapters 1 through 14. Specific applications of these principles and procedures are described in Chapters 15 through 20. These applications include the design of auditoriums, theaters, school buildings, commercial and public buildings, homes, apartments and hotels, churches, and radio, television, and sound-recording studios.

We have collaborated actively over a period of two years in planning and writing this book. Whenever our views or opinions differed, we discussed them with each other and with others until agreement was reached. The manuscript has benefited greatly by the criticism of many of our colleagues. To them we here acknowledge our thanks; a list of these colleagues is too long to include in this preface, but we are especially indebted to H. J. Sabine, Celotex Corporation; W. A. Jack, Johns-Manville Research Laboratory; H. A. Chinn, Columbia Broadcasting System; W. B. Snow, Kellex Corporation; R. K. Cook, National Bureau of Standards; and R. H. Frick, Rand Corporation. Each of the chapters has been read and criticized by several members of the technical staff of the Bell Telephone Laboratories. In the preparation of the manuscript, valuable comments and suggestions were made by C. T. Molloy, R. H. Nichols, Jr., and J. R.

Anderson. Other members of the Laboratories to whom we are indebted are: L. B. Cooke, M. B. Gardner, R. L. Hanson, F. K. Harvey, W. E. Mathews, W. A. Munson, R. K. Potter, and J. C. Steinberg. Figure 11.7 has been reproduced by permission of the Comptroller of His Britannic Majesty's Stationery Office and the Director of Building Research. Finally, we wish to express our appreciation to Miss Judith K. Horton for her help in preparing many of the illustrations.

V. O. K.
C. M. H.

November, 1949

CONTENTS

1 · Properties of Sound

Acoustical designing in architecture begins with the preliminary sketches on the drafting board and continues throughout all stages of planning and construction. Good acoustics will be assured in the buildings an architect designs if he has an understanding of the technological principles of architectural acoustics and knows how to apply them.

Planning for Good Acoustics

Planning for acoustics, like planning for structural strength or the partitioning of a building into serviceable rooms, must be functional. The architect must have adequate comprehension of all functions. To be sure, he must rely upon building experts to some extent, but if he relies wholly on them the outcome may not be a well-integrated and harmonious building; for example, the acoustics, lighting, or air conditioning may be neglected, or possibly gilded, and allowed to depend on the sales ability rather than the competence of the experts. Acoustics, no less than structural strength and partitioning of the building, should be planned for at the beginning. It is not good practice to seek the advice of an acoustical consultant after the plans (or even the building!) have been completed, with the request, "Please prepare recommendations, without of course proposing major structural changes, for providing good acoustics in this building." Sometimes such recommendations can be made but more often they cannot.

There is nothing capricious, mystical, or even unpredictable about the control of sound in buildings. Architectural acoustics is an exact science and a practical art. The architect who has a working knowledge of this subject can plan adequately for the acoustics of the buildings he designs. It is the purpose of this book to present the working principles of this science and art in a simple, useful, and convenient form. Architectural designing based on these principles will assure the construction of rooms and buildings which are free from disturbing noises and which provide the optimum conditions for producing and listening to either speech or music. Functional acoustical design demands scientific, aesthetic, and practical planning.

What Is Sound?

We must know some definitions and properties of sound waves, and the auditory sensations they excite, before we can adequately comprehend even the most elementary rules which govern the acoustics of rooms. First, we must distinguish between objective sound and the sensation it produces by means of the human ear. It is a moot point, for example, whether a sound is produced when a giant tree crashes to the ground in an uninhabited forest. This, of course, is just a matter of definition. In this book, the word *sound* will be used to denote *a physical disturbance,* an alteration or pulsation of pressure, capable of being detected by a normal ear. (In accordance with this definition, the falling tree does generate sound.) In general such a disturbance reaches the ears by traveling through air. In any case, a medium possessing inertia and elasticity is needed to propagate it. Sound waves do not travel through a vacuum.

The auditory sensation produced by sound waves will be called *sound sensation.* The crashing tree produces a sound sensation only when an ear hears it.

Propagation of Sound

Sound has its origin in vibrating bodies. A plucked violin string or a struck tuning fork can actually be seen to vibrate. In the sounding board of a piano and the paper cone of a loudspeaker, as in most other sound sources, the amplitude of vibration is too small to be observed visually but often the vibration can be felt with the finger tips.

Consider a body vibrating in air. As it moves in an outward direction it pushes a "layer" of air along with it; this layer of air is compressed, and its density and temperature are correspondingly increased. Since the pressure in this layer is higher than that in the undisturbed surrounding atmosphere, the particles (that is, the molecules) in it tend to move[1] in the outward direction and transmit their motion to the next layer, and this layer then transmits its motion to the next, and so on. As the vibrating body moves inward, the layer of air adjacent to it is rarefied. This layer of rarefaction follows the layer of compression in the outward direction, and at the same speed; the succession of outwardly traveling layers of compression and rarefaction is called *wave motion.* The speed of propagation is determined by the compressibility and density of the medium—the less the compressibility of the medium and the less its density, the faster will the wave motion be propagated.

The changes in pressure, density, and temperature due to the passage of the sound wave through air are usually extremely small. For example, the effective sound pressure, the root-mean-square of the pressure variations, in the air 3 feet from a trumpet is about 9 dynes per square centimeter. This means that the pressure fluctuations are only about nine millionths of the normal atmospheric pressure, which is 1.01×10^6 dynes per square centimeter, or 14.7 pounds per square inch. (A table of conversion factors for the English and metric systems of units is given in Appendix 3.) A wave of this magnitude would produce variations in the density of air of about 0.001 per cent (the normal density of air is 0.0012 gram per cubic centimeter) and temperature fluctuations of only 0.0008° C.

Displacement Amplitude and Particle Velocity

The individual vibrating particles that transmit a sound wave do not change their average positions if the transmitting medium itself is not in motion. They merely vibrate about their equi-

[1] This movement is added to the random motion of the molecules, but the latter motion can be disregarded in the present discussion.

librium positions. The average maximum distance the individual particles are moved from their equilibrium positions is called the *displacement amplitude.* For the sound waves one ordinarily hears in rooms, the displacements are quite small. Thus the air particles 1 yard away from a struck dinner gong may undergo a maximum displacement of only a few ten thousandths of an inch. The velocity with which the particles move back and forth about their equilibrium positions is called the *particle velocity.* The air particles 1 yard away from the gong would experience a particle velocity of a few inches per second. Particle velocity is directly proportional to the displacement amplitude and the frequency of the sound wave.

Frequency

The number of complete to-and-fro vibrations that the source, and hence the particles in the medium, makes in 1 second is called the *frequency* of vibration. A string that undergoes 256 complete oscillations in 1 second ("middle C") produces a vibration of the same frequency in the surrounding air and in the

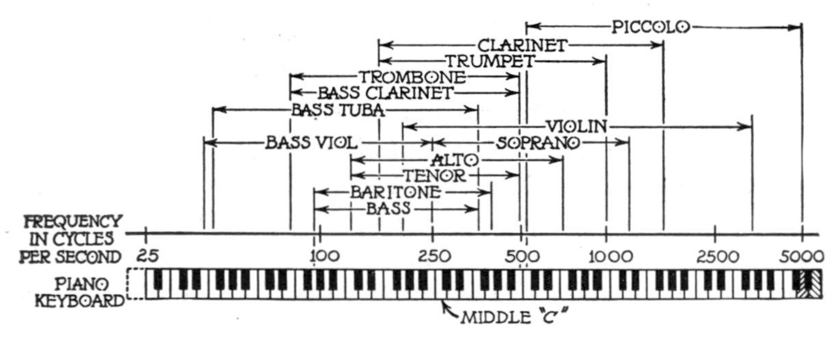

Fig. 1.1 Frequency ranges of several musical instruments and singing voices, shown on a frequency scale which uses the piano keyboard for comparison. (Compiled from *Electronics Spectrum Chart.*)

eardrum of an observer in the sound field. This assumes that the source and the observer are at rest with respect to the medium —the usual assumption in room acoustics. Frequency is a *physical* phenomenon; it can be measured by instruments, and it is closely related to, but not the same as, pitch—a psychological phenomenon. Frequency is usually designated by a number followed by *cycles per second* (abbreviated cps or by the symbol ~) or simply by *cycles*. Thus, the frequency of vibration of the string just referred to is 256 cycles. Figure 1.1 gives the ranges of the fundamental frequencies of some musical instruments and singing voices.

Speed of Sound

If sound traveled with the speed of light, about 186,000 miles per second, many of the causes of poor acoustics in auditoriums would not exist. But sound travels much slower, only about 1130 feet per second (abbreviated ft/sec) in air at room temperature. As a result, echoes and reverberations are likely to be serious defects in many rooms, as we shall show later. Experimental data show that when the reflected sounds which reach an observer are delayed more than about 0.058 second, relative to direct sounds, they are distinguished as echoes. (Sound travels approximately 65 feet in this time interval.) Reverberation, as most simply interpreted, consists of successive reflections of a sound in a room, and since sound travels only about 1130 feet per second there usually will be a rather long succession of these reflections before the sound dies away to inaudibility. Thus we see that the speed of sound plays a significant role in architectural acoustics.

The speed of a sound wave in air does not vary appreciably with frequency in the audible range. Furthermore, the speed does not change with intensity except for very intense waves. For a powerful source of sound such as an air-raid siren, the speed of the sound within a few inches from the source increases slightly with increasing intensity, but, for all practical purposes in architectural acoustics, the speed of sound is independent of frequency, intensity, and changes in atmospheric pressure. Temperature does have a significant effect on the speed, increasing it about 1.1 feet per second per degree Fahrenheit rise in temperature. The dependence of the speed of sound on temperature is one of the prime causes of the bending of sound rays in the atmosphere. This bending (refraction) of sound waves sometimes affects the distribution of sound reaching an audience, especially in open-air theaters.

The speed of sound in air is given by $\sqrt{1.40P_s/\rho}$; P_s is the atmospheric pressure and ρ is its density. For purposes of calculation in this book, the speed of sound in air will be assumed to be 1130 feet per second at a temperature of 72° F. The speed changes only slightly with variations in relative humidity. The actual speed at 72° F for air of different relative humidities does not differ more than 1 or 2 feet per second from the value of 1130 feet per second.

Sound travels much faster in liquids and solids than it does in air. Thus the speed in water is about 5000 feet per second; in hard wood it is about 13,000 feet per second along the fibers and only 4000 feet per second across them; and in stone it is about 12,000 feet per second.

The term *velocity of sound* is often used interchangeably with *speed of sound* although, strictly speaking, the two are not the same. *Velocity* includes both *speed* and *direction* of propagation; velocity is speed in a specified direction; that is, velocity is a vector quantity. The direction of propagation is the direction of the advance of the wave, defined more accurately by the perpendicular to a *wave front* (surface of constant phase) of the advancing wave.

The velocity of sound should not be confused with particle velocity, the speed at which the air particles move back and forth about some equilibrium position; the velocity of sound is a constant under given atmospheric conditions, whereas the particle velocity is proportional to the product of the frequency and displacement amplitude of the sound waves and in general depends on the distance of the waves from the sound source.

Some Characteristics of a Sound Wave

The most elementary type of vibration is that which has a single frequency and is called simple harmonic motion. It is the form of vibration which characterizes a "pure" tone; for example, that given by a good tuning fork which has been struck gently. The form of this vibration and the corresponding form of the pressure variation which is propagated outwardly in the surrounding medium as a sound wave are shown in Fig. 1.2. This is a sine curve; a curve having this shape can be obtained by plotting, on rectangular coordinate paper, the sine of an angle against the angle itself. Thus, a tone produced by a simple harmonic sound source is often called a "pure" tone because it contains only one frequency.

The total pressure in a sound field, at a specified point and instant of time t, is given by the sum of the undisturbed atmos-

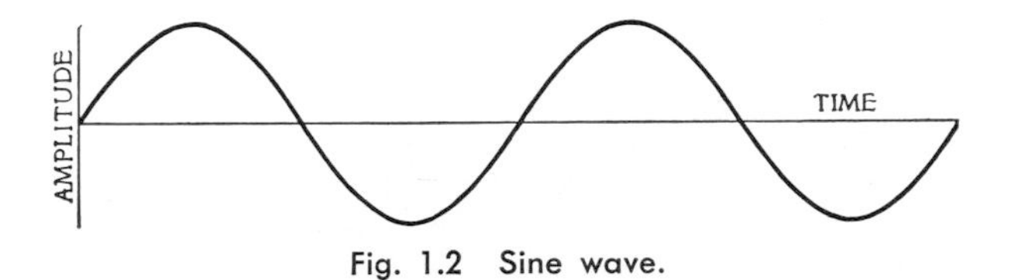

Fig. 1.2 Sine wave.

pheric pressure P_s and the alternating pressure due to the sound wave. The latter is given by

$$P_a \sin (2\pi ft + \theta) \tag{1.1}$$

where P_a is the maximum *pressure amplitude,* f is the frequency of vibration, t is the time, and θ is the phase angle when $t = 0$. This particular (simple harmonic) type of wave motion is important because *all* sound waves can be shown to be made up of a number of different simple harmonic waves. The *effective sound pressure P* is the square root of the time average of the square of $P_a \sin (2\pi ft + \theta)$. The term *sound pressure* is generally used to designate the effective value of the sound pressure. An extraordinarily small sound pressure can be detected by the ear. Figure 1.3 indicates the pressure due to noise in various locations; it shows that at the threshold of audibility the sound pressure is only 0.0000000035 pound per square inch.

The distance that a sound wave travels during each complete cycle of vibration, that is, in $1/f$ second, is called its *wavelength* and is denoted by the Greek letter lambda λ. Wavelength, frequency, and the velocity of sound c are related by the equation

$$\lambda f = c \tag{1.2}$$

Thus on a piano middle C, which has a frequency of 256 cycles, has a wavelength of 1130 feet per second/256 per second, or 4.4 feet. Sounds of higher frequency than a few hundred cycles per second have wavelengths comparable to the dimensions of the architectural embellishments and furnishings of rooms. In later discussions we shall show that these embellishments scatter sound

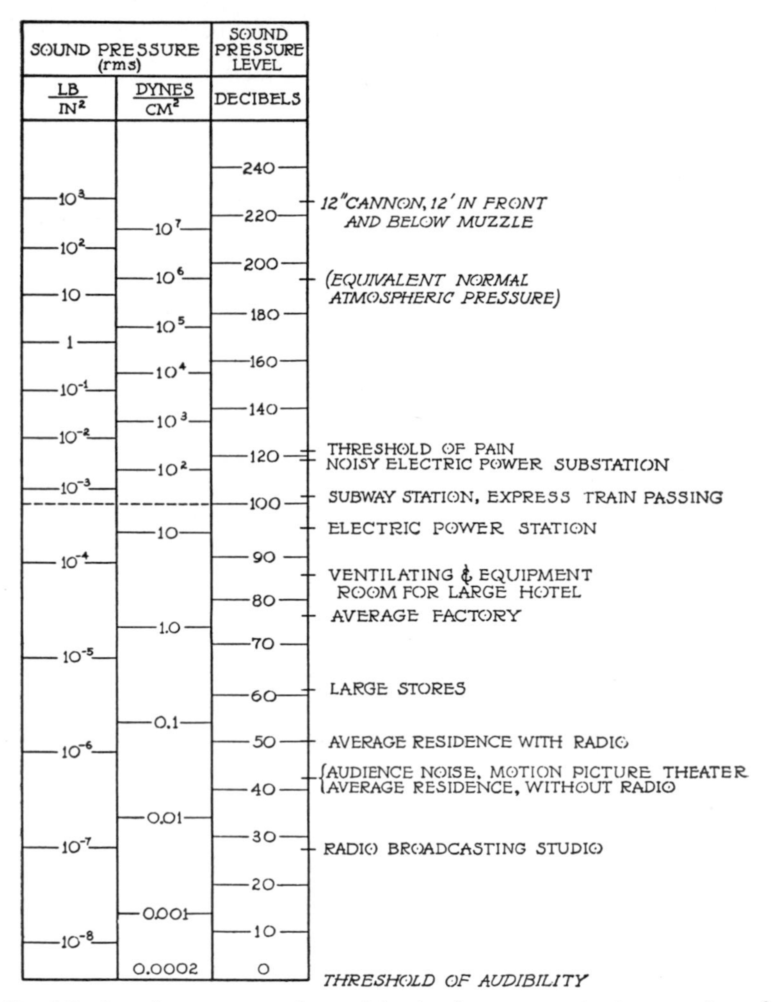

Fig. 1.3 Sound pressures and sound levels of average noise in a number of locations.

and that they aid in providing optimum conditions of diffusion, a prime requirement for ideal acoustics in music rooms, and especially in sound-recording and broadcasting studios.

Wave Form

The *wave form* of a sound wave describes, by means of a graphical representation, the precise nature of a complete to-

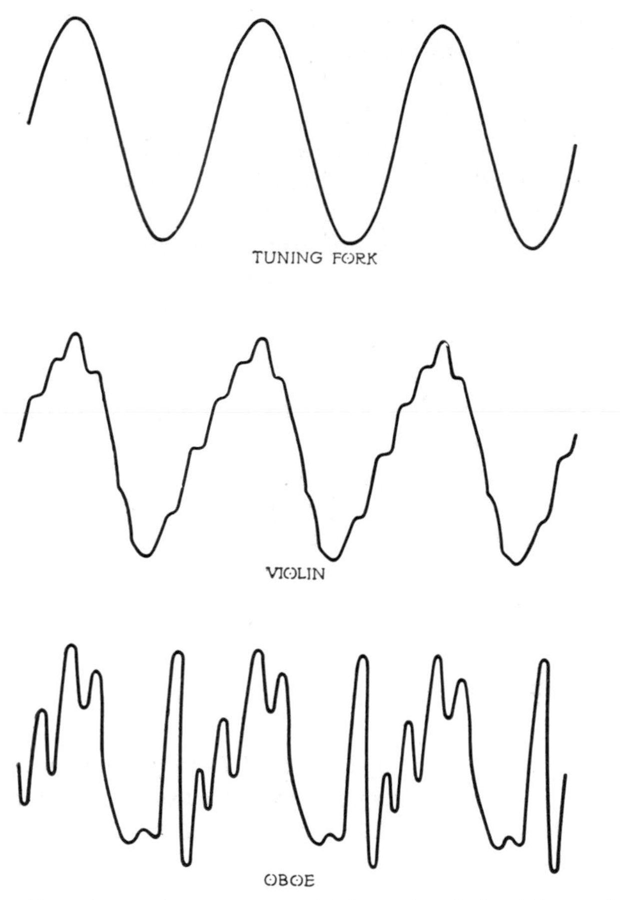

Fig. 1.4 Wave forms of tones produced by tuning fork, violin, and oboe. All three tones are of approximately the same frequency and sound pressure.

and-fro oscillation of the vibrating particles in a sound field. Thus, Fig. 1.2 is a graph of the simple harmonic *wave form* of the sound generated by a gently struck tuning fork; it gives as the function of the time the instantaneous displacement (plotted vertically) of a typical vibrating particle. Each complete cycle in the graph of Fig. 1.2 corresponds to a complete cycle of the tuning fork or of the sound wave it generates. Although the displacements are represented as transverse to the time axis, the actual displacements of the particles in the sound field are parallel to the direction of propagation of the sound wave; that is, the wave motion is *longitudinal.*

The wave forms of musical tones are somewhat more complicated. For example, Fig. 1.4 shows the wave forms of sustained tones produced by a tuning fork, a violin, and an oboe. These records are for sustained musical tones of the same fundamental frequency and approximately the same amplitude of vibration. However, they differ markedly in their wave forms. Although not simple harmonic, the wave forms for these tones are periodic; they repeat at definite intervals. They are called complex waves in contradistinction to simple harmonic waves. It is possible, by mathematical or instrumental means, or both, to analyze complex wave forms, like those characteristic of the oboe or any other instrument, into a series of simple harmonic vibrations. Thus, a complex tone (or its graphical representation as a complex wave form) may be regarded as made up of a series of simple harmonic tones (or waves). Usually the frequencies of these component simple harmonic tones are integral multiples of the frequency of the *fundamental* component, which is sometimes referred to as the gravest component.

Acoustical Power of Average Speakers and Some Musical Instruments

The rate of emission of acoustical energy from most sources of sound, and the corresponding pressures in their resulting sound fields, are very small. For example, the average acoustical power radiated by a person speaking in an auditorium is of the order of 25 to 50 microwatts (a microwatt is one millionth of a watt). It would require, therefore, no fewer than 15,000,000 such speakers to generate a single horsepower of acoustical energy. With such minute amounts of sound power in unamplified speech, the resulting sound pressure in an auditorium is correspondingly small; often the average sound pressure is less than 0.1 dyne per square centimeter. In contrast with the mere 50 microwatts output of a typical speaker, the acoustical power required for good hearing conditions for speech throughout an auditorium, as given by Eq. (14.1), is 10,000 microwatts in a room having a volume of about 100,000 cubic feet.

Most musical instruments radiate a somewhat greater power than does the average human voice. Table 1.1 gives the approximate peak power for a number of typical instruments. These values are small compared to the 37-kilowatt acoustical power output of a large air-raid siren developed during World War II.

TABLE 1.1

THE APPROXIMATE PEAK SOUND POWER OUTPUT OF CONVERSATIONAL SPEECH AND OF SEVERAL MUSICAL INSTRUMENTS

(Bell Telephone Laboratories)

Source	*Peak Power in Watts*
Conversational speech	
Female	0.002
Male	0.004
Clarinet	0.05
Bass viol	0.16
Piano	0.27
Trumpet	0.31
Trombone	6.
Bass drum, 36 in. x 15 in.	25.
Orchestra, 75 pieces	10–70

Sound Intensity

The *sound intensity* [2] in a specified direction at a point in a sound field is defined as the rate of flow of sound energy through a unit area at that point, the unit area being perpendicular to the specified direction. Sound intensity is usually expressed in watts per square centimeter.

As an illustration, we shall calculate the intensity 100 centimeters from the bell of a clarinet. For low-frequency tones, the clarinet approximates a point source; that is, it radiates sound nearly uniformly in all directions. (The sound waves from a perfect point source, which is far from any reflecting surface, are spherical.) Let us assume that the total power output W for a sustained tone from the clarinet is 0.002 watt. Since the

[2] Other definitions have been used in the past, but the one given here is the currently approved definition and it will be used in this book.

area S of a sphere is 4π times the square of the radius, the area of a sphere 100 centimeters in radius is 125,600 square centimeters. Thus the power passing through each square centimeter of this sphere, flowing in the outward direction—the intensity I—is

$$I = \frac{W}{S} = 0.002 \text{ watt}/125{,}600 \text{ cm}^2 = 1.59 \times 10^{-8} \quad \text{watt/cm}^2$$

Since the area of a sphere increases as the square of its radius, we note that the intensity of *free* [3] sound waves originating at a point source diminishes inversely as the square of the distance from the source.

There is a simple relation between intensity I and sound pressure P in *plane* or *spherical free waves* having an effective pressure of P dynes per square centimeter:

$$I = \frac{P^2}{10^7 \rho c} \quad \text{watts/cm}^2 \qquad (1.3)$$

where ρ is the density of air in grams per cubic centimeter, and c is the velocity of sound in centimeters per second. In spherical waves the relation is not valid near the source, but it is a close approximation for most free sound waves. In more complicated sound fields such as occur in most rooms, especially at positions remote from the sound source, the intensity calculated from Eq. (1.3) may differ greatly from the actual value in a specified direction.

Variation of Pressure and Intensity with Distance

If a sound originates at a point in a homogeneous and undisturbed medium, away from all reflecting and diffracting surfaces, the sound is propagated radially in all directions and the wave front is spherical. It was shown in the preceding section that the intensity of such waves falls off inversely as the square of the distance from the source, and hence the sound pressure falls off inversely as the distance in conformity with Eq. (1.3). We have assumed that the sound energy persists as such and is

[3] A free wave is one that is free from the influence of reflective surfaces.

not lost by absorption in air. If sound from a point source is propagated over an absorptive surface, the intensity may fall off even more rapidly, varying, under certain conditions, inversely as the fourth power of the distance (that is, the pressure varies inversely as the square of the distance). In special cases, sound waves having a cylindrical wave front can be obtained by reflection from a very large, hard, cylindrical surface. The intensity of such outgoing waves varies inversely with the distance from the effective source, and the pressure varies inversely as the square root of the distance. In a plane wave there are no losses due to divergence. By a *plane wave* we mean one having everywhere the same direction of propagation. Consequently its points of maximum compression are in planes perpendicular to this direction, and they are separated by a distance equal to the wavelength. Such a wave can be obtained, approximately, by reflecting sound from a point source located at the focus of a large, hard, parabolic surface.

Decibel Scale

The sound pressure near an airplane propeller rotating at top speed is often more than a million times the pressure near the lips of a person producing a faint whisper. In acoustics we must deal with quantities which extend over a wide range. For this and other reasons it is often convenient to describe and measure these quantities on a logarithmic scale called the *decibel* scale.

The decibel (abbreviated db) is a unit which denotes the ratio between two amounts of power, intensity, or sound pressure.[4] It is customary to compare the pressure of all sounds with 0.0002 dyne per square centimeter. This is an arbitrary choice but it approximates rather closely the minimum sound pressure that is audible to the normal human ear. The *sound-pressure level* (usually referred to as *sound level* when measured by a sound-level meter) of a sound wave having a sound pressure of P dynes per square centimeter is therefore defined as

[4] In many sound fields the sound-pressure ratios are not proportional to the square root of corresponding power ratios. Hence, strictly speaking, the term decibel should not be used in such cases. It is common practice, however, to extend the use of this unit to these cases.

$$20 \log_{10} \frac{P}{0.0002} \quad \text{db} \tag{1.4}$$

For example, the sound level of a noise having a pressure of 2 dynes per square centimeter is 20 log (2/0.0002) db = 80 db.

Figure 1.5 shows a graphical relationship between *sound pressure,* in dynes per square centimeter, and *sound level,* in decibels.

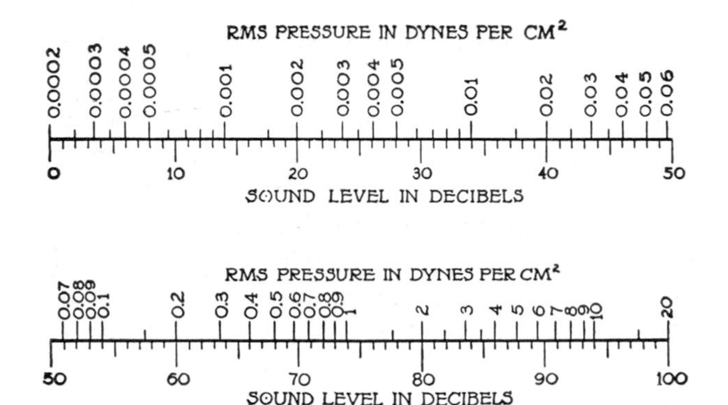

Fig. 1.5 Graphical relationship between sound pressure, in dynes per square centimeter, and sound level, in decibels.

If I_1 and I_2 are the intensities of two sounds, I_1 being greater than I_2, then 10 $\log_{10}$ (I_1/I_2) gives the number of decibels by which I_1 exceeds I_2. The *intensity level* of any sound having an intensity I watts per square centimeter is defined as

$$10 \log_{10} \frac{I}{I_0} \quad \text{db} \tag{1.5}$$

The value of the reference intensity I_0, in watts per square centimeter, must be stated. For most room temperatures an intensity of 10^{-16} watt per square centimeter corresponds closely to a sound pressure of 0.0002 dyne per square centimeter. Therefore it is usual to assign to I_0 a value of 10^{-16} watt per square centimeter.

The acoustical powers of different sound sources similarly can be compared by means of the decibel scale. Thus, two sound sources having powers of W_1 and W_2 differ by 10 $\log_{10}$ (W_1/W_2) db.

Sound-Level Meters

The sound-pressure level in any type of sound field can be measured by a sound-level meter. This device consists of a microphone and appropriate electrical equipment for converting

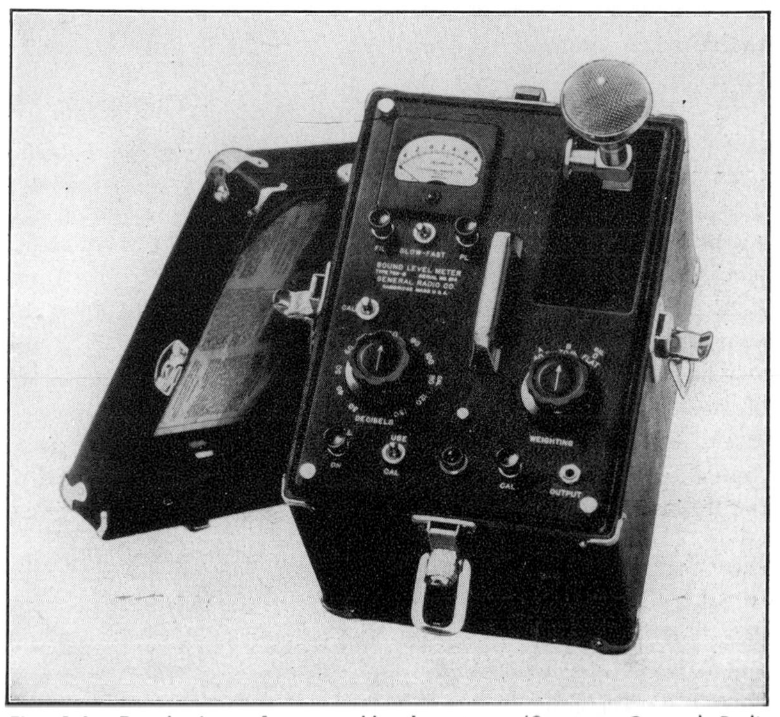

Fig. 1.6 Panel view of a sound-level meter. (Courtesy General Radio Company.)

sound that is incident on the microphone into a meter reading that indicates the sound-pressure level in decibels. Figure 1.6 shows a panel view of one such commercial instrument. The sound-level meter is a physical instrument which, although lacking certain characteristics of the ear, nevertheless is extremely useful for measuring the levels of sounds, including those of interest in architectural acoustics.

In Chapter 2 it will be shown that the ear is not equally sensitive at all frequencies. Hence, even though two noises produce exactly the same sound level, one may be judged to be louder than the other if more of its energy is concentrated in a frequency region where the ear is most sensitive. Sound-pressure level, a physical quantity, is not a direct measure of the loudness of a sound. As we shall see from Fig. 2.5, the sensitivity of the ear is not uniform with respect to frequency. This is illustrated by the curves which represent contours of equal loudness for pure tones.

In order to obtain meter readings which have a closer relationship to what the ear hears than do sound-pressure levels, the frequency response of sound-level meters is modified by the introduction of so-called "40- and 70-db frequency-weighting networks." These networks alter the sensitivity of the sound-level meter with respect to frequency, so that its sensitivity as a function of frequency approximates that of the ear indicated by the contours marked 40 and 70 in Fig. 2.5. The readings of the sound-level meter with either of these networks in use are indicated in terms of decibels and are called "weighted" sound levels. In reporting such readings, the weighting network used should be specified. When the third or "flat" network is used, the meter reads sound-pressure level.[5]

Directionality of Sound Sources

One of the important characteristics of a sound source is its *directionality,* that is, the way in which it distributes sound in a region free from reflecting surfaces. For good listening conditions, this characteristic must receive special consideration in the placement of loudspeakers in all sound-amplification systems. This problem will be considered in greater detail in Chapter 14. Although the radiation patterns of different sound sources vary considerably, most sources have the following properties: (1) When the wavelength of the emitted sound is very large in relation to the dimensions of the source, energy is radiated uniformly in all directions. (2) On the other hand, when the wavelength is small in relation to the dimensions of the source, most

[5] Only under a very special condition do sound-level meters measure intensity level, namely, when the sound field is such that the intensity is proportional to the square of the sound pressure.

of the radiated sound is confined to a relatively narrow beam; the higher the frequency, the sharper the beam. As a result, all the listeners in an auditorium may receive almost the same amount of power for the low frequencies emitted by the loudspeakers of the sound-amplification system. However, those away from the axes of the loudspeakers may not receive an adequate amount of sound at the higher frequencies.

The distribution of sound pressure emitted by an ordinary

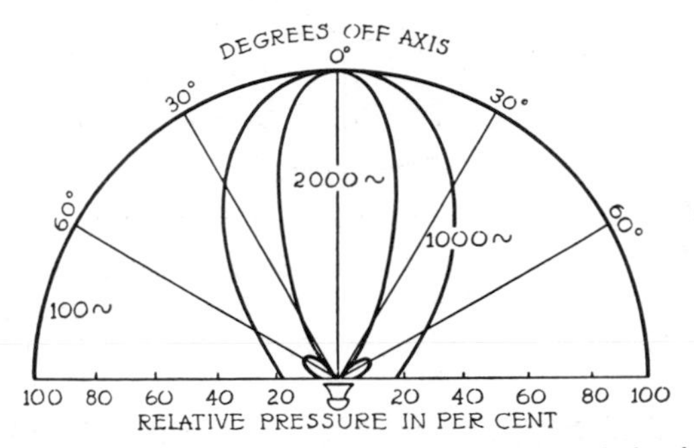

Fig. 1.7 Distribution of pressure about axis of a 12-inch loudspeaker in a large baffle, for frequencies of 100, 1000, and 2000 cycles.

loudspeaker 12 inches in diameter, mounted in a very large baffle, is shown in Fig. 1.7. In this figure the distance from the origin of any point on a curve is a measure of the relative sound pressure in the direction corresponding to that point. Thus a circle, with its center at the speaker, would indicate equal radiation in all directions. Notice that at 100 cycles the sound pattern approximates this condition; at 1000 cycles (where the wavelength is about equal to the diameter of the loudspeaker) the distribution of sound shows a marked directional effect; at 2000 cycles the beam is quite sharp and is centered around the axis. The acoustical radiation pattern becomes more "beam-like" as the ratio of the wavelength to the diameter of the loudspeaker diminishes. Measurements of the sound-pressure field around the human head during speech show a similar effect; Fig. 1.8 gives the sound pattern around the lips of a speaker for three bands of speech centered at frequencies of 200, 600, and 3400 cycles.[6]

[6] Band widths: 125, 200, and 1200 cycles per second, respectively.

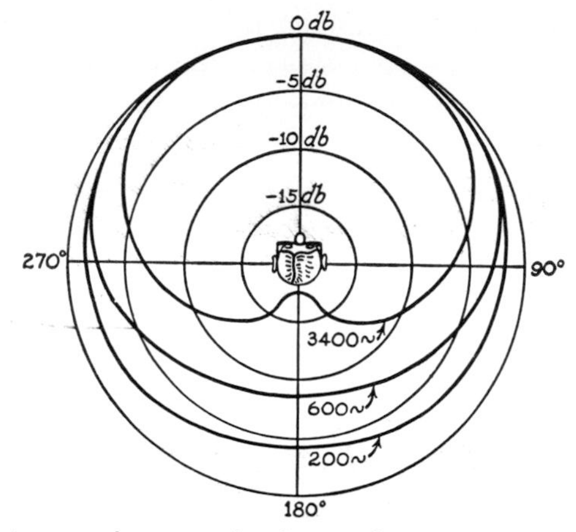

Fig. 1.8 Relative sound-pressure level around the head during speech. Contours represent levels at a distance of about 2 feet from the head in a horizontal plane for three bands of speech centered at the frequencies indicated. (H. K. Dunn and D. W. Farnsworth.)

The low-pitched vowels of speech spread out quite uniformly in all directions around the head of a speaker, but the high-pitched sounds are confined to a narrow beam in front of the mouth of the speaker. It is well known that a "hiss" is very directional.

2 · How We Hear

The hearing mechanism, shown in cross section in Fig. 2.1, can be divided into three parts: the external ear, the middle ear, and the inner ear. The *external ear* consists of an external appendage, called the pinna, and the ear canal, which is closed at the inner end by the eardrum. The *middle ear* contains three tiny bones or ossicles which transmit vibrations from the eardrum to the inner ear. These bones—the hammer, anvil, and stirrup—constitute a lever mechanism that communicates the vibrations of the drum to the membrane of the oval window, which is the entrance to the inner ear. Since the oval window is only about one twentieth as large as the eardrum, the pressure of the vibrations communicated through the oval window to the liquid in the inner ear is increased. Thus the action of the middle ear is that of an efficient mechanical transformer, coupling vibrations in the air to the liquid in the internal ear. The *inner ear* has two distinct functions: (1) the maintenance of body equilibrium, accomplished by the vestibular portion of the ear, which is made up principally of three semicircular canals; and (2) the perception of sound, which is accomplished by the *cochlea* and its associated apparatus.

The cochlea is a liquid-filled spiral canal, subdivided along its length into two canals by a bony structure and a tough membrane. The end of one of these canals is closed by the oval window. It is through this window that the vibration of the ossicles is transmitted to the liquid in the cochlea. Sensitive nerve endings, associated with tiny hair cells in a third canal, are excited by the vibration set up in the cochlear liquid, and they send impulses to the brain by way of the nerve fibers. It is believed that the rate at which the total number of these nervous

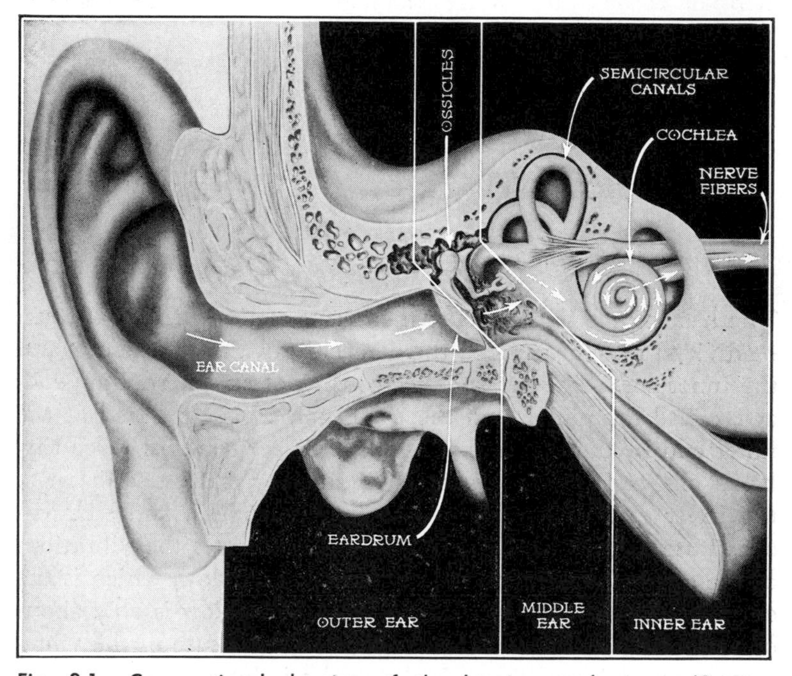

Fig. 2.1 Cross-sectional drawing of the hearing mechanism. (Courtesy Western Electric Co.)

impulses are communicated to the brain determines the *loudness*. This rate depends on the number and activity (pulsing rate) of the nerve endings stimulated. It increases with the sound pressure of the wave striking the ear. The *pitch* of the sound sensation is determined principally by the location of the nerve endings that are most excited by the resonant vibration of various sections of the basilar membrane; however, at low pitch the frequency of arrival of the nervous impulses at the brain may be the chief determinant. *Tonal quality* is determined largely by the number, location, and extent of the excited nerve endings and is complexly related to the wave form and pressure of the sound wave striking the ear.

Sensitivity of the Ear

A sound wave must have a certain minimum value of pressure in order to be heard by an observer. This value for selected

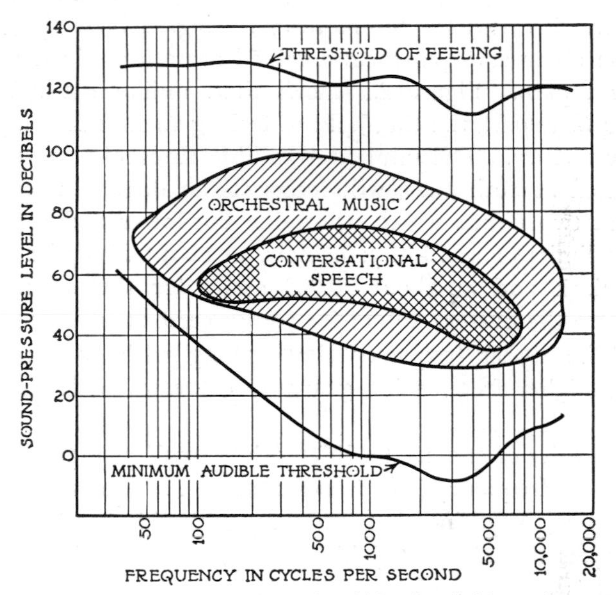

Fig. 2.2 Chart showing the minimum audible threshold vs. frequency, and the threshold of feeling. (Bell Telephone Laboratories.)

observers, who have good hearing, who are facing the source of plane progressive waves and listening with both ears, is called the *minimum audible threshold* for a free field. It is shown in the lower curve of Fig. 2.2. Frequency is indicated along the horizontal axis; and the pressure level of the plane progressive sound wave that is just barely audible is indicated along the vertical axis. One notes that the sensitivity of hearing varies enormously for sounds of different frequencies. Fortunately, the ear is most sensitive in the frequency range that is most important for the intelligibility of speech sounds. Since, in the evolution of man, speech and music were developed later than the sense of hearing, it appears that speech and music have developed in such a manner as to be well adapted to the sensitivity characteristics of the ear.

An observer in the field of a free plane progressive wave will notice that, as the pressure of the wave is increased, the resulting sound becomes louder and louder until it attains a level at which the sound can be *felt* (a sort of tingling sensation) as well as heard. This level is called the *threshold of feeling.* Above this threshold, the observer experiences a mixed sensation of sound, feeling, and pain. Figure 2.2 shows that, unlike the *minimum audible threshold,* the threshold of feeling varies relatively little with frequency. The minimum audible threshold curve, if extrapolated at both ends, will intersect the threshold of feeling curve at two points which determine the lower and upper frequency limits of audibility; namely, at about 20 cycles for the lower limit and at about 20,000 cycles for the upper limit. These are the *average* values for young persons with good hearing. The upper frequency limit, along with the sensitivity for the higher frequencies, generally decreases with increasing age.

The ability of the ear to differentiate small changes of sound pressure or frequency is of importance in the hearing of speech and music. Anything that interferes with this function of the ear renders the understandability of speech or music more difficult. It is therefore of interest to inquire about the ear's capability in this respect. The minimum perceptible increment of sound-pressure level of a pure tone varies with both pressure and frequency, but, for levels greater than about 40 db above the threshold of audibility and for frequencies between 200 and 7000 cycles, the minimum perceptible increment in pressure level varies from one quarter to three quarters of a decibel. The smallest perceptible change in frequency that the ear can detect is different for different pressure levels and frequencies, but, for pure tones more than 40 db above threshold and for frequencies greater than 500 cycles, it is of the order of 0.3 per cent for *monaural* listening with an earphone. In the discussion of the acoustics of rooms in Chapter 8, we shall see that a variation of the frequency of a sound source in a room may produce a marked variation in the pressure distribution within the room. Therefore we should expect, and rightly, that an observer could detect a smaller change in the frequency of a source in a room than in the open air.

Impaired Hearing

A person with normal hearing ordinarily experiences some difficulty in understanding speech in a large auditorium, even if the room has been well designed. The reason is that the unamplified power of the voice of the average person is generally inadequate to "fill" the auditorium. Therefore a person with minor hearing impairment may experience considerable difficulty in listening to speech in large rooms. For this reason, complaints concerning the poor acoustics of auditoriums may come from persons who have slightly impaired hearing of which they are not aware.

A loss in acuity of hearing is manifested by the required increase in the sound level at which a pure tone is just audible. Furthermore, it has been shown that this shift in threshold is a fairly good measure of the loss of ability to understand speech. By definition, the *hearing loss* of an ear at a given frequency, expressed in decibels, is the difference between the threshold of audibility for that ear and the normal threshold of audibility at the same frequency. The *normal threshold of audibility* is the modal value of the minimum audible threshold of a large sample of the general population in the United States.

The relationship between the hearing loss and various inabilities to hear has been investigated by the U. S. Public Health Service. In its survey, nearly 9000 persons were asked to classify their hearing ability into one of the five following groups:

(1) Normal hearing; no noticeable difficulty.

(2) Unable to understand speech in a public place such as a church or theater.

(3) Unable to understand speech from a person speaking 2 or 3 feet away.

(4) Unable to understand speech from a telephone.

(5) Total deafness; unable to understand speech under any condition.

The hearing loss in the speech-frequency range of those in each of the above groups was measured. The average results are shown in Fig. 2.3. It will be seen, for example, that the group reporting an inability to understand speech in a typical church or theater shows an average hearing loss of about 25 db.

It is of interest to know what percentages of a group of people

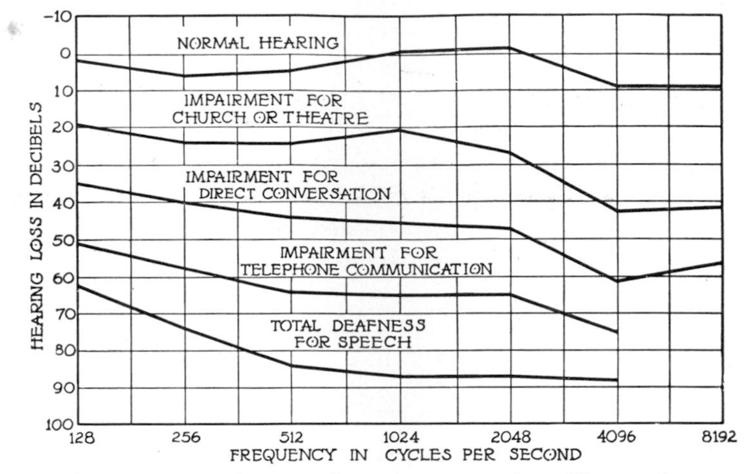

Fig. 2.3 Average audiograms for groups reporting different degrees of hearing. (U. S. Public Health Service Survey.)

have various degrees of hearing loss. Such data may be obtained from the comprehensive survey conducted on a large sample of the population—visitors to the 1939–1940 World's Fairs in San Francisco and New York. A series of curves (Fig. 2.4) has been constructed from these data by Steinberg, Montgomery, and Gardner.[1] The curves give the percentage of the population having a hearing loss at least as great as that indicated by the contour lines. For example, 5 per cent of the population have a hearing loss of at least 20 db for frequencies below 1000 cycles. In other words, for a pure tone to be just audible for this 5 per cent of the population, it would be necessary to raise the level of that tone at least 20 db above the level required for a person with "normal" hearing.

[1] J. C. Steinberg, H. C. Montgomery, and M. B. Gardner, *J. Acoust. Soc. Am.*, **12**, 291 (1940).

The loss in the acuity of hearing generally increases with age. This is shown in Table 2.1, which gives the percentages of the population, according to age groups, having a hearing loss of at least 45 db at the several frequencies indicated; these frequencies are representative of the important frequency range for the proper reception of speech. For example, 16 per cent of the male

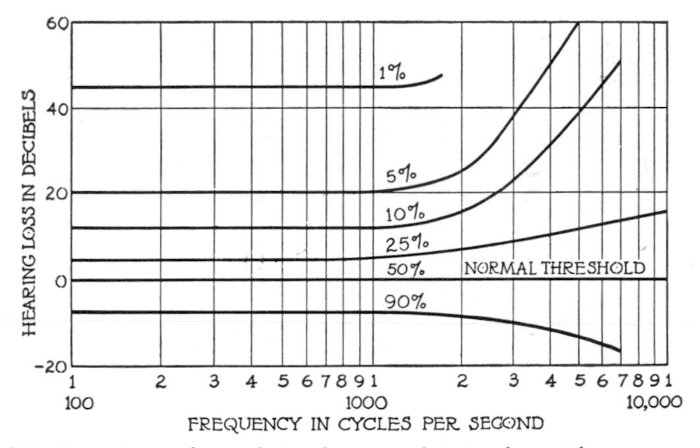

Fig. 2.4 Percentage of population having a hearing loss at least as great as that indicated by the contour lines. (Based on data of J. C. Steinberg, H. C. Montgomery, and M. B. Gardner.)

population between the ages of 40 and 49 have a loss at 3520 cycles of 45 db or more. Note that, in a given age group, the loss at the higher frequencies is greater for men than it is for women.

Loudness and Loudness Level

The loudness of a sound (that is, the magnitude of its sensation) depends not only on the pressure of the sound but also on its frequency spectrum. Loudness can be described quantitatively in terms of another subjective characteristic of sound, the so-called loudness level, which itself is defined in terms of the sound pressure and frequency of a pure tone; see Fig. 2.5. These curves are the well-known Fletcher-Munson contours of equal loudness.[2] They were obtained by employing a pure 1000-cycle tone as a

[2] H. Fletcher and W. A. Munson, *J. Acoust. Soc. Am.*, **5**, 82 (1933).

TABLE 2.1

PERCENTAGE OF THE POPULATION HAVING A HEARING LOSS OF AT LEAST 45 DB

(J. C. Steinberg, H. C. Montgomery, and M. B. Gardner)

Age Group	Loss in Decibels		
	Frequency		
	880 cycles	1760 cycles	3520 cycles
10–19			
Men	0.6	0.6	1.8
Women	0.6	0.4	0.3
20–29			
Men	0.1	0.3	2.7
Women	0.4	0.3	0.7
30–39			
Men	0.3	0.6	6.0
Women	1.2	0.8	1.6
40–49			
Men	1.4	2.6	16.0
Women	2.1	1.5	3.0
50–59			
Men	2.6	6.0	27.0
Women	4.0	3.0	7.0

reference tone and adjusting the sound-pressure level of tones of other frequencies until they were judged to be of the same loudness as that of certain arbitrarily chosen pressure levels of the reference tone. Thus, by definition, the *loudness level,* in *phons,* of a sound is numerically equal to the sound-pressure level, in decibels, of the 1000-cycle reference tone which is judged by listeners to be equal in loudness. For example, Fig. 2.5 shows that a 500-cycle tone having a sound level of only 25 db sounds equally as loud as a 50-cycle tone having a sound level of 64 db. Both have a loudness level of 20 phons. Notice that at low frequencies a given change in sound level produces a much larger change in apparent loudness than does the same change in sound level at higher frequencies. It should be emphasized that the curves in Fig. 2.5 are contours of equal loudness for *pure tones* and may not apply to continuous sound spectra like room noise. This is one of the reasons why sound-level meters do not measure the loudness level of such sounds correctly. A method for ob-

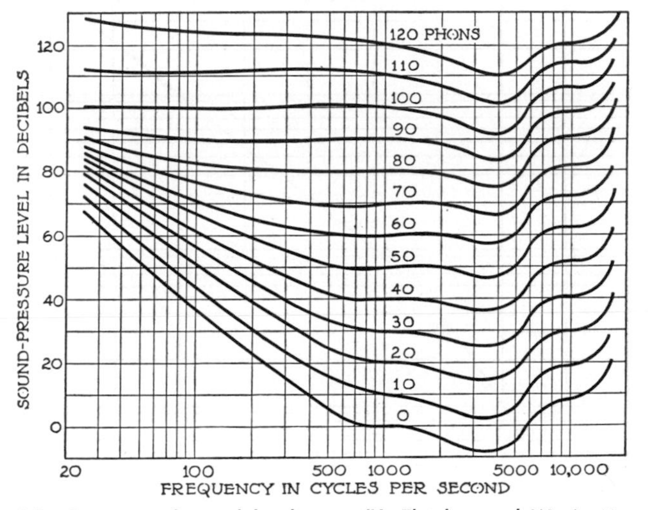

Fig. 2.5 Contours of equal loudness. (H. Fletcher and W. A. Munson.)

taining the loudness level of room noise will be discussed later in this chapter.

The loudness of a sound is related to the total nerve energy produced by the sound in the ear, whence it is sent to the brain. It is measured in *sones, millisones,* or *loudness units* (1 sone = 1000 millisones = 1000 loudness units). A loudness of 1 millisone corresponds to the threshold of hearing, a loudness of 10 sones is twice as loud as a loudness of 5 sones and 10 times as loud as 1 sone, etc. The relationship between loudness and loudness level is given in Fig. 2.6. This curve applies to sounds of any frequency, or any combination of frequencies.[3] For example, if the

[3] In general, tones of very short duration are judged to be less loud than equally intense tones of longer duration. See W. A. Munson, *J. Acoust. Soc. Am.,* **19**, 584 (1947).

loudness level of a sound is 40 phons, it will have a loudness of 1000 millisones or 1 sone; if the loudness level is 70 phons, it will have a loudness of 10 sones. In the range above 40 phons (the range of chief interest in architectural acoustics), a change in

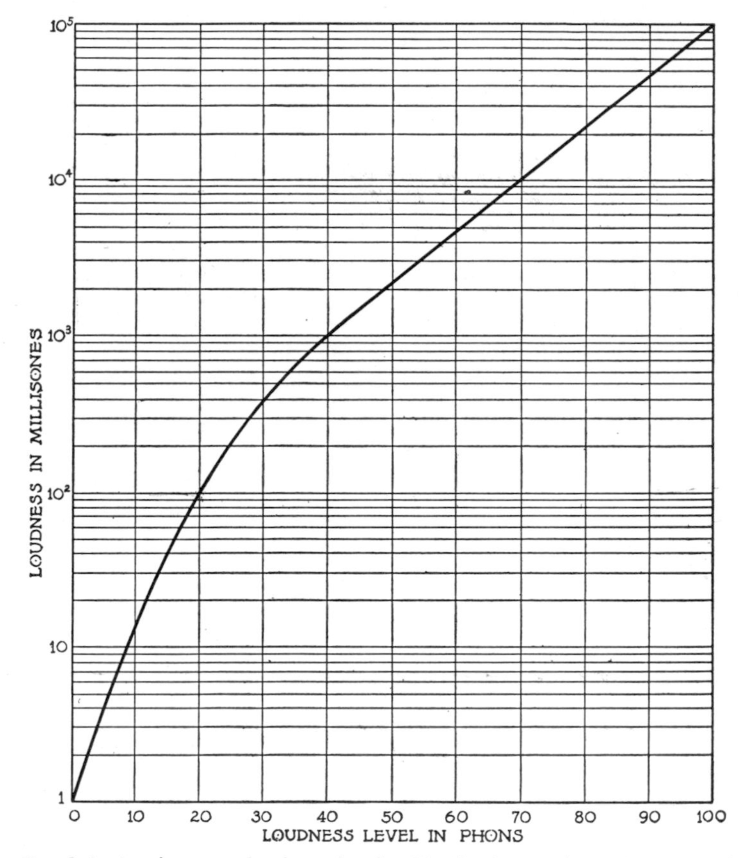

Fig. 2.6 Loudness *vs.* loudness level. (H. Fletcher and W. A. Munson.)

loudness level of 30 phons corresponds to a tenfold change in loudness, and a change in loudness level of 9 phons corresponds to a twofold change in loudness. This relationship is of considerable interest in regard to the reduction of noise in rooms by sound absorption or by sound insulation. Thus, a reduction of loudness level in a room from 60 to 51 phons will be judged by the average listener as a reduction in loudness of one half.

Effect of Noise on Hearing

Many auditoriums and theaters are not adequately insulated against noise. Everyone has probably had the experience of being unable to hear some critical lines in a play because noises from the foyer or street often occur just as these lines are spoken. In spite of the apparent correlation, no one has demonstrated the existence of a "masking demon" that knows the play and takes delight in making noise at these most crucial moments. However, one may legitimately conclude that, aside from the annoyance that it causes, noise has the effect of reducing the acuity of hearing; that is, it elevates the threshold of audibility. This shift in threshold of audibility is called *masking,* and the shift, in decibels, defines the amount of masking. Unless the loudness of speech or music is sufficiently above the level of the surrounding noise, the speech or music cannot be fully recognized or appreciated because of the masking effect of the noise; it is impossible to ignore completely a loud noise and listen only to the wanted sound.

The subject of the interfering effect of noise is so pertinent to the hearing of speech and music in auditoriums that considerable space will be devoted to it in subsequent chapters. Only a few of the fundamental properties of auditory masking are presented in this section.

Masking data are generally represented in the form of a curve called a *masking spectrum* (sometimes called a masking audiogram) which gives the number of decibels at each frequency that the threshold level of a pure tone is shifted when heard by a normal observer in the presence of masking sounds. As an illustration, the masking spectrum due to "average room noise" is given in Fig. 2.7. For instance, a tone of 1000 cycles would have to be raised 25 db above the *minimum audible threshold* to be heard in the presence of this average room noise. The masking spectrum in this case, and in general, is not constant with frequency. It depends on the pressure level and the nature of the masking sounds. Here we shall discuss two types of masking sounds: first, a sustained pure tone; and, second, a continuous noise spectrum typical of those which occur in auditoriums.

Experiments indicate that low-pitched tones, especially if they are of considerable loudness, produce a marked masking effect upon high-pitched tones, whereas high-pitched tones produce only little masking upon low-pitched tones. The auditory masking of one tone upon another is greatest when the masking tone is almost identical with the masked tone. In general, all tones, especially if they are loud, offer considerable masking for all

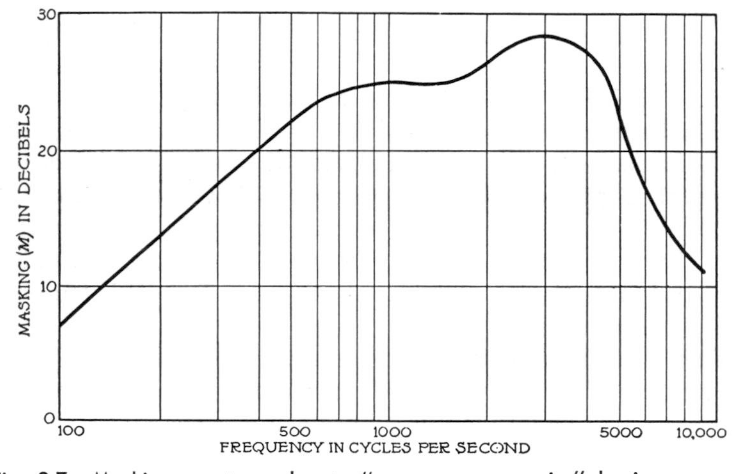

Fig. 2.7 Masking spectrum due to "average room noise" having a sound level of 43 db.

tones of higher frequency than the masking tone. Therefore, very intense low-frequency hums or noises are especially troublesome sources of interference for the hearing of speech or music since they mask nearly the entire audible range of frequencies.

Calculation of Masking Spectra from Sound-Pressure Spectra

The level of the noise surrounding a listener—not his threshold of audibility—generally determines the lowest value of sound pressure he can hear; that is, the noise elevates his apparent threshold of audibility. The effect of this increase in the threshold of audibility is the reduction of the intelligibility of speech for him, especially when the speech is at low levels. It is important, therefore, that we should be able to predict, at least approximately, the masking effect in the average room or auditorium.

The masking effect of noise depends not only on its total sound pressure, but also on its frequency or *spectral* distribution. This distribution can be obtained by the use of a *sound analyzer,* a device which measures the sound pressure within a limited frequency range called a *band.* By "sweeping" this band across the entire audible range it is possible to determine how the sound pressure is distributed with respect to frequency. Not all analyzers have the same band width; in order to make sound frequency analyses comparable, sound pressures P_w dynes/cm² for a band having a frequency width of w cycles are frequently converted to pressure-spectrum levels L_{ps} (that is, the pressure level for a band 1 cycle wide) by the equation

$$L_{ps} = 10 \log_{10} \frac{(P_w/0.0002)^2}{w} \text{ db} \tag{2.1}$$

A graph of pressure-spectrum level *vs.* frequency is called a *sound spectrum* or a *noise spectrum.* The spectrum for average room noise is shown in Fig. 2.8.

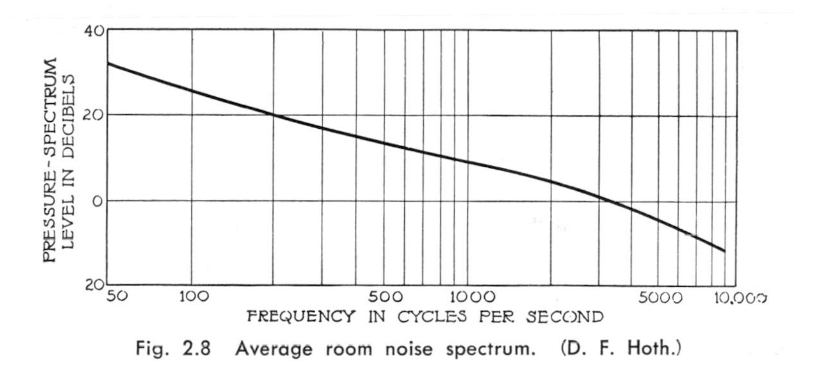

Fig. 2.8 Average room noise spectrum. (D. F. Hoth.)

The first step in computing the masking effect of noise is to determine the spectrum level of the noise, using Eq. (2.1). Having obtained the sound spectrum of the noise, the masking M at every frequency is determined from Fig. 2.9.

As an illustration we shall calculate the masking spectrum due to "typical room noise." Surveys of room noise in telephone subscribers' locations have been made by the Bell Telephone Laboratories (see Figs. 10.3 and 10.5). The average sound-pressure level in residences with radios turned off, as measured with a sound-level meter (using the 40-db frequency-weighting network), was found to be 43 db. About half of the residences had sound levels between 38 and 47 db, and 90 per cent had levels between 33 and 52 db. The noise spectra for all types of rooms are similar, and they have the shape shown in Fig. 2.8. We have chosen the ordinates of this curve so that the total sound level (that is, the level of the integrated value of L_{ps}) corresponds to 43 db, the average sound level in residences. The masking spectrum (Fig. 2.7) for this noise distribution is then obtained from Fig. 2.9.

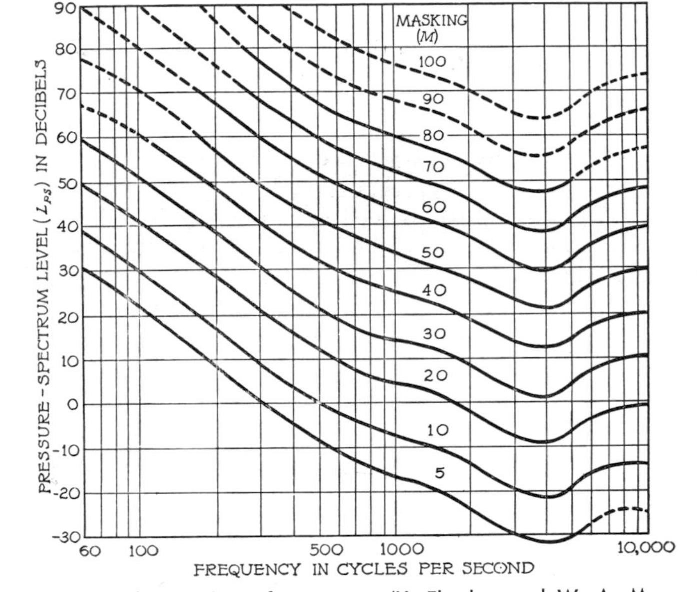

Fig. 2.9 Masking (M) vs. frequency. (H. Fletcher and W. A. Munson.)

For example, in Fig. 2.8 at 200 cycles the pressure spectrum level is 20 db. Then in Fig. 2.9 the value of M for L_{ps} equal to 20 db at 200 cycles is 13 db. Other points for deriving the curve in Fig. 2.7 were obtained similarly. The application of these data is limited to portions of the noise spectrum that do not exhibit abrupt changes with frequency.

Loudness Calculations for a Case of Typical Room Noise

The reduction of room noise is a routine task in architectural acoustics. It is most advisable to stop the noise at its source, but frequently this procedure is impossible or impractical. Then the room noise is usually decreased by the installation of sound-absorptive material. The over-all reduction in the level of the noise can be given in decibels, but in order to express the decrease in terms which correspond more closely to what is recognized by the ear, physiological

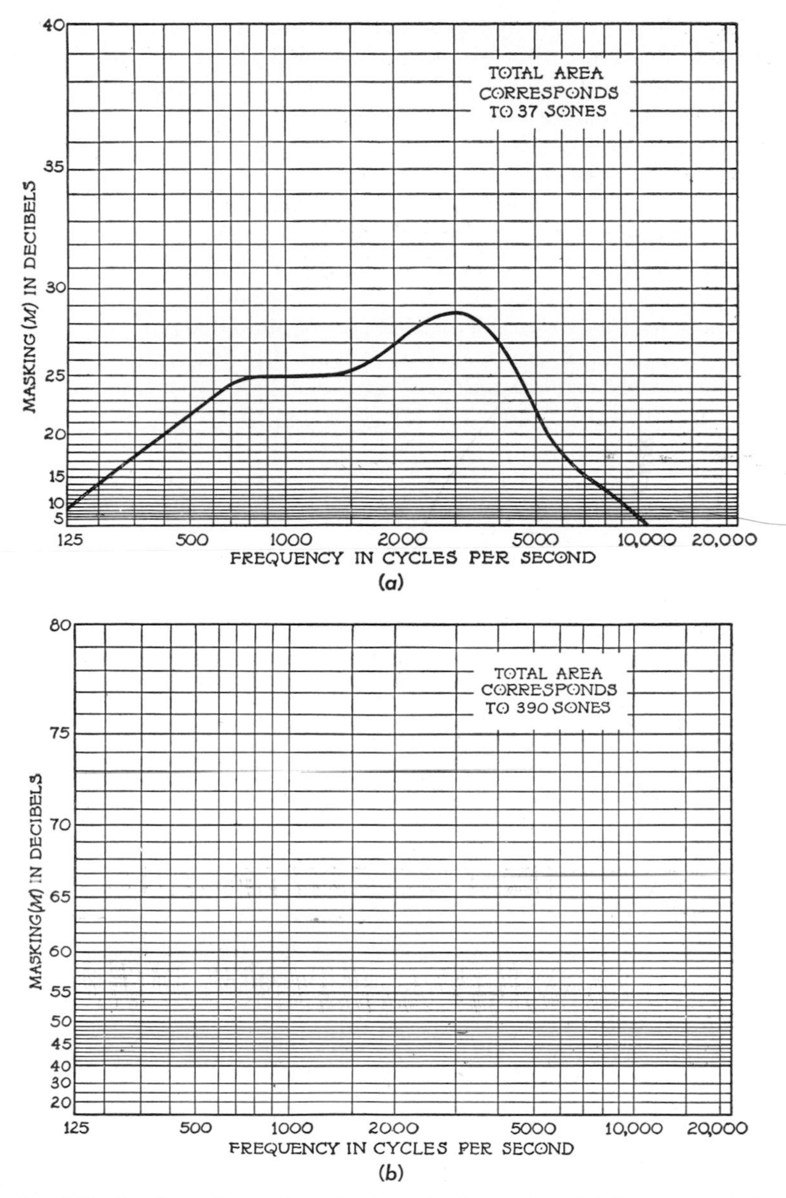

Fig. 2.10 Loudness Computing Charts. The total area of (*a*) corresponds to a loudness of 37 sones; that of (*b*) to a loudness of 390 sones. (W. A. Munson.) The curve which has been added to (*a*) is for the room noise spectrum of Fig. 2.8.

units, such as loudness units, should be used. Hence, the following method is presented for computing the loudness of a sound whose spectrum does not change abruptly in magnitude from one frequency region to another, such as average room noise: [4]

(1) Determine the *masking spectrum* from the *sound-pressure spectrum* by the method given above.

(2) Plot the *masking* M on the special Loudness Computing Chart, Fig. 2.10(*a*) or 2.10(*b*).

(3) Determine the area under this curve; the area is directly proportional to the loudness, the constant of proportionality depending on the unit of area used. The total area of the Loudness Computing Chart in Fig. 2.10(*a*) corresponds to 37 sones. Thus, if the area under a curve plotted on this chart is 10 per cent of the total area of the chart, it represents a loudness of 3.7 sones. The total area of the chart in Fig. 2.10(*b*) corresponds to 390 sones; this chart should be used when the other one is not large enough to accommodate the masking spectrum.

As an illustration, we shall calculate the loudness of the average room noise spectrum shown in Fig. 2.8. Its masking spectrum was determined and plotted in Fig. 2.7. Replotting these data on Fig. 2.10(*a*), we find that the area under this curve is 21 per cent of the total area of the Loudness Computing Chart. Hence the loudness due to typical room noise is 21 per cent of 37 sones, or 7.8 sones. A method is given in Chapter 11 for the calculation of the loudness reduction of room noise resulting from the installation of sound-absorptive material.

[4] Note that the loudness of a pure tone can be determined by the use of equal loudness contours (Fig. 2.5) and the loudness *vs.* loudness-level curve (Fig. 2.6). However, for sounds such as average room noise (Fig. 2.8) which are distributed in frequency the method of this section should be employed.

3 · Speech and Music

In order to predict and control the behavior of speech and music in auditoriums, theaters, music rooms, and lecture halls, it is, of course, necessary to know the physical properties of speech and music. Thus, the characteristics which distinguish speech sounds must be known so that rooms in which these sounds are to be heard can be designed in such a way that their essential characteristics will be preserved as they are transmitted from their source to the listeners. Similarly, if a room is to be designed to enhance the beauty of music, the properties of the sounds which issue from musical instruments must be known. These sources of sound will be considered in this chapter. As we shall see, the physical properties of speech differ considerably from those of music; hence, it is to be expected that the acoustical properties of speech rooms should differ appreciably from those of music rooms.

Noise, Music, and Speech

Sounds are frequently classified into three types: noise, music, and speech. However, this classification is not always clear-cut. It is sometimes questionable, for example, whether a sound should be classified as music or noise. In general, *noise* may be defined as *unwanted sound.* Thus, if one is listening to a concert in an auditorium, a conversation in the next row may be regarded as noise. On the other hand, if one is trying to converse on the telephone while "Junior's Dixieland Four" is holding forth in the living room, this music, as far as the person on the telephone is concerned, very definitely falls under the classification of noise.

Sound may also be classified as *ordered* or *disordered*. In an ordered sound the instantaneous pressure follows a regular pattern. Furthermore, a frequency analysis of such sound will show a definite overtone structure; that is, the sound can generally be resolved into a fundamental frequency and a series of overtones, the latter having frequencies that often are integral multiples of the fundamental frequency. Overtones possessing this simple relationship of frequencies are called *harmonics*. On the other hand, the peaks of acoustic power in disordered sound (for example, the background noise in a large auditorium) occur more or less *at random*. In such sound, practically all audible frequencies, from the lowest to the highest, are present. The periodic qualities of ordered sound are lacking.

Street noise is an example of disordered sound. A spectrum analysis of this noise will show that practically all frequencies are present and that it is highly irregular in nature. This irregularity can be illustrated by means of sound spectrograms. These visual records, obtained with a sound spectrograph [1]—an instrument developed by the Bell Telephone Laboratories—provide a frequency analysis of a sound source as a function of time. As an example, a sound spectrogram of street noise in a busy city is given in Fig. 3.1. The frequency scale of the spectrogram is linear, and it covers 3500 cycles, as shown by the vertical scale to the left of the figure. The time scale is also linear and is marked in seconds along the horizontal axis. Thus, on the spectrograms, a sustained pure tone of 1750 cycles produces a single dark horizontal line midway between the top and bottom of the record. The greater the pressure of the sound, the darker is the line.[2] The spectrogram of street noise shows that the

[1] For a more complete discussion of sound spectrograms see *Visible Speech* by R. K. Potter, G. A. Kopp and H. C. Green, D. Van Nostrand Co., Inc., 1947.

[2] The width of the analyzing band-pass filter used in obtaining the spectrograms presented in this chapter was 300 cycles per second, with the exception of the spectrogram of Fig. 3.2, for which it was 45 cycles per second.

peaks of power occur at random. Figure 3.2 is the spectrogram of noise from a ventilating fan. Note the regularity of the pattern and also the predominance of certain frequencies.

Music is generally, though not always, made up of ordered sound. The power peaks come at periodic intervals, as illustrated

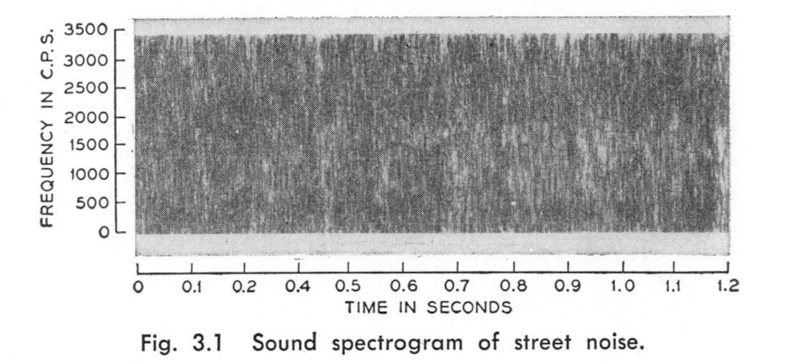

Fig. 3.1 Sound spectrogram of street noise.

in Fig. 3.3, which is a spectrogram of a portion of a clarinet solo. This record also indicates another characteristic of most musical tones: the overtone structure is harmonic. The component frequencies in Fig. 3.3 are integral multiples of the fundamental frequency.

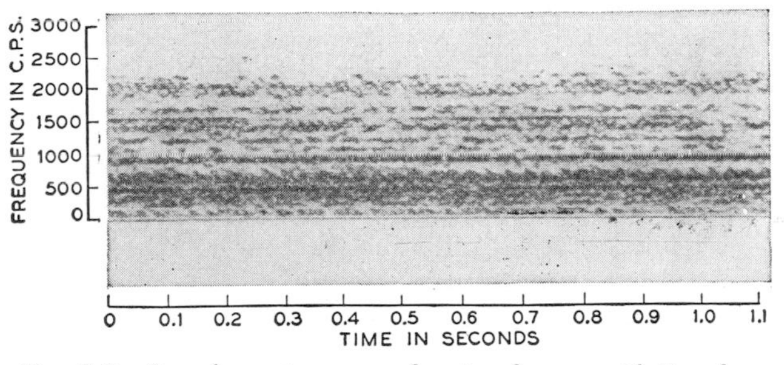

Fig. 3.2 Sound spectrogram of noise from ventilating fan.

Speech consists of both ordered and disordered sound. The spectrograms in Fig. 3.4 illustrate the spoken words, *Acoustical Designing in Architecture.* Note that the hiss "s" in the word "acoustical" produces much the same record as street noise does. This "s" sound is non-periodic as contrasted with the vowel sounds, which show a definite overtone structure with the bursts of peak power coming at regular intervals of time, as shown by the vertical striations.

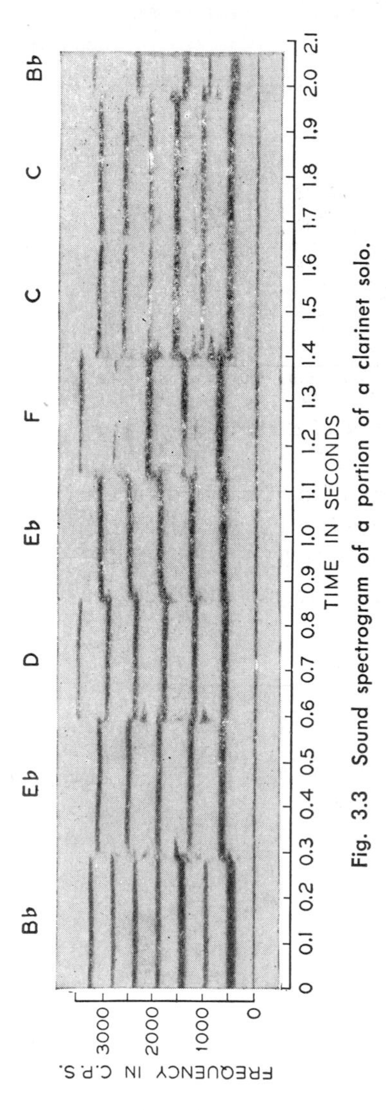

Fig. 3.3 Sound spectrogram of a portion of a clarinet solo.

Articulated speech consists of a flow of various combinations of consonants and vowels. The nature of the articulation of the separate syllables and words in speech, and the rapidity with which the separate syllables follow one another, have a great bearing upon how well the speech is heard. If the separate syllables are inaccurately formed, and if they follow each other in rapid succession, they may not be heard distinctly. The effect on speech of an enclosure, such as a room, is very significant. It is of such great importance that a separate section is devoted to it in this chapter, and this effect will be considered again after the acoustics of rooms has been discussed.

Nearly all the difficulties which arise in the hearing of speech in auditoriums are attributable to errors in the recognition of the consonants. This is due partly to the very small energy in the consonants compared to the energy in the vowels and partly to the differences in the detailed structures of consonants and vowels. Figure 3.4 shows that below 3500 cycles the vowels are characterized by three or more characteristic "bars" or regions of resonance. These "bars" aid in the identification of the speech sounds.

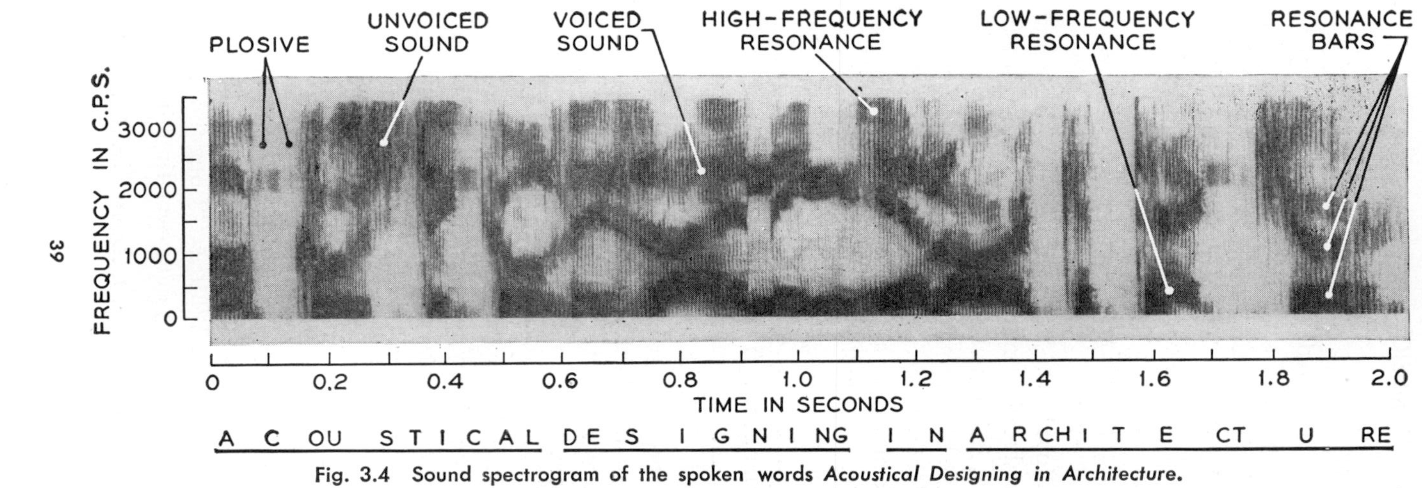

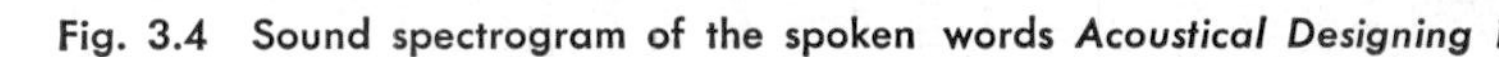

Fig. 3.4 Sound spectrogram of the spoken words *Acoustical Designing in Architecture*.

Speech Power

In this section we shall discuss the acoustical power output of speech. Later we shall show how such data may be applied to a room in order to calculate the average sound-pressure level throughout the room. The average person is surprised at the exceedingly minute amount of energy contained in speech. As mentioned in Chapter 1, approximately 15,000,000 lecturers speaking at the same time generate acoustical energy at a rate of only 1 horsepower. When the speech power of a single speaker is diffused in a large auditorium, the sound pressure in the room is reduced to extraordinarily small values. Under such circumstances, it is easy to understand why it is difficult to hear well in a large room, and why very feeble sources of extraneous noise may produce serious interference with the speech. For example, the noise of a distant ventilating fan or motor, the shuffling of feet on the floor, the jarring of a near-by door, or the whispering or coughing of inconsiderate "spectators" may be sufficient to mask many of the speech sounds, and especially the feeble consonants, which reach an auditor in a large auditorium.

Since the amounts of acoustical power generated in speech are very small, the acoustics of auditoriums used primarily for unamplified speech must be carefully controlled to make the best possible use of the usually inadequate speech power. In large auditoriums, as might be expected, the amplification of speech is an indispensable requirement.

The *instantaneous speech power,* the rate at which sound energy is radiated by a speaker, varies considerably with time. Its maximum value in any given time interval is the *peak speech power.* The *average speech power* has, in general, a very much lower value than the peak value and depends on the method of averaging (that is, on the length of time over which the average is taken) and on the inclusion or omission of the pauses between syllables and sentences in this time interval.

An extensive investigation of the conversational speech power output of individuals of two groups, 6 men and 5 women, was conducted by Dunn and White.[3] "Long-time-interval averages" were obtained by averaging data over time intervals of a minute

[3] H. K. Dunn and S. D. White, *J. Acoust. Soc. Am.,* **11**, 278 (1940).

or more of continuous speech, including all natural pauses between syllables and sentences. Their results show that the long-time-interval average power output varied from one individual to another within the group of 6 men, ranging from 10 to 91 microwatts, with an average for the group of 34 microwatts. The extremes for the women were 8 and 55 microwatts, and the average was 18 microwatts. The short-time-interval average and peak power outputs of typical speakers, speaking at a conversational level, can be and often are much higher. Calculations of these quantities were made for $\frac{1}{8}$-second intervals, a length of time of the order of magnitude of the duration of a syllable. At least 1 per cent of the $\frac{1}{8}$-second intervals had an average power in excess of 230 microwatts for men and 150 microwatts for women, and a peak power in excess of 3600 microwatts for men and 1800 microwatts for women. This study indicates that the average male person produces a long-time-interval average sound-pressure level of about 64 db at a distance of 1 meter, directly in front of him, when he talks in a normal conversational voice; the average for women, as shown by this study, is about 61 db at a distance of 1 meter.

The above data are for conversational speech in a quiet location in the absence of reflecting surfaces. Noise, the size of the room in which a person is speaking, his distance from the auditor, the acoustical conditions of the room, and other factors affect the power output of his speech, and especially the sound-pressure distribution throughout the room. If a noisy condition prevails, he will raise his voice in order to "override" the noise. He will, in general, increase his power output as his distance from an auditor is increased. Furthermore, it is well known that a speaker attempts to raise the power output of his voice when he is speaking in an auditorium, and the larger the auditorium the more he exerts himself. Tests conducted in a small auditorium (27,000 cubic feet) indicate that the long-interval average speech power of the speakers in a test group of typical university lecturers was about 27 microwatts. Similar tests conducted in a larger auditorium (240,000 cubic feet) indicate that the average speech power in this large auditorium was approximately 50 microwatts. These results confirm a reasonable expectation based upon everyday observations, namely, that a speaker increases the power of his voice in his attempt to discount the effect of the size of the auditorium in which he is speaking. He attempts to speak so that he will be heard by all auditors in the room. That he falls short of the requirements for good hearing in large auditoriums will be made manifest in Chapter 9.

The percentage of the speech power lying below a given frequency, for the average speaker, is given in Fig. 3.5. There

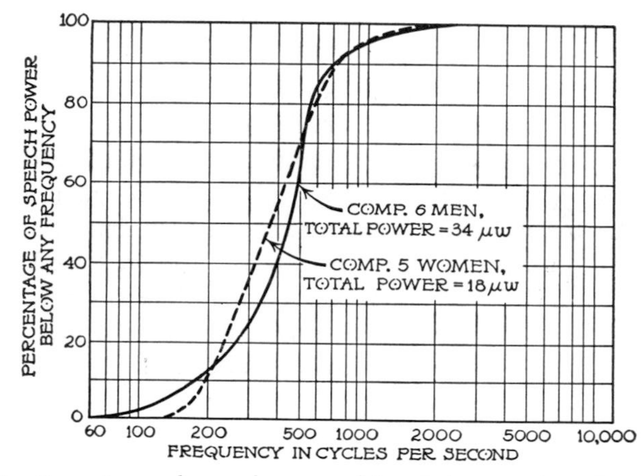

Fig. 3.5 Percentage of speech power lying below a given frequency *vs.* frequency. (H. K. Dunn and S. D. White.)

is relatively little power in the frequencies above 1000 cycles, the frequency range that characterizes most consonants. Figure 3.6 shows how the total power of average conversational speech is distributed in frequency. The level of speech power per cycle is plotted as a function of frequency. Since these curves represent data averaged over a long time interval, their shapes are affected by the frequency of occurrence of the speech components as well as by their acoustic power. If these curves were "corrected" for the sensitivity of the ear so that the ordinates represented the loudness of the various frequency components as heard by the ear, the maximum would occur between 500 and 1000 cycles. This will be more fully appreciated if one recalls the nature of the hearing sensitivity curve, Fig. 2.2. All portions of the frequency range do *not* contribute equally to the intelligibility of speech; this is illustrated in Fig. 3.6, where the speech spectrum is divided into ten bands that make approximately equal contributions to the intelligibility of speech. Although frequencies below 250 cycles and above 7000 cycles contribute to the "naturalness" of speech, they add little to its intelligibility.

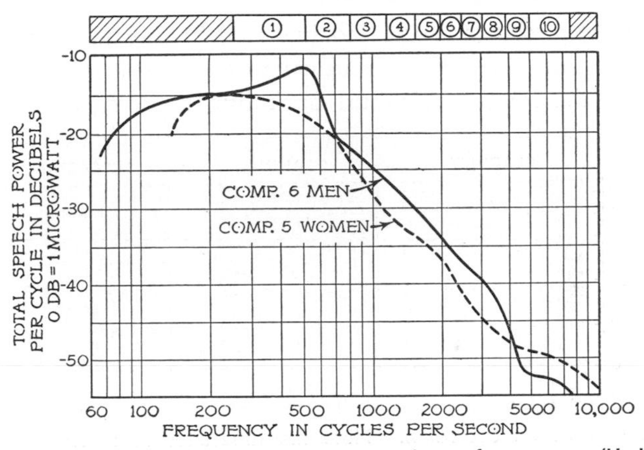

Fig. 3.6 Total radiated speech power per cycle vs. frequency. (H. K. Dunn and S. D. White.) All portions of the frequency spectrum do not contribute equally to the intelligibility of speech, as is indicated by 10 bands that provide equal contributions.

Properties of Musical Sounds

The physical characteristics of musical sounds differ from those of speech in several important respects. In general, they are not so transient in nature. The separate tones of music often are sustained for an appreciable fraction of a second or longer, and the change in frequency is nearly always ordered in conformity with the relations among the frequencies which make up the musical scale. This is illustrated by a comparison of the speech spectrogram, Fig. 3.4, with one of a portion of a clarinet solo, Fig. 3.3.

The separate tones that comprise music are in general made up not of a single simple harmonic vibration, but of long and complex series of such vibrations. In some instances the overtones may be much more prominent than the fundamental. The number and prominence of these overtones, together with the differences in their rates of build-up and decay, are the chief determinants of the tonal characteristics of various musical instruments. The overtones from most string and pipe instruments are, at least very approximately, harmonic.[4] The overtones from reeds, bars (vibrating transversely as in the xylophone), disks, bells, etc., are, in general, not harmonic. The differences in the overtone structure of different musical tones of the same pitch are illustrated in Fig. 3.7. Portions of spectrograms of sound from three musical instruments are shown. In each case the lower horizontal dark line represents the funda-

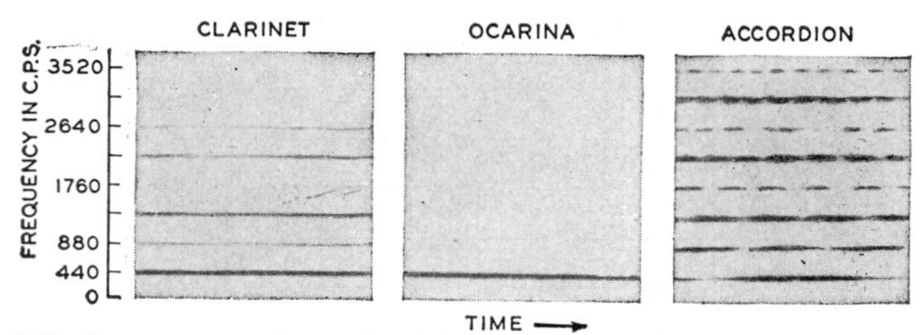

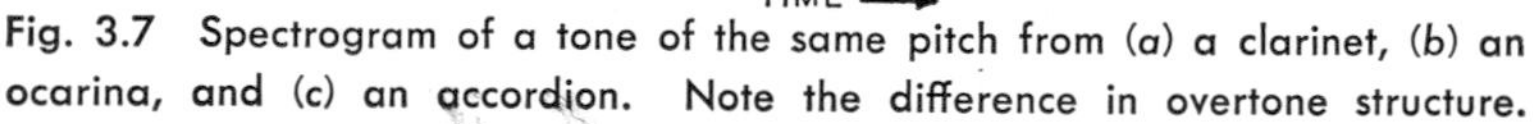
Fig. 3.7 Spectrogram of a tone of the same pitch from (*a*) a clarinet, (*b*) an ocarina, and (*c*) an accordion. Note the difference in overtone structure.

mental component—a frequency of about 440 cycles. Dark bands above it correspond to the harmonic overtones. Some of the harmonics are "stronger" than others, and some do not exist at all. Thus, in the clarinet tone the fourth and sixth harmonics are particularly weak, and in the ocarina tone all higher harmonics appear to be absent. The spectrogram of the accordion tone shows that its even harmonics are very weak. This record is especially interesting because it furnishes an excellent illustration of "amplitude vibrato," a rapid periodic variation in the acoustic output as a function of time.

The acoustical power generated by musical instruments, including the singing voice, is in general considerably greater than

[4] An overtone that has twice the frequency of the fundamental is called the second harmonic, one that has three times the frequency of the fundamental is called the third harmonic, etc. The second, fourth, and all even-numbered harmonics are called even harmonics; the first (or fundamental), third, etc., are called odd harmonics. Closed organ pipes produce tones that have only odd harmonics; tones from open pipes have both odd and even harmonics.

that generated in speaking. Thus, whereas the average speaker generates an average power of about 50 microwatts in a fairly large auditorium, a vocalist or a musical instrument generates in the same auditorium about 100 microwatts, and often 500 to 5000 microwatts (see Table 1.1 for *peak* power values); therefore, the sound-pressure level of music in a room is usually several decibels higher than the average pressure level of speech. For this reason, less difficulty is encountered in hearing music than in hearing speech. The power generated in singing, or in the playing of musical instruments, is usually adequate for satisfactory hearing, even in auditoriums considerably larger than those in which the listening conditions for unamplified speech are just barely tolerable. The frequency range over which this power is distributed is considerably wider than that used in speech.

Some Effects of a Room on Speech and Music

When sound waves strike the boundaries of an enclosed space they are reflected back and forth until their energy is finally dissipated. The persistence of sound in an enclosure as a result of these repeated reflections is known as *reverberation*. This phenomenon has a very pronounced effect on both speech and music. For example, the sound pressure at a given distance from a source in a room is, in general, greater than it would be at this same distance from the source in the open air. This increase in level is quite helpful where sound sources, such as the voice, have relatively weak outputs. Furthermore, a certain amount of reverberation will contribute to the acoustical quality of a room intended for music. Indeed, listener (and especially performer) preference has shown that, properly controlled, it is a desirable property.

(*a*) Effects on Speech. The normal rate of speech is about 10 individual sounds per second. Thus, each sound has about one tenth of a second in which to make its impression upon the auditory mechanism. Since the time of reverberation in a room is nearly always in excess of 1 second, a number of sounds preceding the one upon which attention is focused will yet remain audible and will produce a masking effect that is dependent on their loudness and frequency composition and similar to that of noise. This is illustrated by the sound spectrograms in Fig. 3.8. All three spectrograms were made from the same magnetic-tape recording of the words "reverberation analysis." The record shown in Fig. 3.8(*a*) is a direct analysis of the original record. The next two records show the effects of adding reverberation.[5] As a little reverberation is added—see Fig. 3.8(*b*)—the individual sounds begin to overlap. After still more reverberation is added, namely, an amount corresponding to a reverberation time of about 3 seconds—the condition for the record in Fig. 3.8(*c*)—only a few of the more prominent identifying bars of the vowels can be recognized. This record has a greater resemblance to that of street noise than it does to the original record! It is indeed remarkable that the ear can resolve, as well as it does, such a jumble of sound. But even an instrument as extraordinary as the human ear is not infallible in resolving the confusions of sounds in excessively reverberant rooms.

(*b*) Effects on Music. It is obvious that in order to preserve the original quality of a musical tone it is necessary that the relative magnitudes of all harmonic components be unaltered as they are transmitted from their source to the listeners in an auditorium. This is strictly impossible in an auditorium or in any other enclosed space. The air within the enclosure acts as if it were an assemblage of resonators that respond to certain (resonant) frequencies more than to others. Furthermore, the rate of absorption of sound by the boundaries and contents of the room, and even by the air in the room, may be greater for certain frequency components than for others. Many acoustical materials are of such a nature that the high-frequency components will be absorbed more rapidly than the low-frequency ones,

[5] The words "reverberation analysis" were recorded on magnetic paper tape so that the same record could be reproduced many times. The output voltage from the reproducer was split into two separate channels. One channel was used as a source of voltage for driving a loudspeaker in a highly reverberant room, called an "echo" chamber. The sound in this chamber was then picked up by a microphone and fed into a "mixer." The second channel went directly from the output of the reproducer to the mixer. By combining the output from this direct channel with different amounts of the output from the echo chamber, various ratios of reverberant to direct sound energy could be obtained.

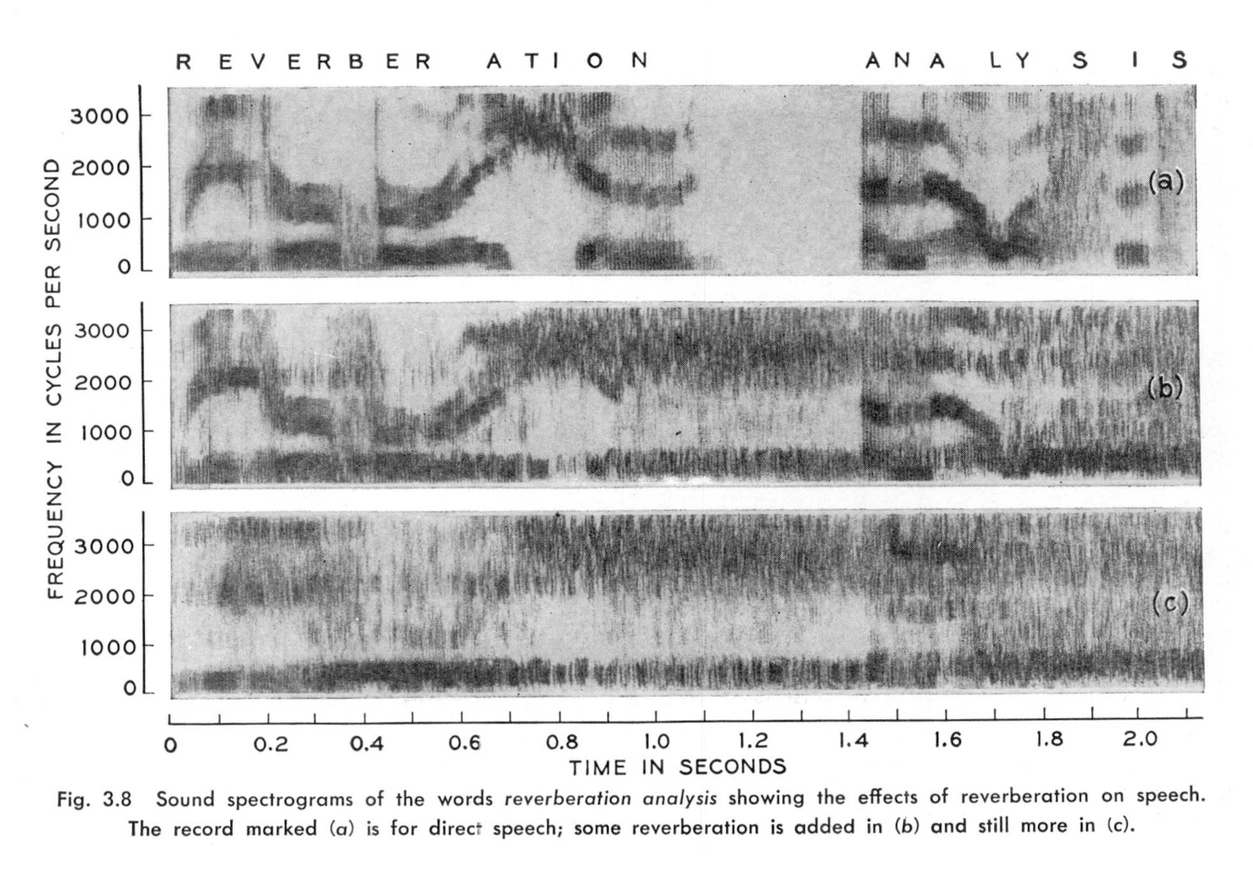

Fig. 3.8 Sound spectrograms of the words *reverberation analysis* showing the effects of reverberation on speech. The record marked (*a*) is for direct speech; some reverberation is added in (*b*) and still more in (*c*).

and the air itself is highly absorptive at very high frequencies. For these reasons sounds at some frequencies are enhanced while others are suppressed.

It is most fortunate for the hearing of both speech and music in auditoriums that these frequency distortions can be tolerated to a considerable degree without sacrificing the characteristics that are essential for the correct auditory recognition of sounds of speech or for the enjoyment of music. Thus, one is able to recognize a given sound when it is spoken by different men and women even though the frequency spectra of this sound, as spoken by these individuals, may be quite different. Except for the absorption in air, frequency distortion of the above type is not encountered outdoors. Music in the open, however, lacks the beneficial effects of reverberation; also, the loudness level may be too low for optimum listening conditions. It is apparent from the foregoing qualitative considerations of the effects of enclosures on speech and music that the acoustical properties of rooms are greatly dependent on such factors as noise, reverberation, loudness, and room resonances. These factors will be considered quantitatively in subsequent chapters, where general principles will be described to guide the architect and the engineer in the acoustical designing of all speech and music rooms.

4 · Reflection and Diffraction of Sound in Rooms

When a sound wave impinges on a non-yielding wall,[1] part of the incident sound is reflected from the wall; another part is transmitted into the wall, where some of it is dissipated as heat; and the rest is transmitted through the wall. For example, if the wave encounters a wall of very porous material, such as mineral wool, that portion of the wave which is transmitted into the wall suffers considerable attenuation as it is propagated through the material. Although this reduction in sound intensity is due largely to the viscous losses within the capillary pores of the material, the vibration of the fibers of the material often contributes to the attenuation.

Most wall structures are *not* non-yielding; they vibrate as a whole, or in parts, under the action of the pressure pulsations of incident sound waves: the wood, plaster, or even masonry walls of a room are set into vibration, like diaphragms, and hence radiate sound energy. Since most of the sound that is communicated from one room to an adjacent one is transmitted in this manner, rigid, heavy walls should be better insulators of sound than flexible, light ones; and experience gives abundant evidence that they are. One of the most effective means for

[1] *Wall* refers, here, to all the boundaries of a room and includes floor, ceiling, windows, doors.

providing a high degree of sound insulation makes use of a combination of rigid partitions and porous materials.

The relative magnitudes of the absorbed, transmitted, and reflected components of a sound wave incident on a wall depend on such factors as the frequency and angle of incidence of the sound wave, the nature of the wall material, and the manner in which the wall is supported, reinforced, backed. The present chapter will deal primarily with the reflection and diffraction of sound by the boundaries of a room. Absorption and transmission will be treated in subsequent chapters.

The Reflection of Sound

When a "free" sound wave (one free from the influence of reflective surfaces) strikes a uniform surface that is large compared to the wavelength of the sound, the reflection of the wave is

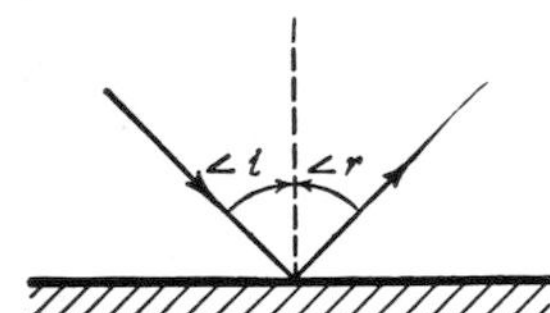

Fig. 4.1 Reflection from a plane surface.

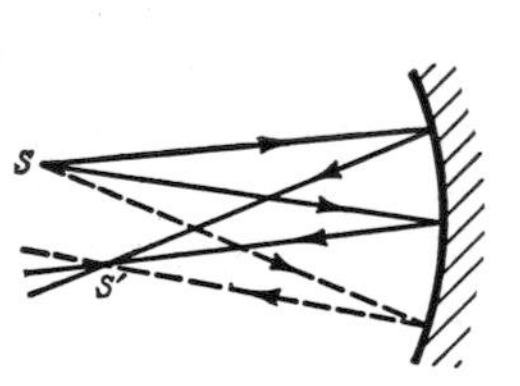

Fig. 4.2 Reflection from a concave spherical surface.

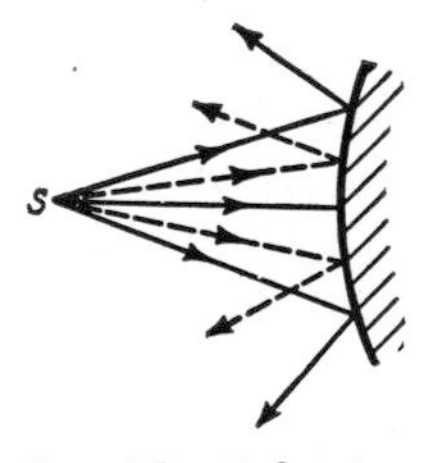

Fig. 4.3 Reflection from a convex surface.

similar to the familiar reflection of light. Suppose that the path of this sound wave is represented by a *ray* along which the wave advances, that is, by a line perpendicular to the advancing wave front. Then, by the *law of reflection,* the angle of reflection for this ray equals the angle of incidence, and the reflected ray lies in the plane of incidence. Thus in Fig. 4.1, which represents reflection from a plane surface, $\angle i = \angle r$. When this law is applied to the analysis of reflection from curved surfaces, the curved surface being regarded as made up of many small plane surfaces, the character of the reflections from a concave or a convex surface can be derived as shown in Figs. 4.2 and 4.3.

Evidently, as shown in Fig. 4.2, a concave surface tends to concentrate the reflected waves. Large concave surfaces may be used to advantage as reflectors, but if used indiscriminately they may ruin the acoustics of a room. In Chapter 9, we shall discuss the reflective properties of concave surfaces. Occasionally the reflections from such surfaces are beneficial, but more frequently they are deleterious. On the other hand, since a convex reflector (Fig. 4.3) tends to "spread" the reflected waves, convex surfaces at the boundaries of a room tend to diffuse the sound throughout the room. For this reason a number of radiobroadcasting studios and other special rooms have been constructed with cylindrical convex panels as part of the wall construction. Their action in dispersing sound waves is illustrated in Fig. 20.6.

Acoustical engineers make frequent use of the law of reflection for investigating the effects of various shapes of a proposed room on the distribution of sound in that room. Such studies can lead to the design of interior surfaces that will give beneficial reflections, or to the elimination or modification of surfaces that otherwise would give rise to echoes or other harmful reflections. Caution must be exercised, however, in such applications of the law of reflection; many mistakes are made in the acoustical design of buildings because it is assumed that sound is reflected in the same manner as light. It will be shown in Chapter 8 that this simplification of the behavior of sound in rooms has *limited validity* and use.

Diffraction of Sound

In the preceding sections there have been discussed many of the basic principles of sound, such as reflection, transmission, and absorption, all of which have analogies in the subject of light. The assumption that light is propagated in straight lines gives rise to that branch of optics called geometrical optics, and this "rectilinear" propagation accounts for the sharp shadows and images that can be formed by light. For example, light coming through a small opening such as a crack in a door is confined to a narrow beam of about the same shape and cross-sectional size as the opening. *Geometrical optics* corresponds to *geometrical acoustics,* which assumes that sound is propagated in straight lines; it is valid only for wavelengths that are short compared to the dimensions of rooms and the openings and reflecting surfaces in them. It should be remembered that many surfaces in rooms are *not* large in comparison with the wavelengths of low-pitched sounds. Windows, doors, pilasters, beams, coffers, any form of relief ornamentation, and patches of absorptive material—all introduce *diffraction* which greatly alters the direction and magnitude of the reflected sound. *Physical optics* and *physical acoustics* are based upon wave properties and thus describe many aspects of light and sound that cannot be handled by a geometrical treatment; among these, diffraction is the most important.

When sound comes through a crack in the door, it spreads out almost uniformly; or when sound encounters the corner of a building, it bends around the corner. In such instances, we say the sound has been diffracted or bent. Diffraction is the change in direction of propagation of sound waves due to their passage around an obstacle. The extent of the diffraction depends on the relationship between the wavelength of the sound and the size of the obstacle. This is illustrated in Fig. 1.8, which gives the sound-pressure distribution around a person's head while he is speaking. It will be seen that at low frequencies there is approximately equal radiation in all directions, whereas at higher frequencies, where the wavelength is much shorter and the bending is much less, the distribution is fairly directional. Similarly, sound from a point source reflected from a hard parabolic surface can produce a concentrated "beam" which converges to a focus or which diverges very little if the wavelength of the sound is small compared to the dimensions of the parabolic reflector. On the other hand, if the wavelength is large compared to the dimensions of the reflector, it will be relatively ineffective in its influence on the sound waves. As a further illustration of diffraction, Fig. 4.4 represents plane waves striking the edge of a partition. Notice how the waves, traveling from left to right, are bent around the edge of the obstacle.

It would appear that, in regard to diffraction, light and sound behave very differently. However, both theory and experiment show that sound and light behave very much alike if the openings and obstacles in the sound field are in the same proportions to the wavelengths of sound as the openings and obstacles in the light field are to the wavelengths of light. Visible light has wavelengths of the order of 0.000015 to 0.000030 inch, whereas audible sound has wavelengths of the order of 0.06 to 60 feet. It is principally because of this great disparity in the wavelengths of the sound and the light that we usually observe that light travels in straight lines through openings and past obstacles, whereas sound spreads out very considerably under similar circumstances.

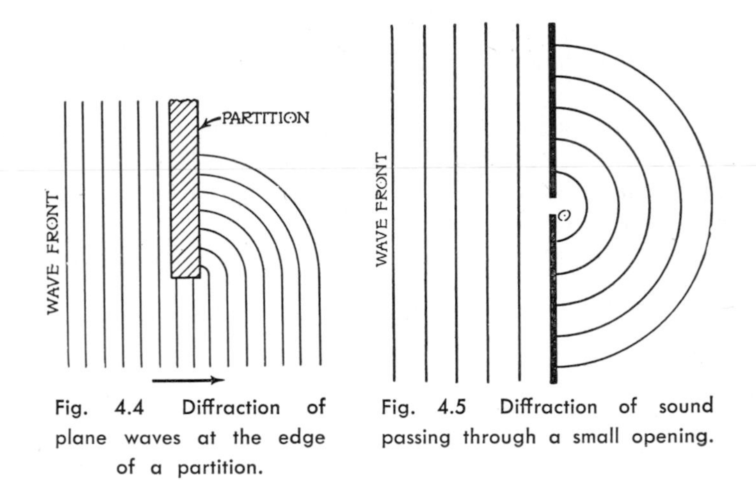

Fig. 4.4 Diffraction of plane waves at the edge of a partition.

Fig. 4.5 Diffraction of sound passing through a small opening.

Diffraction of Sound Transmitted through Openings. An elementary principle of physical acoustics, known as Huygens' principle, enables us to determine the extent of spreading of a sound wave when it is transmitted through an opening of known size. Consider a plane wave falling upon a very large surface in which there is an opening that is small in comparison with the wavelength of the sound. A familiar example is a small hole in a door, as illustrated in Fig. 4.5. According to Huygens' principle, which states that each point of the wave front at any instant may be regarded as a source of secondary waves, the opening *O* may be regarded as such a source from which the sound spreads out as a spherical wave. Hence, whereas the supposed wave was plane before it reached the small opening, it emerges approximately as a spherical wave and thus diverges widely. If the opening is large compared to the wavelength of the sound propagated through it (for example, a large proscenium), there is only slight bending near the edges.

Since the wavelengths of sound vary from about 0.06 to 60 feet, the diffraction may be pronounced for some frequencies and negligible for others. For example, a 3-foot door opening would be small compared to a wavelength of 60 feet. Therefore, such a low-frequency sound would be very much diffracted in going through the door, the emergent sound spreading out almost uniformly in all directions. In contrast, this same door opening would be very large compared to a wavelength of 0.06 foot (a frequency of about 19,000 cycles), and therefore a sound having such a wavelength would be transmitted through the door with little diffraction. Obviously, sounds such as those in speech and music, which are made up of a wide range of frequencies, are selectively diffracted because the low-frequency components will diverge widely while the high-frequency components will continue in a relatively narrow beam.

Diffraction of Sound from Reflective and Absorptive Surfaces. In architectural acoustics the diffraction effects that accompany the reflection of sound are even more important than those that accompany the transmission through openings. The architectural and decorative treatment of rooms, such as beams, columns, and ornamental plaster, results in regular or irregular breaks or discontinuities in the boundaries of the room, and consequently the interiors of most rooms are of such a nature as to introduce complicated diffraction phenomena. As an instance of how sound is reflected and diffracted from a broken surface, such as the coffers in a ceiling, a spark photograph obtained from a model of an auditorium is shown in Fig. 4.6.[2] An inspection of this photograph will show that the many wavelets originating at the ribs of the coffers cannot be explained on the basis of the simple law of reflection; they are, however, readily accounted for by diffraction. The diffraction of sound from the projecting ribs is similar to that already described for the transmission of sound through small openings: the edges of the ribs of the coffers are small in comparison with the wavelength, and

[2] This and other methods of investigating diffraction effects are described on p. 202.

in accordance with Huygens' principle these edges diffract and thus diffuse the sound.

Discontinuities in the sound-absorptive treatment of a wall (such as "patches" of absorptive material), as well as irregularities in the shape of the wall (such as "bumps"), will diffract sound waves that strike them. Thus, patches of absorptive material on

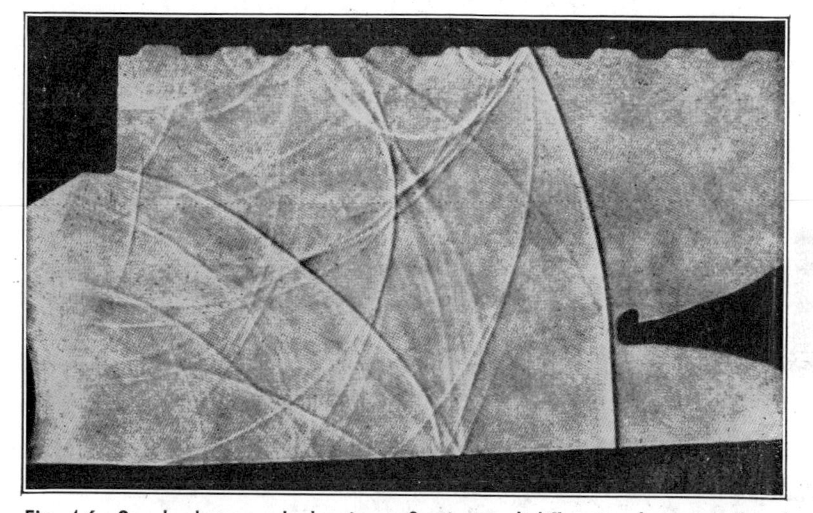

Fig. 4.6 Spark photograph showing reflection and diffraction from a coffered ceiling. Cross section, Royce Auditorium, University of California at Los Angeles.

the walls of a room diffract or scatter incident sound waves and aid in diffusing the sound throughout the room. Since a certain degree of diffusion of sound within a room is a desired condition for good acoustics, this subject will receive more detailed discussion in Chapter 9.

Diffraction from an obstacle in the path of a sound wave is frequently referred to as scattering.[3] The obstacle alters the sound field in its immediate neighborhood; the alteration (that is, the difference between the existing wave and the wave which would exist if the obstacle were absent) is called the *scattered*

[3] A comprehensive mathematical treatment of this phenomenon is given in Sec. 29 of *Vibration and Sound* by P. M. Morse, McGraw-Hill Book Co., 1948.

wave. Despite the complicated nature of this type of diffraction, certain properties of diffraction are worth remembering: (1) when an obstacle is large in relation to the wavelength of the incident sound, a sharp "shadow," similar to a light shadow, is cast; (2) when an obstacle is small in comparison to the wavelength of the incident wave, the sound is scattered in all

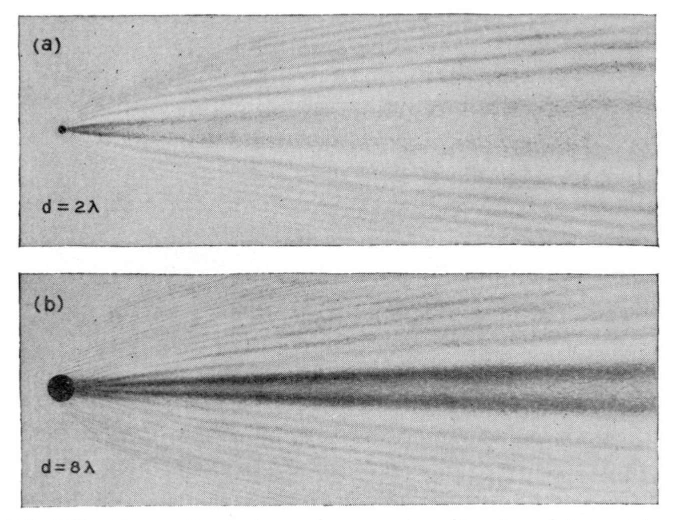

Fig. 4.7 Diffraction pattern of sound waves, traveling from left to right, striking a solid cylinder. In (*a*) the cylinder has a diameter equal to twice the wavelength of the sound waves; in (*b*) the diameter is equal to 8 wavelengths. (G. W. Willard.)

directions; (3) when the size of the obstacle is comparable to the wavelength, the sound is scattered in a complex but regular pattern, which depends on such factors as the shape, size, and absorptive properties of the obstacle and the wavelength of the sound and its direction of propagation with respect to the obstacle. A number of scattering effects are illustrated in the photographs shown in Fig. 4.7. These pictures show plane sound waves of a very high frequency traveling from left to right in water, striking a solid steel cylinder whose cross section appears as a dark circle. In (*a*) the cylinder has a diameter equal to twice the wavelength of the sound waves; in (*b*) the diameter is equal to 8 wavelengths.. The larger cylinder casts a much more pronounced shadow than does the smaller one.[4] Scattering due to obstacles within a room and to wall surface irregularities is very important in contributing to the uniformity of the sound field within the room.

Enough has been said concerning diffraction to demonstrate that the simple laws of reflection are wholly inadequate to predict the complete behavior of sound in architectural interiors. If the surfaces are large, uniform, and plane, the laws of reflection will give a fair approximation to the actual behavior. The architectural treatment of a room, however, often introduces broken surfaces which greatly affect the distribution of the reflected sound. One of the most satisfactory methods of investigating these diffraction effects consists in obtaining spark photographs of sound waves traveling through model sections of the auditorium (see Fig. 4.6). It is usually possible, however, to determine the principal effects of reflection and diffraction from a study of the plans and sections of a room, since most of the surfaces comprising the boundaries of a room are large compared to the wavelengths of the sounds they reflect. Very helpful in such a study is the formulation of the simple laws of reflection in the method of *acoustical images.*

Acoustical Images

If a point source of sound is placed on one side of an extended plane reflecting surface, it may be considered to have an "image" at an equivalent distance on the other side of the reflecting surface, along the perpendicular projection from the source to the plane, analogous to the familiar optical image. The effects of reflecting surfaces on sound waves in rooms can frequently be predicted by the use of such images. Although this method is precise only when the reflecting wall is non-absorptive (for example, if it is absorptive, a spherical wave will not be reflected as such), the method is often of considerable help in investigating the action of sound waves in rooms. A source in front of a

[4] It is interesting to note that in the central region of the shadow there is a small intense beam—a phenomenon well known in the corresponding optical case. These photographs were made by the Debye-Sears method. See G. W. Willard, *Bell Lab. Record,* **25**, 194 (1947).

very reflective wall—hard plaster, concrete, and masonry reflect 97 to 99 per cent of incident sound energy—will have an image of "strength" almost equal to that of the source, and the image will "vibrate" in phase with the source.[5] Thus, to obtain the total effect at any point due to the combined action of a source and such a reflecting surface, the effects due to the source and its image are simply added (that is, the effects due to the source itself and another source of equal strength placed at the image point). If there is another reflecting surface present, this "first-order image" will have an image which is called the second-order image of the source. An image of the second-order image is said to be a third-order image, etc.

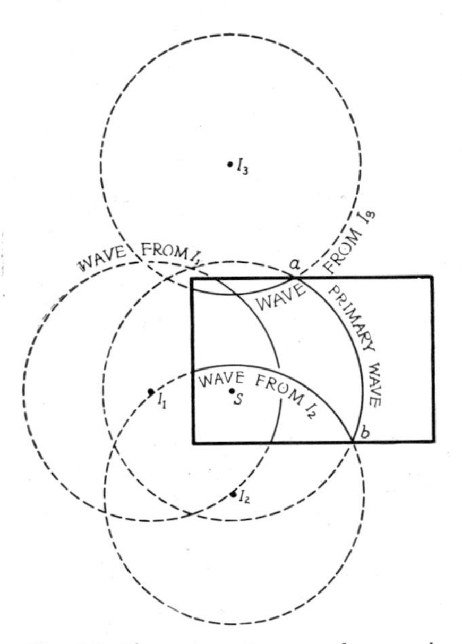

Fig. 4.8 The nearest images of a sound source S near the three walls of a rectangular room, and the reflections resulting from these three images.

In Fig. 4.8 are shown the three first-order images from three near-by walls for a point source of sound at S in a rectangular room having nonabsorptive boundaries. The primary and first-order reflected wave fronts are shown as they would be, with diffraction neglected, at the instant the primary wave has advanced to the position ab. The wave fronts are arcs of circles having their centers at S (for the primary wave), and at I_1, I_2, and I_3 for the reflected waves.

[5] Suppose that a point source is placed infinitesimally close to a perfectly reflective wall of infinite extent and that the *volume-velocity* of the source is maintained constant. Under these conditions the *radiation resistance* of the source and its image will be doubled, so that the power output of each will be twice as great as that of the source when it is in a free field. However, twice as much power is required to maintain the constant volume-velocity. Hence, at any point on the source side of the reflecting plane, the intensity will be increased by a factor of 4 and the pressure will be increased

There will be three similar first-order images from the other end wall, the floor, and the ceiling; these first-order images will then have second-, third-, and higher order images, which recede farther and farther from the walls of the room as the order of the images increases, and all of which contribute to the sound energy in the room. Usually, however, in architectural acoustics, the first-order images are the ones of prime importance, and it is rarely necessary to consider orders higher than the second in determining the most favorable reflecting surfaces for architectural interiors. If the reflected wave from I_1 is sufficiently close behind the primary wave *ab*—closer than 55 feet—it will provide a beneficial reinforcement of sound to an auditor to the right of *S;* but if it is delayed more than about 55 feet it will produce an interfering or "blurring" effect, and if it is delayed as much as 65 feet it will be heard as an echo. Similar effects result from all other reflected waves. Thus, the reflected wave that appears to come from I_2 will be beneficial for auditors seated near the side wall adjacent to I_2, and, similarly, auditors seated near the opposite side wall will receive beneficial reinforcement from I_3. In very large auditoriums the delayed reflections in the central part of the auditorium may cause considerable interference, and therefore the central seats on the main floor of an auditorium may be relatively poor seats for hearing.

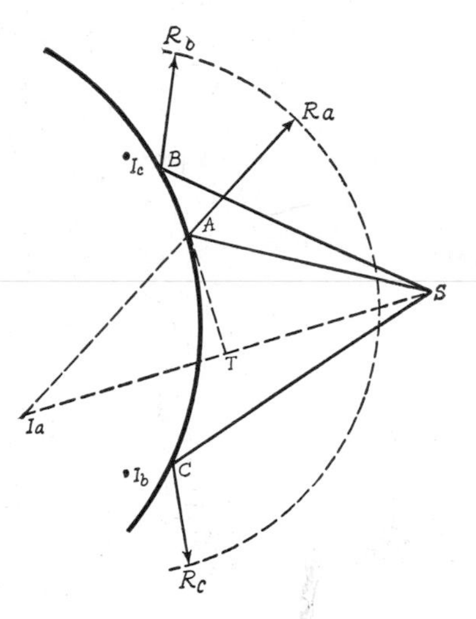

Fig. 4.9 Reflection of a spherical wave from a convex surface.

by a factor of 2. As the source recedes from the wall the radiation resistance varies as $1 + [(\sin x)/x]$ times its value in a free field. Here, x is 4π times the distance from the wall divided by the wavelength. At large distances from the reflector, $(\sin x)/x$ approaches zero so that the radiation resistance of the source is the same as it would be in a free field. Thus the effect of the wall is merely to double the intensity. This is the case we consider in the *method of images.*

When conditions permit the application of the method of images, the wave front of a reflected sound wave from a curved surface may be found by an extension of the method used for plane surfaces. A spherical wave emanating from a point source S and striking a convex surface, as shown in Fig. 4.9, exemplifies the method. Each ray will be reflected at an angle equal to its angle of incidence. The reflected ray AR_a will appear to come from an image at I_a. In order to determine this point, a tangent to the surface at A is drawn. Then I_a will lie on a line perpendicular to the tangent AT, so that $I_aT = ST$. The images I_b and I_c for the rays BR_b and CR_c can be found in a similar manner. The reflected wave front will then be the locus of points along the reflected rays such that their distances, measured along each ray back to the reflector and then to the source, are equal. For the reflected wave shown, $SAR_a = SBR_b = SCR_c$, and thus $R_bR_aR_c$ is the wave front; that is, it is the front of the reflected wave at the time $t = SAR_a/c$ seconds after it has left the source S. Figure 4.9 clearly reveals the diffusing action of a convex surface. A similar graphical construction can be applied to other curved surfaces. As already shown in Fig. 4.2, if the reflecting surface is concave toward the source S, the reflected wave front will be convergent. By the method of acoustical images the manner of convergence can be precisely displayed. From most concave surfaces the wave will converge toward some region in space, approximately a point or a line in many cases. When the surface is so shaped that the wave converges to a point S' (Fig. 4.2) that point is called the conjugate focus of S.

In Chapter 9 a number of room shapes will be considered that possess either good or poor acoustical features. The method of acoustical images, discussed in this section, is helpful in analyzing the acoustical problems of room shape, such as those considered in Chapter 9. Use of the method of images will be made in the following chapter for investigating the influence of reflective surfaces of the shell for an open-air theater.

5 · Open-Air Theaters

The first Greek theater was little more than a marked-out place in a hollow at the base of a hillside. As the name implies, the theater (derived from the Greek word *θεατρον, a place for seeing*) was initially a place for seeing rather than a place for hearing.[1] The spectators stood on the hillside and watched the action, usually dancing, which took place on the cleared space or stage. Later this marked-out space developed into a circular area called the orchestra, with banks of benches extending about two thirds of the way around it; see Fig. 5.1. A skene or platform behind the orchestra, a later addition, was originally only a place for utility, rest, and recreation of the actors; all action occurred on the circular orchestra. The skene developed into the logeion, which was gradually deepened and elevated to form the type of stage developed in the Roman theater; see Fig. 5.3.

The Roman theater was not located in a hollow but was usually on a level plain outside the city. It was erected as a single unit—the auditorium—in the form of semicircular banks of benches, connecting directly to the stage structure. The orchestra was reduced to a semicircle, or less, and became a part of the auditorium. The skene became a large platform or stage, well elevated, and enclosed by side and rear reflective walls, shown

[1] The earlier Egyptian "theaters" apparently were open-air inner courts of temples, with a level floor for standees and a slightly elevated section for the "stage."

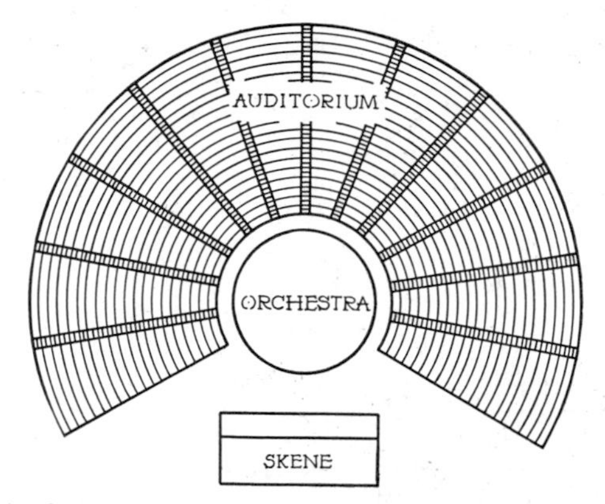

Fig. 5.1 Early form of Greek theater showing a separate *skene* erected behind the *orchestra*.

Fig. 5.2 Photograph of early Greek theater.

at *a* and *b*. The walls were permanently ornamented with all manner of relief work and were pierced with five large doorways, three in the rear and one on each side. Although these reflective surfaces contributed to the acoustical merit of the theater, they did not completely eliminate one of the major shortcomings of early Greek and Roman theaters—the lack of effective reflectors for reinforcing the voices of the performers. A practice of the ancient Greeks indicates that they realized that the power of

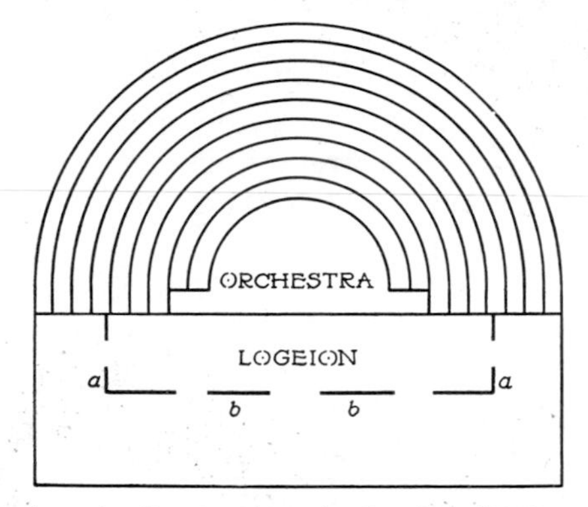

Fig. 5.3 Plan of early Roman theater showing *logeion* and semicircular *orchestra*.

the average voice is inadequate to provide distinct hearing in the more distant parts of a large open-air theater. Actors used very large masks not only to exaggerate their facial expressions so they could be seen from the most remote seats, but also to enhance the power of their voices by reason of the action of the masks as megaphones.

Nearly all theaters of the Greek and Roman types are subject to a peculiar acoustical defect which results from the shape of the auditorium and the location of the stage. In these classical theaters the action takes place on the stage or platform, which is located near the center of curvature of the regularly spaced and elevated rows of seats and benches. As a result, the speakers or actors are frequently disturbed by the converging reflections from these circular rows. Not only do these reflective surfaces return the sound and converge it to a focus near the source of the action, but also the uniform (echelon-like) spacing between the successive backs of the benches causes the selection and enhancement of a single frequency which gives rise to a tonal echo, the pitch of which is determined by the spacing between the seats. This "musical" but monotonous echo is most disagreeable to speakers. If a complex sound, such as clapping or shouting or even talking, is generated at the center of curvature of the terraced seats, the tonal echo from these rows of seats has a wavelength equal to twice the distance between successive rows of seats. Thus, when the rows are spaced 30 inches apart, the reflected sound will be preponderantly composed of sound waves having a length equal to two times 30 inches, or 5 feet. From Eq. (1.2) we find that the corresponding frequency in this case is about 226 cycles; the pitch of the echo is therefore slightly below middle C. When all the seats are occupied, this peculiar selective reflection of sound is diminished, but even under the most favorable circumstances (a full house) the echo may be annoying. In the design of open-air theaters such a condition should be avoided. Forms other than the circle should be adopted for the arrangement of benches; or the center of curvature should be well removed from the stage or scene of action. The defect is not so troublesome if the backs or risers of the successive rows of seats are inclined backward at a small angle (about 10°) from the vertical instead of rising vertically, and if the risers (as well as the concave surfaces of retaining walls) are covered with absorptive material such as climbing vines, shrubs, plants, or bushes. One concludes that, although early Greek and Roman theaters had many good features that should be preserved and perpetuated, such as the steeply sloping floor, they also had many defects. *Today, an open-air theater should be designed in the light of our present-day knowledge of acoustics.* The principles underlying such design will be described in this chapter.

Propagation of Sound in the Open Air

The acoustical problems associated with the design of open-air theaters can be more fully appreciated, and can be better solved, if one has an understanding of the factors affecting the propagation of sound in the atmosphere. In Chapter 1 we considered a few elementary properties of sound waves traveling in a free, homogeneous, undisturbed medium. In that case, we noted that the sound pressure of spherical waves originating at a point source decreases inversely with the distance from the source, so that the difference in sound-pressure level between two points whose distances from the source are D_1 and D_2 is given by

$$\text{Difference} = 20 \log \frac{D_1}{D_2} \quad \text{db}$$

Thus, there is a drop of 6 db with each doubling of the distance from the source, or a drop of 20 db for each tenfold increase of distance.

The wind and temperature variations in the atmosphere may greatly modify the distribution of energy about a sound source by bending the sound rays from their usual rectilinear paths. These effects on the propagation of sound in the atmosphere, as well as the absorptive properties of the air itself and the influence of sound-absorptive surfaces in the sound field, are discussed in the following paragraphs of this section.

EFFECT OF WIND UPON THE PROPAGATION OF SOUND. In Chapter 1 it was shown that the speed of sound in still air, at a given temperature, is constant, and equal to about 1130 feet per second.

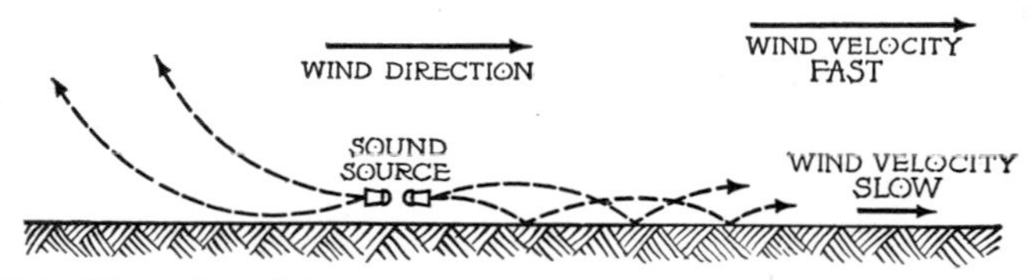

Fig. 5.4 Effect of wind direction on propagation of sound rays. (The vertical scale is exaggerated.)

If the air is in motion, or if the temperature changes, the sound speed will be altered. The speed of sound in the direction of the wind is equal to the speed of the wind plus the speed of sound in still air. In all cases of the propagation of sound in moving air, the vector velocity of the sound, with respect to an object at rest on the earth, is equal to the vector sum of the velocity of the sound in still air and the velocity of the wind.

Suppose that the wind is blowing past a source of sound as shown in Fig. 5.4. Then, since the speed of the wind is generally slowest at the surface of the earth and increases at higher elevations above the surface, the normal to the wave front of the sound that travels with the wind is bent more and more toward the earth, whereas the normal to the wave front of the sound that travels against the wind is bent more and more away from the earth.[2] Consequently, the upper portions of the sound waves that travel with the wind are deflected downward and they contribute to the flow of sound energy near the earth's surface, thus intensifying the sound near the earth and facilitating the propagation of sound to great distances in the direction of the wind. On the other hand, the upper portions of waves that travel against the wind are relatively retarded so that these waves are deflected upward from the level plane, thus making impossible the propagation of sound to great distances in the direction against the wind. The wind has a marked effect upon the distribution of sound; the pressure of the sound wave in the direction of the wind, at a given distance over a level plane, amounts to several times the pressure at the same distance but in the direction against the wind. If the wind has approximately a constant direction in a region where an open-air theater is to be constructed, preference should be given, other things being equal, to that site for which the wind generally blows from the stage to the audience. This preference increases in importance with increase in size of the open-air theater.

Effect of Temperature Differences in the Air upon the Propagation of Sound. It has been seen that in the presence of a wind the best condition for sound propagation is one where the upper portion of the waves travel faster than the lower portion, thus bending the wave front downward and augmenting the flow of energy along the earth's surface. A similar type of bending may take place which is due to a vertical temperature gradient of the air. Since the speed of sound in still air is given by $\sqrt{1.40P_s/\rho}$ (Chapter 1), the speed is inversely proportional to the square root of the density of the air and therefore directly proportional to the square root of the absolute temperature of the air. Thus the speed of the upper portion of sound waves may

[2] See *Acoustics* by G. W. Stewart and R. W. Lindsay, D. Van Nostrand Company, 1930.

be increased or decreased with respect to the lower portion as a result of temperature differences in the atmosphere. Suppose that the temperature of the air decreases with the altitude above the earth's surface, as it most commonly does. Then the upper portions of sound waves originating at a sound source will be retarded in relation to the lower portions, and consequently the

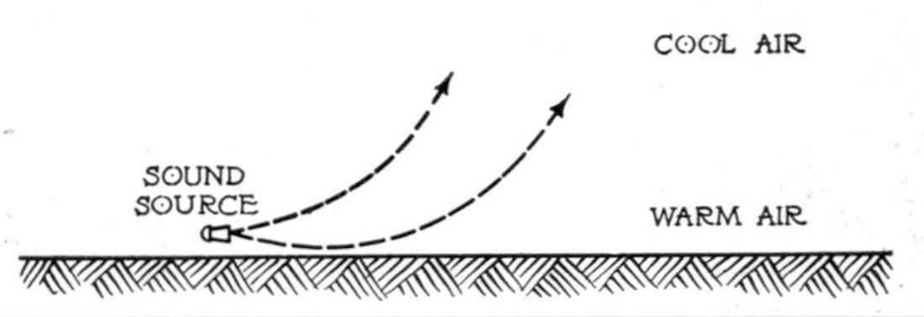

Fig. 5.5 Effect of temperature gradient on progagation of sound rays—decreasing temperature with increasing altitude. (The vertical scale is exaggerated.)

wave front will be bent upward, as shown in Fig. 5.5. On the other hand, suppose that the air temperature increases with the altitude, as it frequently does over land surfaces just after sunset or whenever meteorological conditions give rise to an "inverted temperature gradient." Then the upper waves travel faster than the lower ones do, and consequently the wave front will be bent downward, as shown in Fig. 5.6. Under cer-

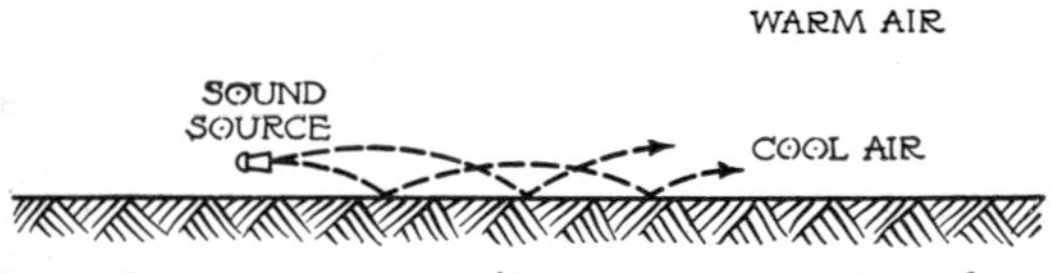

Fig. 5.6 Effect of temperature gradient on propagation of sound rays—increasing temperature with increasing altitude. (The vertical scale is exaggerated.)

tain conditions of increasing temperature with altitude, an appreciable portion of the sound originating at a point source will be totally reflected by the upper and warmer layers of air. When these circumstances prevail there will be repeated reflections between the earth and the upper layers of air. Therefore the pressure of the sound waves decreases only as the inverse square root of the distance instead of as the inverse of the distance, the usual decrease for a spherical wave in free space. These conditions often are approximated in the air over a frozen lake when it is possible on a quiet day to hear and understand ordinary conversation at a distance of a half mile or even more.

If an open-air theater is located where the temperature decreases with a rise in altitude, as such theaters usually are, the slope of the seating area should be slightly steeper than that required in a homogeneous air or in a region where the temperature of the air increases with altitude. If the slope of the seating area rises more rapidly than do the advancing wave fronts, all auditors in the theater will be well elevated into the main flow of sound energy and will receive a relatively large amount of the sound energy coming from the stage. *It is desirable, at least from the standpoint of possible atmospheric effects, to grade the seating area of an open-air theater so that the slope, except for the seating area near the stage, is at least 8 degrees above the horizontal.*

ABSORPTION OF SOUND IN AIR. It is well known that every type of wave motion, including sound, loses part of its energy as it is propagated through a ponderable medium such as air. The attenuation of sound is due to viscosity, heat conduction, radiation, scattering, and molecular absorption.[3,4] The attenuation of sound waves having pressures ordinarily associated with speech and music depends principally on the frequency of the sound

[3] V. O. Knudsen, *J. Acoust. Soc. Am.*, **5**, 112 (1933); also **18**, 90 (1946).

[4] Owing to this dissipation of energy, the pressure and intensity of a *plane* wave in a homogeneous medium diminish exponentially with distance. Thus in a distance x, in the direction of propagation, the pressure amplitude is reduced by a factor $e^{-\alpha x}$ and the intensity is reduced by a factor e^{-mx} where $m = 2\alpha$. Alternate expressions for the pressure-reduction factor $e^{-\alpha x}$, frequently found in the literature, are $e^{-mx/2}$ or $10^{-0.05Ax}$, where α, A, and m have the following relative values: $A = 4.343m = 8.686\alpha$. The unit of attenuation coefficient α is the *neper per unit distance,* and the unit of the equivalent coefficient A is the decibel per unit distance, the distance unit in either case being the unit chosen for x. Thus, if $m = 0.02$ per foot (which is about the maximum value of m in air for a sound wave of 10,000 cycles), then $A = 4.343 \times 0.02$, that is, about 0.087 decibel per foot, or 8.7 decibels per 100 feet.

wave, the relative humidity, and the temperature. The curves of Fig. 5.7 are data on the attenuation (in decibels per hundred feet) of a *plane* sound wave in the atmosphere, as a function of the relative humidity at a temperature of 68° F for various frequencies. The attenuation increases with the temperature; it becomes as much as 16 db per 100 feet at 10,000 cycles in hot desert air and almost vanishes in cold winter air. For air of the same *absolute* humidity, and for temperatures between 60° F and 80° F, the temperatures usually encountered in open-air theaters, the attenuation increases about 8 per cent for each 5° F rise in temperature. In a *spherical* wave, the decrease of pressure with the increase of distance from the source is due to a combination of *attenuation* and *divergence* of the sound waves from the source. If r is the distance from the source, the pressure amplitude is proportional to $(1/r)\, e^{-ar}$. The loss due to divergence alone is 6 db for each doubling of the distance.

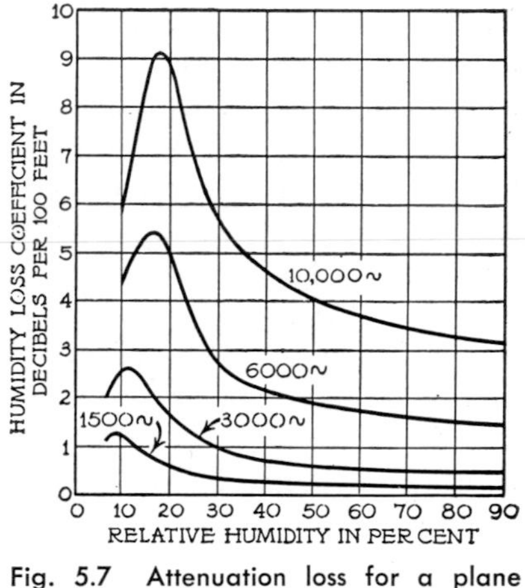

Fig. 5.7 Attenuation loss for a plane wave propagated in the atmosphere. The loss is given in decibels per hundred feet as a function of the relative humidity in per cent.

As an example of the use of the curves in Fig. 5.7, let us calculate the difference in sound level due to attenuation in the atmosphere between the front and back rows of an open-air theater that are separated by a distance of 200 feet. Suppose that the temperature of the air is 68° F and the relative humidity is 20 per cent. Under these conditions, the loss for a tone of 6000 cycles is 5 db per 100 feet of transmission; or, for a distance of 200 feet the loss due to attenuation alone is 10 db. In a similar manner, calculations may be made for other frequencies, and for other humidities, temperatures, and distances. In order to find the total difference in sound level between the front and back rows of an open-air theater this loss must be added to the losses due to the divergence of the sound waves, the bending of the waves, and the absorption by neighboring surfaces. These cumulative losses (especially those from divergence) often are quite large, and they place limitations on the size of open-air theaters if certain specified standards of acoustics are to be maintained.

Closely associated with the absorption and scattering of sound in the atmosphere is the phenomenon of "fluctuations." The slow but sometimes large fluctuations in the loudness of the sound coming from a distant airplane is a familiar observation. A study of the micrometeorological properties of the atmosphere reveals great turbulence, especially near the surface of ground which has been heated by the sun. Temperature changes of 5° F or more, occurring several times a second, are not uncommon; the wind is ever-changing; convection currents keep the air in a state of agitation. The motion in the air of smoke particles or of small bits of paper reveals the turbulent nature of the atmosphere. Sounds of long wavelength are not greatly influenced by the micrometeorological properties of the atmosphere, but sounds of short wavelength are subject to violent fluctuations. Consideration should be given to this phenomenon of fluctuations when the site for an open-air theater is selected. In general, sites which have relatively uniform terrain (at least in respect to heat absorption and radiation) and which are not subject to excessive heating by the sun should be given preference, all other things being equal.

Propagation of Sound over Absorptive Surfaces. It is well known that the application of sound-absorptive material to the ceiling or walls of a long corridor, as in a hospital, is an effective means of impeding the transmission of noise along that corridor. Similarly, the application of absorptive material to the ceiling or walls of a long auditorium, or the presence of an audience, or both, impede the transmission of sound to the rear of the auditorium. A similar effect can be observed by listening to sound that grazes over a field covered with tall grass, or freshly fallen snow, both of which are very absorptive. Sound that has traveled an appreciable distance over such an absorptive surface is much more diminished near the absorptive surface than it is a few feet above the surface. In a large open-air theater with a level "floor" and a sound source located only 2 or 3 feet above the floor, the sound pressure at 50 feet or more from the source is much less at the level of an audience than it is 3 or 4 feet above the audience; the absorptive audience "extracts" sound energy from the sound wave.

This effect, which is important acoustically for enclosed as well as for open-air theaters, has been studied quantitatively by Békésy [5] and by Rudnick [6] who have measured the pressure of a sound wave propagated over a highly absorptive surface as a function of frequency, distance, and the height of the source and microphone above the surface. Békésy made his measurements over an absorptive sheet of wood fiber 1½ inches thick. His data show, for example, that at a distance of 39 feet from an 800-cycle source, with both the source and microphone 1 foot above the absorptive sheet, the sound level is about 21 db less than it would be if the absorptive sheet were removed from the sound field. When the microphone was elevated to 4 feet above the sheet, the loss due to the presence of the absorptive sheet was reduced to about 7 db.

It is clear that the level of the sound reaching the audience will be greatly influenced by the location of the sound source and the slope of seating area. In order to minimize the sound-transmission losses of the type here considered, the source should be well elevated above the audience and the seating area should rise steeply toward the rear.

Effects of Clouds and Fogs on the Propagation of Sound. When a sound wave strikes a cloud or a fog bank, most of the sound energy usually is refracted (with a very small change of direction) into the cloud or fog, and only a small portion of the sound energy is reflected. If, however, the sound wave strikes the cloud or fog bank at nearly grazing incidence, the sound wave may be totally reflected, in which case the direction of propagation of the sound wave may be appreciably altered. It is not often, however, that such reflections become a factor in the acoustics of open-air theaters.

[5] Georg v. Békésy, *Z. tech. Physik,* **14,** [1], 6 (1933).

[6] I. Rudnick, *J. Acoust. Soc. Am.,* **19,** 348 (1947). In this paper is given a theoretical treatment of the propagation of sound waves over absorptive surfaces.

Speech-Articulation Tests in the Open Air

Speech-articulation tests are designed to measure the recognizability of speech sounds under a given set of conditions. Here they are employed as a measure of how well speech can be heard in the open air in order to determine the limiting dimensions of open-air theaters.

Percentage syllable articulation, usually referred to as "percentage articulation," signifies the percentage of meaningless syllabic sounds which are heard correctly. Thus, if a speaker calls out 1000 meaningless speech sounds and an observer hears 850 of them correctly, the speech articulation is said to be 85 per cent. If the syllable articulation is 85 per cent, the conditions for the hearing of speech are very good; if it is 75 per cent, the conditions are satisfactory; and, if it is 65 per cent, the conditions are barely acceptable.

Tests have been conducted on a level site to determine the percentage articulation of speech in the open air as a function of distance from the speaker. The listeners were stationed at distances in front of the speaker, to the right and left of the speaker, and behind the speaker. All the speakers were instructed to speak as though they were addressing an audience. The results of these tests are shown in Fig. 5.8. The short arrow shows the direction the speaker was facing. The curves indicate the distances from a speaker at which the speech articulation was 75 per cent. Under quiet conditions these distances were found to be approximately 140 feet in front of the speaker, 100 feet to the side of the speaker, and 55 feet behind him. In a wind which varied from 20 to 25 miles per hour, the corresponding distances were 85 feet, 52 feet, and 26 feet. Since the action of the wind decreased the intelligibility of speech in all directions around the speaker, it is apparent that the wind interfered with hearing primarily because it introduced a masking noise in the ears of the listener. (For the distances used in these tests, the effects of refraction apparently are small compared to these masking effects; otherwise the curves would be lopsided.) Other speech-articulation tests have been made with the wind velocity as low as 5 to 10 miles an hour. Even such gentle winds were found to interfere appreciably with the hearing of speech. These tests indicate that an open-air theater should be located in a site which is free from winds. If such a site cannot be found, the theater should be so oriented that the prevailing wind will blow from the stage toward the audience. If the prevailing wind exceeds about 10 miles an hour, and if the open-air theater is to

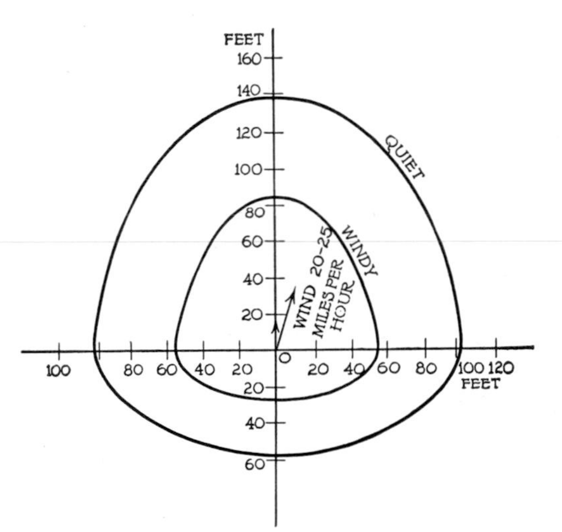

Fig. 5.8 Curves showing how the hearing of speech depends on the distance and direction from the speaker. The speaker was located at 0 and facing in the direction of the short arrow. The outer contour represents the limiting distances at which normal speech can be heard satisfactorily on a quiet, level plain. The inner contour represents similar data taken in the presence of wind blowing in the direction indicated.

be used without the benefit of a sound-amplification system, the size of the theater in such a site should be reduced by an amount which is roughly indicated by a comparison of the contours in Fig. 5.8.

The noise level prevailing during the above tests was unusually low. A higher noise level would reduce these distances considerably. Some articulation tests conducted in Hollywood Bowl show the effect of noise on the hearing of speech. These tests were made, for the most part, during the early part of the night, at a time when disturbances from city traffic and other sources were comparable with the noise present during programs in the

Bowl. In these tests the speakers were aided by a reflective shell. The percentage of articulation obtained at different positions in the Bowl is indicated in Fig. 5.9. It will be noted that the syllable articulation decreases from about 89 per cent near the front central part (where the mean background noise level is 40

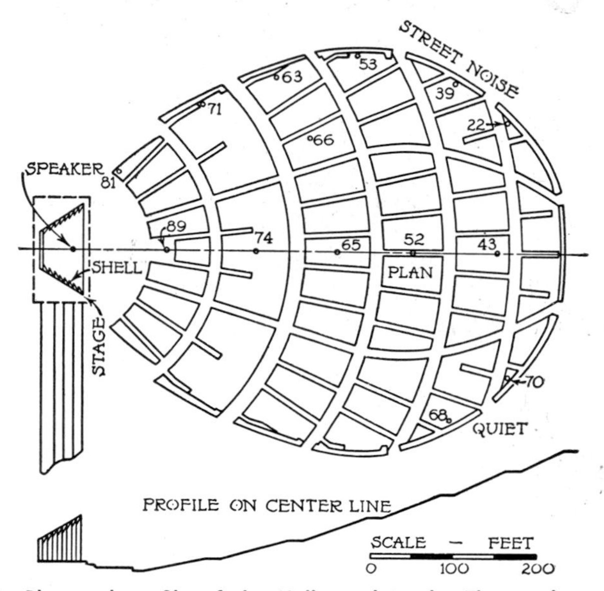

Fig. 5.9 Plan and profile of the Hollywood Bowl. The numbers give the percentage syllable articulation in various locations with a speaker located in the orchestra shell.

to 43 db) to a value as low as 22 per cent at the upper corner of the Bowl (where the background noise is 50 to 58 db, owing to the proximity of two busy boulevards and the inadequate embankment surrounding this corner of the Bowl). Speech-articulation tests conducted at about 3:00 A.M., when there were practically no disturbances from noise, gave articulations as high as 65 to 70 per cent in the most remote seats.

The Design of Open-Air Theaters

The design of an open-air theater should include (*a*) careful selection and grading of the theater site, (*b*) the design of an appropriate orchestra shell (to be discussed in the next section), and (*c*) provision for a sound-amplification system (see Chapter 14), especially where there will be 600 or more auditors. It is recommended that theaters used principally for dramatic productions be provided with a stereophonic sound system in order to reproduce the sound with its proper spatial distribution and hence with greater naturalness than would be attained with an ordinary single-channel sound-amplification system.

The selection of the site for an open-air theater should be based upon a thorough survey of the topographical, meteorological, and acoustical properties of all available locations for the theater. *Quietness is the most important of all acoustical considerations in the selection of the site.* It should be well removed from all traffic arteries, both on the ground and in the air; it should be shielded on all sides by the natural slopes of surrounding hills, by artificial embankments, and by a dense growth of trees; and it should be free from winds which have velocities of more than 10 miles per hour. Noise surveys should be made on all proposed sites to determine not only the average noise level but also the standard deviation, maximum and minimum noise levels, and the frequency and times of occurrence of the maxima. In order that a site be a satisfactory one, the average noise level should not exceed 40 db, the standard deviation should be small, and the site should be free from occasional loud noises.[7] If these occasional disturbances have sound levels as high as 60 db, as they frequently do, and if they occur two or three times an hour, the site is undesirable. A cove on a gently sloping hillside, well removed and shielded from automobile, rail, and airplane traffic, is usually a good site. The slope of the seating area should not be less than about 12 degrees in order to provide all auditors with an abundant flow of sound.

The distribution of sound due to a source of sound on the stage can be calculated approximately for the seating area of an open-air theater by the method of acoustical images. Suppose

[7] Unless otherwise stated, the noise levels given in this chapter refer to measurements made with a sound-level meter using the 40-db frequency-weighting network (see Chapter 1).

that the stage is enclosed by a rear vertical wall, two diverging side walls, and a sloping ceiling (see Fig. 5.10). Then assume that the sound energy spreads out from the source and from the images of the source, such as I_1, I_2, and I_3, so that the pressure falls off inversely as the distance, and assume that the intensity at any point is the sum of the intensities radiated by the source and its effective images. In general, the greatest contributions

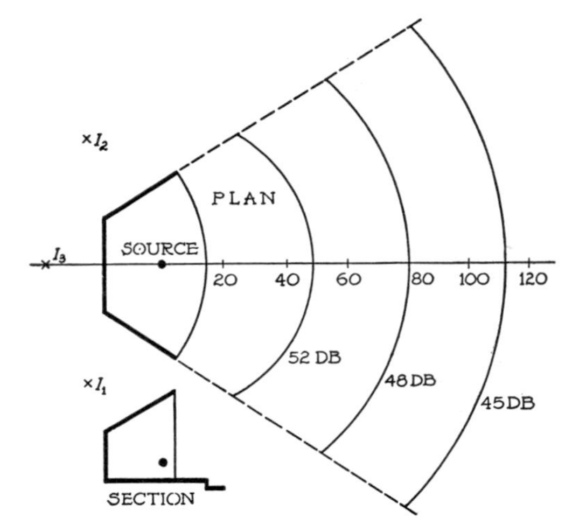

Fig. 5.10. Distribution of sound level due to a 100-microwatt source in front of a stage enclosed by a rear wall, diverging side walls, and a sloping ceiling.

will be due to the three images from the rear and the side walls, the image from the floor, and the two images from the ceiling (one the ceiling image of the source, and the other the ceiling image of the floor image of the source). For example, Fig. 5.10 gives the calculated values of sound level due to a source of sound having a power of 100 microwatts and located at the position indicated. This output corresponds, approximately, to the "long-interval" average speech power of a person speaking in a loud voice. The use of scale models to obtain the distribution of reflected sound will be discussed in Chapter 9.

A seating capacity of about 600 should be regarded as the upper limit of size when the theater, not equipped with a sound-amplification system, is to be used principally for spoken drama, if all auditors are to hear without undue strain and if the actors are to speak without undue effort. This corresponds to a maximum depth of about 75 feet and a maximum width of about 85 feet. If open-air theaters are designed to accommodate a larger audience than this, it is probable that auditors in an area beyond these dimensions will experience difficulty in hearing the performance unless the speakers raise their voices and speak with deliberate clarity. On the other hand, open-air theaters to be used principally for music may be somewhat larger than those planned primarily for drama and unamplified speech, partly because of the greater acoustical power of musical instruments. In addition, when listening to speech, for good intelligibility one must be able to hear even the weakest portions of the sounds (the unvoiced consonants). This is not a prime requirement for the enjoyment of a musical program. Of course, if a high-quality sound-amplification system is employed, the theater is not subject to the above limitations in size.

A disturbance that often masks the hearing of speech or music is the noise of footfalls or the scuffing of feet. This nuisance can be alleviated by the proper treatment of the aisles of the theater. In mild climates a good grade of fiberboard may be found suitable. It wears for several years, and the cost of replacement is small. Furthermore, it provides a comfortable and safe surface upon which to walk. For installations where fiberboard would be impractical, other compliant surfacing, such as asphalt tile, should be utilized.

Figures 5.11 and 5.12 show a design of an open-air theater based on a consideration of the principles presented in this chapter. The shape of the auditorium was determined from the results of speech-articulation tests in the open. The theater is designed to seat approximately 2000, which is above the practical limit in which speech can be heard satisfactorily without the use of a sound-amplification system. It is assumed that this theater will be used to a large extent for dramatic productions, and therefore a stereophonic sound system has been provided; note the two loudspeakers, one on either side of the orchestra shell in Fig. 5.12.

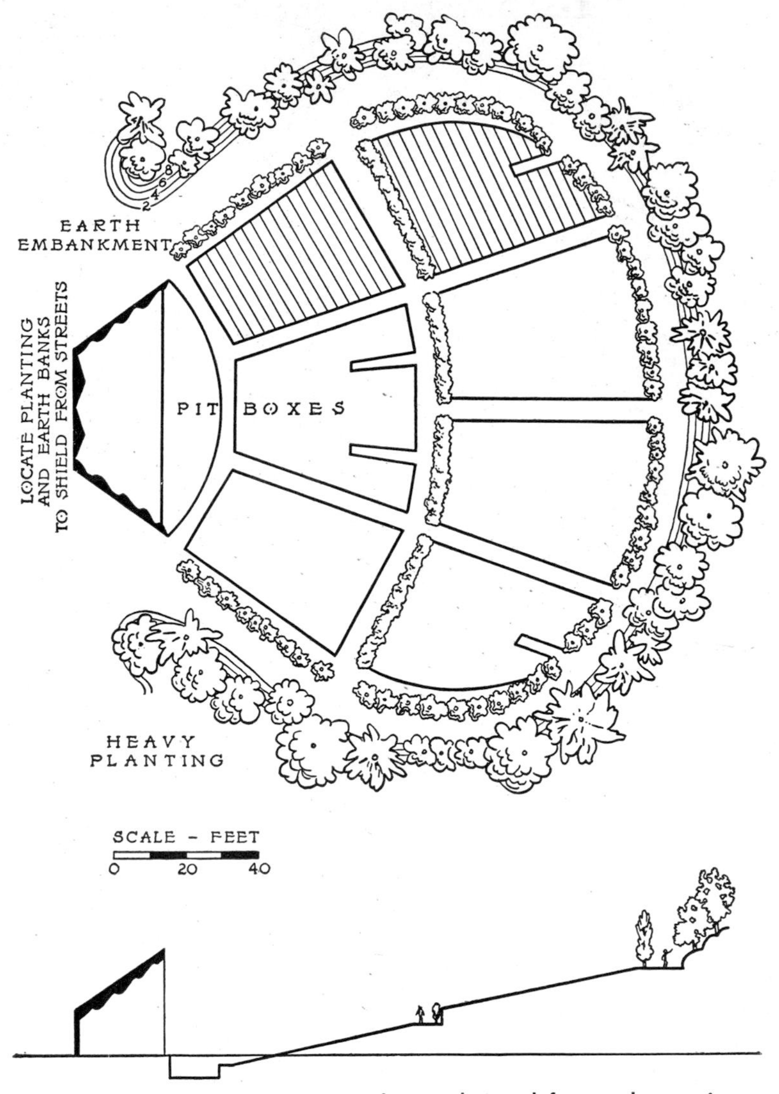

Fig. 5.11 Plan for an open-air theater designed for good acoustics.

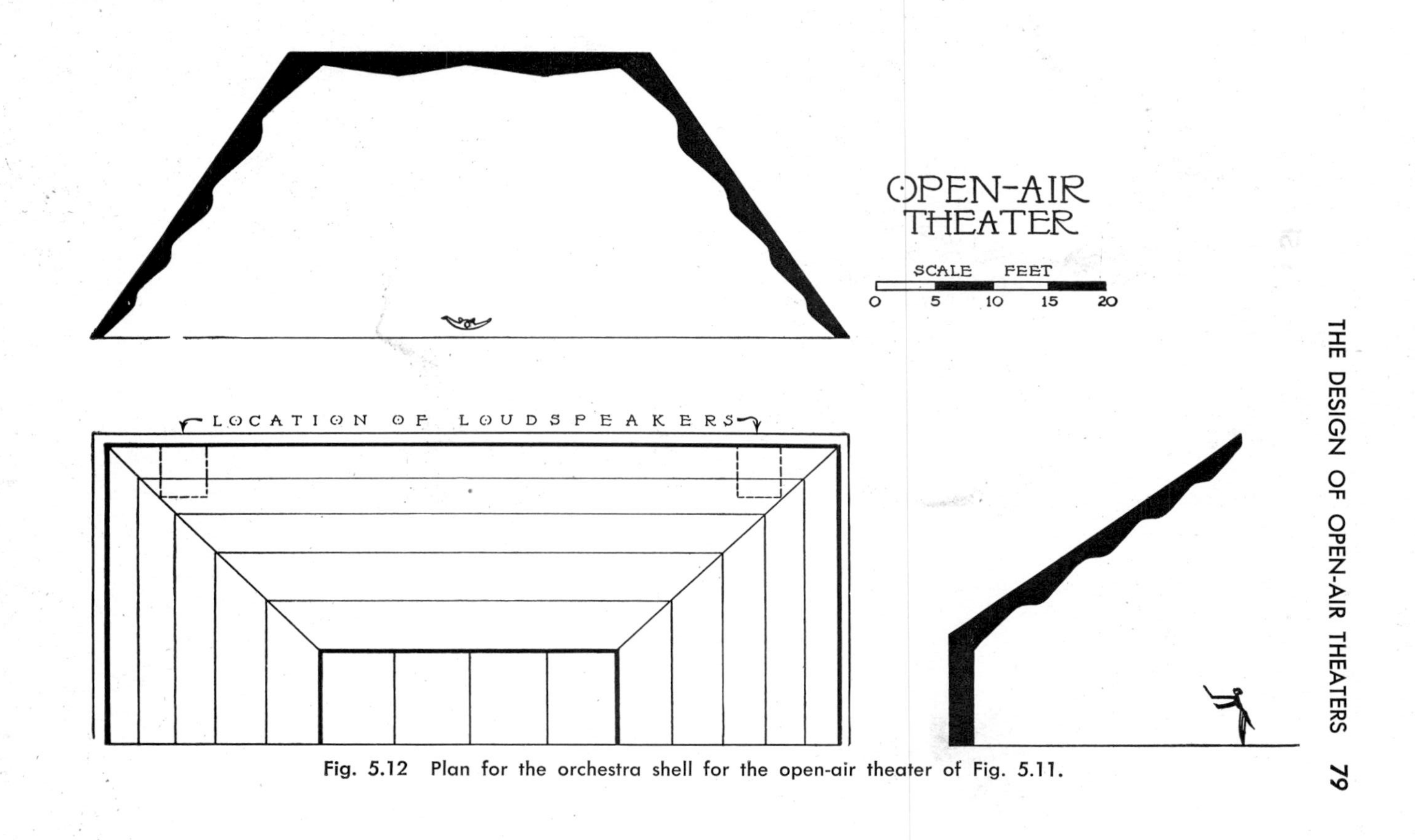

Fig. 5.12 Plan for the orchestra shell for the open-air theater of Fig. 5.11.

Orchestra Shells

There are two principal reasons for providing an orchestra shell in an open-air theater. First, the reflective power of a well-designed shell increases the average sound level throughout the auditorium and provides a more uniform distribution of sound pressure over the entire seating area. It is possible to reflect sound downward on the audience in such a manner as to compensate largely for the losses which normally occur if no shell is present. Second, the shell enables the performers on the stage to hear each other more clearly. Musicians find this quality a very desirable one; in fact, without it, they cannot play in unison or in proper loudness balance (without an exceptionally good conductor). Of course, the shell design should be coordinated with the other needs, such as space requirements, lighting, and sound-amplification equipment.

The simplest type of orchestra "shell" is a highly reflective vertical wall directly behind the stage. Such a wall would approximately double the intensity of the sound projected to the audience. There is much to be said in favor of this type of shell (especially for very small theaters where little reinforcement is needed), principally because it is almost free from directional effects and is entirely free from focusing or converging action. Consequently, a musical instrument that acts as a point source will be almost equally reinforced in all directions in front of the wall and a nearly uniform flow of sound energy will be directed to all parts of the audience. The stage floor should be well elevated, and the auditorium floor should be sloped up toward the rear as steeply as other requirements will allow. The addition of vertical side walls to the shell will give some additional reinforcement but also will make the shell more directive. If side walls are employed, they should be non-parallel since parallel surfaces (such as those used in the classical Greek theaters) permit sound waves to reflect back and forth between them. These multiple reflections are undesirable for they give rise to a raucous "flutter echo" when the parallel walls are smooth; they are a source of annoyance to the performers; and they impose serious limitations on the placement of microphones. By splaying the side walls outward, these difficulties are avoided to a great extent. The use of splays or convex surfaces incorporated in the shell helps to give a non-directional distribution of diffuse sound throughout the seating area; they are especially desirable when sound amplification is to be used.

An overhead inclined reflector is a valuable addition to a vertical wall behind the stage. An example of a shell of this type, shown in Fig. 5.13, is especially attractive because of its simplicity and its freedom from focusing effects. The angle made by the overhead surface with the horizontal is important. It should impart to the reflected sound a direction which is approximately parallel to the slope of the seating area. Hence, if the angle of elevation of the profile of the center line of the auditorium is $\theta°$, the angle of the reflecting ceiling, with respect to the horizontal, should be $45° + (\theta°/2)$. For example, if the seating area of a bowl is inclined at an angle of about 12° above the horizontal, the overhead reflective surface should be pitched to an angle of 51° above the horizontal. If the seating area is steeply inclined or if the stage is very large, portions of the shell might be at a relatively great distance from some of the performers. Then delayed reflections resulting in echoes are possible. This difficulty can be met, by adding to the ceiling, interior reflecting bands which make the required angle of $45° + (\theta°/2)$ with the horizontal and then pitching the overhead reflector to a smaller angle. A photograph of a shell of this type is shown in Fig. 5.14. Application of the principles of Chapter 4 will show that sound having wavelengths small compared to the widths of the reflective bands will be reflected at an angle determined by the pitch of the bands; on the other hand, sound having wavelengths large compared to the widths of the bands will be relatively unaffected by them, the angle of reflection being determined principally by the pitch of the entire ceiling. Hence, in order for the reflective bands to be effective over the important frequency range of speech, they should be at least 4 feet in width. Furthermore, in order to avoid selective reflection (with respect to frequency), it is advisable to use bands which are non-uniform in width.

In very large open-air theaters, there occasionally is need for more reinforcement of sound than can be obtained with plane reflective surfaces. In these cases, if no sound-amplification sys-

Fig. 5.13 Band shell. (Wm. A. Ganster, Architect.)

Fig. 5.14 Band shell. (Alden B. Dow, Architect.)

tem is to be used, it may be feasible to design shells that are somewhat more directive then those we have considered thus far.

In general, *concave surfaces such as sections of cylinders, ellipsoids, spheres, and paraboloids should be avoided.* Curved shapes of this kind usually are expensive to build; besides, and even more important, they nearly always are unsatisfactory acoustically, since they overemphasize those instruments or voices which are located near the focal points or lines of the curved surfaces and they generally provide a non-uniform distribution of reflected sound throughout the seating area.

6 · Sound-Absorptive Materials

The rate at which sound is absorbed in a room is a prime factor in reducing noise and controlling reverberation. All materials used in the construction of buildings absorb some sound, but proper acoustical control often requires the use of materials that have been especially designed to function primarily as sound absorbers. Such materials are popularly known as "acoustical" materials. Within the past few decades there has been a tremendous increase in the use of these products, especially for the reduction of noise in office buildings, hospitals, and restaurants. Many people tolerate noise, but most people do not like it. This probably is the reason why many business establishments have found that the cost of acoustical treatment usually is more than offset by the profits resulting from the increase in patronage after the installation of the absorptive material. Similarly, the value of such treatment has been amply demonstrated by the resultant improvement of the acoustics in innumerable schools, theaters, and churches, and in places where people gather for work or play or for listening to speech or music. When properly planned, the absorptive treatment of rooms contributes to good acoustics, making it possible for speech to be heard clearly and for music to be enjoyed to the fullest extent.

Many different acoustical materials are now available for use in buildings. In addition, many new and diverse absorptive treatments of rooms can be devised by architects and builders who have imagination and who are familiar with the basic processes by which sound is absorbed. There is no one "universal" material that is best suited for all installations. Each job requires separate and painstaking consideration. It is always necessary to choose materials with the proper acoustical characteristics, but this is not enough. All other physical and decorative properties of the materials must be given proper attention. Having determined which materials have the required absorptive properties, the architect must raise about each material such questions as the following: Is it combustible or fire-resistant? How much light will it reflect? What about its structural strength, absorption of water, and attraction for vermin? How foolproof is it? Can its application be entrusted to the average journeyman? What is its appearance, and what are its decorative possibilities? How much does the material cost? Will it be expensive to install and maintain?

In this chapter we shall consider the acoustical properties of various types of materials and combinations of materials, and point out the advantages of each type. Other forms of absorptive treatment are described in the following chapter.

How Sound Is Absorbed

Sound is absorbed by a mechanism which converts the sound into other forms of energy and ultimately into heat. Most manufactured materials depend largely on their porosity for their absorptivity. Many materials, such as mineral wools, pads, and blankets, have a multitude of small deeply penetrating intercommunicating pores. The sound waves can readily propagate themselves into these interstices, where a portion of the sound energy is converted into heat by frictional and viscous resistance within the pores and by vibration of the small fibers of the material. If the material is sufficiently porous, and of appropriate thickness, as much as 95 per cent of the energy of an incident sound wave may be absorbed in this manner.

When sound waves strike a panel, the alternating pressure of these waves against the panel may force it into vibration. The resulting flexural vibrations use up a certain amount of the incident sound energy by converting it into heat. If the panel is massive and stiff, the amount of acoustical energy converted into mechanical vibrations of the panel is exceedingly small; on the other hand, if the panel is light and flexible, the amount of energy absorbed may be very large, especially at low frequencies. For example, Fig. 6.1 shows that fiberboards such as Masonite, and acoustical tile such as Acousti-Celotex are much more absorptive at frequencies of 128 and 256 cycles when they are nailed to wood strips—and can vibrate as panels—than when they are cemented or otherwise fastened against a rigid surface. Plaster on lath over studs provides much more absorption at low frequencies than does the same type of plaster applied directly to solid masonry walls. In general, the rate at which a flexible panel absorbs acoustical energy is proportional to the product of the amplitude and frequency of vibration, to its internal damping coefficient, and to the frictional losses at the edges of its mounting.

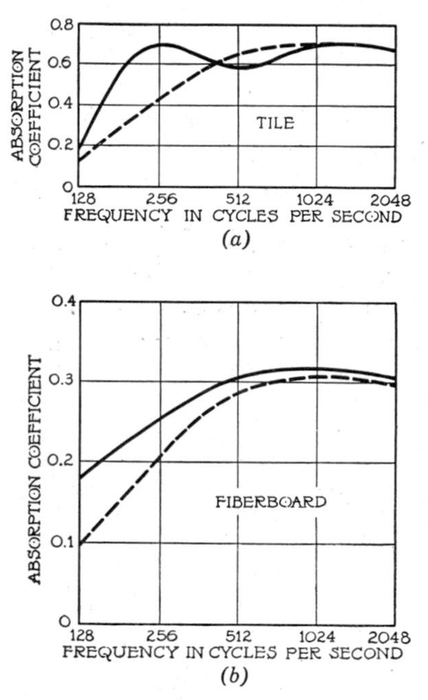

Fig. 6.1 Absorption coefficient *vs.* frequency of (*a*) an acoustical tile and (*b*) a fiberboard panel. The dashed line is the coefficient with the material mounted directly against concrete; the solid line represents data for the same material on 1-inch furring strips.

Absorption by porous materials normally is large at high frequencies and small at low frequencies. Absorption by panel vibration is small at high frequencies but may be large at low frequencies. Both of these types of absorption are important in the control of sound in rooms. By using them in the proper proportions, it is possible to control the absorption of sound throughout the audible range of frequencies. This becomes a necessity in sound-recording and radio studios and is often desirable elsewhere.

Rating of Acoustical Absorptivity of Materials

The efficiency of a material in absorbing acoustical energy at a specified frequency is given by its *absorption coefficient* at that frequency. This quantity is the fractional part of the energy of an incident sound wave that is absorbed (not reflected) by the material. Thus if sound waves strike a material, and if 55 per cent of the incident acoustical energy is absorbed and 45 per cent is reflected, the absorption coefficient of the material is 0.55. Each square foot of this material is the absorptive equivalent of 0.55 square foot of a perfectly absorptive surface. A *sabin* is a measure of the sound absorption of a surface; it is the equivalent of 1 square foot of perfectly absorptive surface. A surface of area S having an absorption coefficient α has a total absorption of $S\alpha$ sabins. A sabin is often referred to as a *square-foot-unit* or a *unit* of absorption. The coefficient of absorption varies with the angle at which the wave strikes the material. In this book, the absorption coefficient at a given frequency will be understood to be the value of the coefficient (at that frequency) averaged over all angles of incidence. The coefficient will be denoted by the Greek letter alpha, α. Since α varies with frequency, it has been common practice to list values at specific frequencies in tables of absorption coefficients, usually at 128, 256, 512, 1024, 2048, and 4096 cycles. There is a trend among many acoustical investigators toward the use of 125, 250, 500, 1000, 2000, and 4000 cycles. For all practical purposes, these two series of frequencies can be regarded as identical. Therefore the coefficient obtained for 500 cycles, for example, can be regarded also as the coefficient for 512 cycles. The absorption coefficients depend not only on the nature of the material but also on other factors such as its thickness, on the way in which it is mounted, and on the depth of the air space behind it.

Acoustical impedance is a more fundamental quantity describing the acoustical properties of an absorptive material than is the absorption coefficient; it is defined as the complex ratio of the sound pressure to the corresponding particle velocity at the surface of the material. Recent work has indicated that the acoustical properties of a room can be predicted somewhat more accurately in terms of the impedance than the absorption coefficients of the boundaries of the room. However, in practical problems in room acoustics, the mathematical labor involved in a detailed analysis is so great as to make such use of acoustical impedance impractical for the architect or engineer.

NOISE-REDUCTION COEFFICIENT. This coefficient of a material is the average, to the nearest multiple of 0.05, of the absorption coefficients at 256, 512, 1024, and 2048 cycles. This number is sometimes used in comparing materials for noise-reduction applications in offices, hospitals, banks, and corridors. It is better practice, however, to choose materials on the basis of the absorptive characteristics best suited to suppress the type of noise to be absorbed. For example, one should use materials which have relatively high absorption at the low frequencies for applications where the noise to be reduced is made up predominantly of low-frequency sounds.

The Revised U. S. Federal Specifications (SS-A-118-a) for prefabricated acoustical units classify noise-reduction coefficients by the use of "grades."[1] Noise-reduction coefficients are specified as follows:

Grade	*Noise-Reduction Coefficients*	*Grade*	*Noise-Reduction Coefficients*
1	0.90 or over	7	0.60
2	0.85	8	0.55
3	0.80	9	0.50
4	0.75	10	0.45
5	0.70	11	0.40
6	0.65	12	0.35

For example, a grade 5 material has a noise-reduction coefficient of 0.70. Another classification is also given in these specifications. It is based on the absorption coefficients at the frequency of 512 cycles. The sound-absorption coefficients for auditorium treatment are specified as follows: grade 101 (0.90 or over), grade 102 (0.89 to 0.85), grade 103 (0.84 to 0.80), grade 104 (0.79 to 0.75), etc.[2] Thus, if a material has an absorption coefficient of 0.83 at 512 cycles, it would be in grade 103. *Since the absorption coefficients of materials may vary markedly with*

[1] Printed copies of these specifications are sold by the U. S. Government Printing Office, Washington, D. C., for 5 cents each.

[2] Subject to the further requirement that the coefficient of sound absorption at 128 cycles shall be not less than 1/8 of the coefficient at 512, and the coefficient at 2048 cycles not less than 3/4 of the coefficient at 512 cycles.

frequency, the use of this means for representing the absorptivity of a material is not recommended.

The combustibility of acoustical materials differs widely. *Fire-resistance ratings* are given in Federal Specification SS-A-118-a. When materials are subjected to the test prescribed in this specification, their ratings are defined and abbreviated as follows:

"i. *Incombustible Material.* . . . no flame shall issue from the specimen during or after flame application. Glow shall not progress beyond the fire-exposed area.

"r. *Fire-Retardant Material.* . . . no sustained flaming shall issue from the specimen. Any flame which occurs shall be limited to intermittent short flames from the area directly exposed to the test flame. . . . no flaming shall occur more than 2 minutes after the test flame is discontinued.

"s. *Slow-Burning Material.* . . . all flaming shall cease within 5 minutes after the test flame is discontinued.

"c. *Combustible Material.*. Material not conforming to any of the above requirements."

Types of Acoustical Materials

Most commercially available acoustical materials are included in one of the three following categories:

(1) *Prefabricated Units.* These include acoustical tile, which is the principal type of material available for acoustical treatment; mechanically perforated units backed with absorbent material; and certain wall boards, tile boards, and absorbent sheets.

(2) *Acoustical Plaster and Sprayed-On Materials.* These materials comprise plastic and porous materials applied with a trowel; and fibrous materials, combined with binder agents, which are applied (sprayed on) with an air gun or blower.

(3) *Acoustical Blankets.* Blankets are made up chiefly of mineral or wood wool, glass fibers, kapok batts, and hair felt. The physical characteristics of the materials in each of these categories will now be considered.

Prefabricated Acoustical Units

Prefabricated acoustical materials have been subclassified in order that similar products may be grouped together. The classification given here is that of the Revised U. S. Federal Specification for "Acoustical Units; Prefabricated," SS-A-118-a, which furnishes a convenient means for grouping materials having similar properties. Prefabricated units are separated into three types described in detail below. These groups include tile, absorbent material covered by mechanically perforated units, and certain building boards and sheets.

Perhaps the most outstanding feature of an acoustical tile is its "built-in" absorptive value. The tile is a factory-made product; the absorptivity is relatively uniform from tile to tile of the same kind. This makes it foolproof, a highly desirable characteristic. The amount of absorption added to a room by acoustical tile therefore is quite independent of the skill, or lack of skill, of the persons who install the material. Another merit possessed by acoustical tile is its relatively high absorptivity. In a factory-made product it is possible to control such factors as porosity (including the number and size of pores), flexibility, density, and the punching or drilling of holes—factors which are paramount in determining the absorptivity of materials, and factors which often are difficult to control in certain types of acoustical plasters. In addition, tile can be given structural and decorative properties which usually are well adapted to the requirements for artistic interiors. Because of its high absorptivity, acoustical tile is well adapted to rooms in which a relatively small surface is available for acoustical treatment.

Several acoustical units on the market (for example, Acousti-Celotex, Fibretone, Cushiontone, and Sanacoustic Tile) have the advantage that they can be decorated with oil-base paint without having their high absorptivity impaired. This property is due to the mechanically made holes which permit the sound waves to reach the interior of the tile and be absorbed as a result of viscous forces in the tiny pores of the material. Laboratory and field tests have shown that Sanacoustic Tile and Acousti-Celotex are as absorptive when decorated with an oil-base paint as when they are unpainted. Consequently these materials provide a high degree of assurance against loss of absorption through repeated decoration.

The principal disadvantages of an acoustical tile are its limitations for architectural treatment and its cost compared with that of other acoustical materials. It is quite impossible to conceal entirely the joints between adjacent tiles, and for this reason such treatments should be limited to rooms or surfaces where a tile or ashlar effect is not objectionable. With certain types of tile it is possible to secure the appearance of a continuous or monolithic surface by using tight, unbeveled joints and by decorating an entire surface. But in rooms with low ceilings, or in other rooms with tile on the walls, the ashlar effect is noticeable with any type of decoration. For this reason, the edge is frequently beveled around the tile to emphasize, rather than attempt to conceal, its masonry effect. The bevels also serve to "conceal" slight irregularities in the fitting of the tiles.

Most types of acoustical tile on the market are relatively costly. In comparing the cost of acoustical tile with that of other types of acoustical treatment it should be borne in mind that the cost per square foot should not be considered alone. Acoustical tiles often are two or three times more absorptive than acoustical plasters, and for this reason as much absorption may be attained with one square foot of tile as with two or three square feet of plaster.

The U. S. Federal Specification SS-A-118-a (revised) classifies prefabricated units into four types. The absorption coefficients given in Table A.1, Appendix 1, use this classification. These types and their subclassifications are listed below, together with the name of one or more representative commercial products.[3] Figure 6.2 shows the surface appearance of the different types of materials.

"Type I. Cast units having a pitted or granular-appearing surface.

"Class A. All-mineral units composed of small granules or finely divided particles with portland cement binder."

The masonry-like surface appearance of the units makes them particularly suited for installation in buildings of the monumental type and in some churches. These tiles are rated as incombustible.

[3] The lists of commercial products given under each type of unit may not be complete, nor are they necessarily products recommended by the authors.

Type I-A (Akoustolith Tile, R. Guastavino Co.)

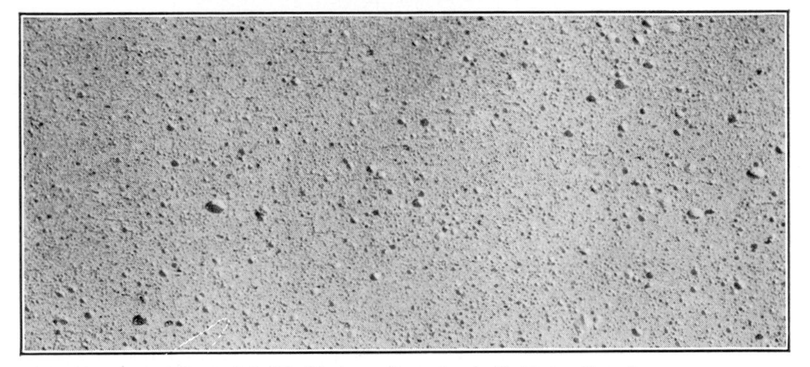

Type I-B (Muffletone, Standard, Celotex Corp.)

Type I-C (Softone, American Acoustics, Inc.)

Fig. 6.2 Half-scale photographs showing surface appearance of the various types and classes of prefabricated acoustical units. (The type is specified by the Roman numeral, the class by the letter.)

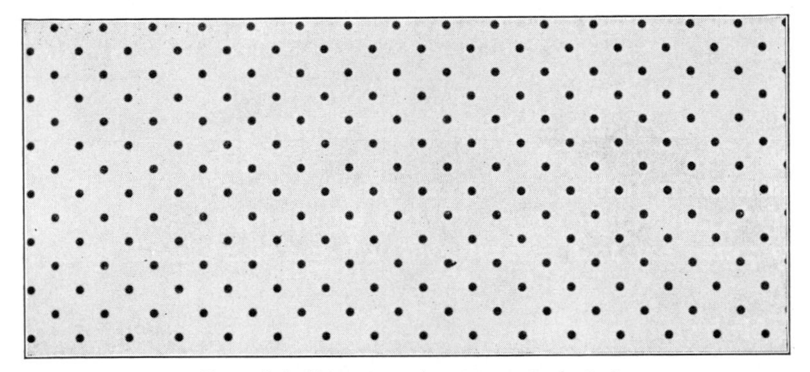

Type II-A (Arrestone, Armstrong Cork Co.)

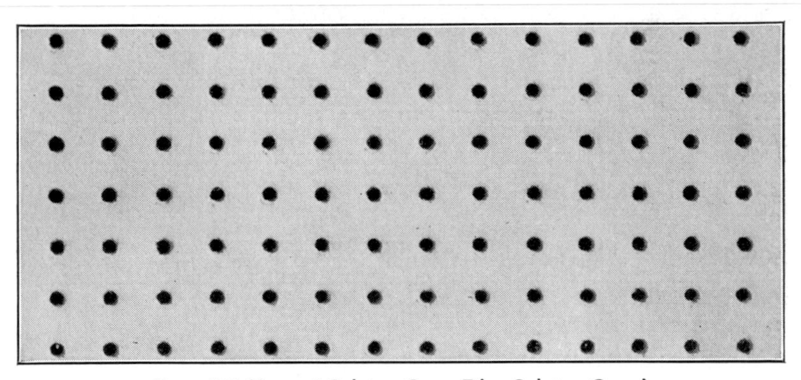

Type II-B (Acousti-Celotex Cane Tile, Celotex Corp.)

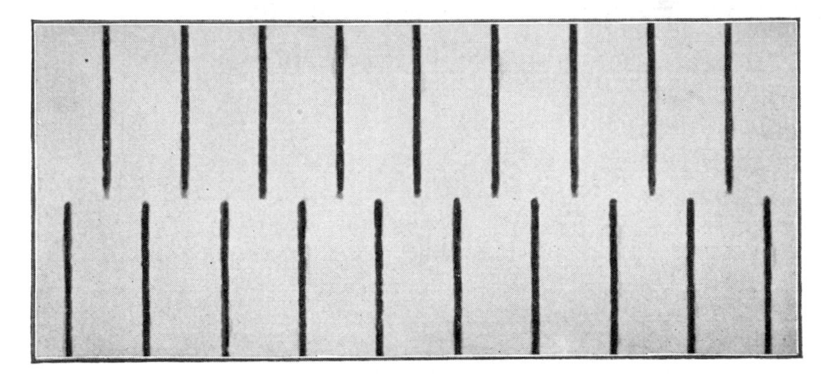

Type II-C (Auditone, U. S. Gypsum Co.)

Fig. 6.2 (*Continued*)

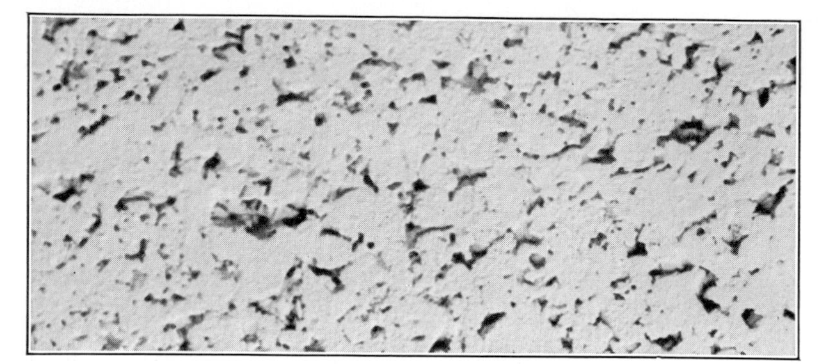

Type III (Corkoustic, Armstrong Cork Co.)

Type IV-A (Absorb-A-Noise, Luse-Stevenson Co.)

Type IV-B (Econacoustic, National Gypsum Co.)

Type IV-C (Q-T Ductliner, Celotex Corp.)

Fig. 6.2 (*Continued*)

Paint normally reduces their sound-absorptive properties, but decoration is seldom required. The surfaces of materials in this class are reasonably smooth.

AKOUSTOLITH TILE R. Guastavino Co.

"Class B. All-mineral units composed of small granules or finely divided particles with lime or gypsum binder."

MUFFLETONE Celotex Corp.

"Class C. Units composed of small granules or finely divided particles of mineral or vegetable origin with incombustible mineral binder."

SOFTONE American Acoustics, Inc.

"Type II. Units having mechanically perforated surface; the perforations to be arranged in a regular pattern.

"Class A. Units having a perforated surface which acts as a covering and support for the sound absorbent material. The facing material to be strong and durable and substantially rigid."

In this type of unit an absorptive pad, blanket, or rigid element (frequently consisting of compressed mineral wool) is covered by perforated sheet metal or board. The perforated covering does *not* reduce the absorption in proportion to the area covered. For example, the absorption coefficient of a blanket covered with perforated sheet steel which exposes only 15 per cent of the absorptive material may have a coefficient, up to 4000 cycles, almost as high as if the covering were not there at all! This is due to diffraction, which is discussed in Chapter 4.

Type II-A prefabricated units can be painted repeatedly without impairing their absorption, if reasonable care is taken not to fill or bridge the holes with paint. If the holes are ⅛ inch in diameter or larger, it is highly improbable that they will ever become bridged by painting. Since the perforated coverings offer good mechanical protection for the absorptive material, the units can be installed in locations where they will be subject to considerable wear and tear. Most units of this class are incombustible.

Many are moisture-resistant and hence find application in swimming pools, kitchens, etc. Some offer interesting possibilities for combining acoustical construction with air conditioning and lighting control. For example, there are metal pan units and special

Fig. 6.3 Acoustical ceiling construction using Perfatone metal pan units which are interchangeable with flush-type fluorescent lights. (Courtesy U. S. Gypsum Co.)

flush-type fluorescent lighting fixtures that can be interchanged. Figure 6.3 shows an office employing such a ceiling installation.

Acousteel	Celotex Corp.
Acoustimetal	National Gypsum Co.
Arphon	A. B. Arki (Stockholm, Sweden)
Arrestone	Armstrong Cork Co.
Perfatone	United States Gypsum Co.
Sanacoustic Unit	Johns-Manville
Transite Acoustical Unit	Johns-Manville

"Class B. Units having circular perforations extending into the sound absorbent material."

Prefabricated units of this class usually have large perforations and therefore are especially serviceable in installations that require frequent redecoration. Laboratory and field tests show that these tiles may be painted repeatedly without noticeable reduction of their sound-absorptive properties.

The presence of holes in porous materials, as in Acousti-Celotex or Cushiontone, has the effect of greatly increasing the absorptivity of the material. The holes increase both the superficial area and the effective porosity of the material. The perforations can be used to conceal the heads of nails or screws when used for attaching the units to wood furring strips or wood decking.

Acousti-Celotex Cane Tile	Celotex Corp.
Acousti-Celotex Mineral Tile	Celotex Corp.
Acoustifibre	National Gypsum Co.
Cushiontone	Armstrong Cork Co.
Fibretone	Johns-Manville
Paxtiles	Newalls Insulation Co., Ltd. (England)
Stenitplatta	A. B. Arki (Stockholm, Sweden)
Simpson Acoustical Tile	Simpson Industries

"Class C. Units having slots or grooves extending into the sound absorbent material."

The action of the slots or grooves is similar to that of the holes in the tiles of the preceding classification.

Auditone	United States Gypsum Co.
Treetex (Type B)	Treetex, Ltd. (Sweden and England)

"Type III. Units having a fissured surface."

This type, illustrated in Fig. 6.2, includes tiles differing widely in composition. Some consist largely of filaments or mineral wool granules; in others, vermiculite or cork is the principal ingredient. The action of the fissures in causing absorption of sound by the units is very similar to that of the perforations in type II-B. These tiles have surfaces that are sanded or planed smooth. They may be painted without loss of absorption if the fissures are numerous and are not filled with paint.

ACOUSTONE	United States Gypsum Co.
CORKOUSTIC	Armstrong Cork Co.
FISSURETONE	Celotex Corp.
TRAVERTONE	Armstrong Cork Co.

"Type IV. Units having a felted fiber surface.

"Class A. Units composed of long wood fibers."

Units of this class are made of wood shavings or excelsior, generally pressed together with a mineral binder. The wood fibers may be fine, medium, or coarse.

ABSORB-A-TONE	Luse-Stevenson Co.
ABSORB-A-NOISE	Luse-Stevenson Co.
L. W. INSULATION BOARD	Brown and Tawse, Ltd. (England)
POREX	Porete Manufacturing Co.
SONO-THERM	SONO-THERM Co.

"Class B. Units composed of fine felted vegetable fibers or wood pulp."

Included in this class are small tiles and also acoustical fiberboards. In general, these materials are not fireproof. The fiberboards provide a means of obtaining absorption at relatively low cost. They are commonly manufactured in large panels, 4 feet wide, and 8, 10, or 12 feet long. The use of fiberboards presents a difficulty in the matter of decoration and redecoration. Oil, lead, and other non-porous paints will close the surface pores of the material and hence destroy the absorptive value. On the other hand, thin dyes and stains, stencil designs with heavier paint, or dry paint dusted on with a pounce-bag can be used without impairing the acoustical value of the material. In spite of these limitations, certain acoustical fiberboards are useful for the control of noise and reverberation in buildings. There are many school and industrial jobs, where cost is an important consideration, in which fiberboards may be used to advantage.

ACOUSTILITE	Insulite Co.
ECONACOUSTIC	National Gypsum Co.
FIBRACOUSTIC	Johns-Manville
LLOYD BOARD	Lloyd Boards, Ltd. (England)
NUWOOD BEVEL LAP TILE	Wood Conversion Co.

"Class C. Units composed of mineral fibers."

AIRACOUSTIC SHEETS	Johns-Manville
FIBERGLAS ACOUSTICAL TILE	Owens-Corning Fiberglas Corp.
PAXFELT	Newalls Insulation Co., Ltd. (England)
Q-T DUCTLINER	Celotex Corp.

Acoustical Plaster and Sprayed-On Materials

The use of selected types of acoustical plastic materials has proved highly satisfactory for the treatment of offices, school rooms, corridors, and many public buildings. They can be used in most places where ordinary lime or gypsum plaster can be used without altering the architectural effects. Two coats of acoustical plaster may be applied instead of the finish coat in the ordinary plaster treatment for an added cost of about one dollar per square yard. These materials have deficiencies in regard to cleaning and decorating. Although these shortcomings are not serious in localities where the air is relatively clean, they are an important consideration where air is laden with smoke or dust. As plastic materials are improved, and as the correct manner of their application is more fully understood and practiced by plasterers, their use may be extended to more and more buildings.

The absorptivity of such material as acoustical plaster is dependent on its thickness and composition, and on the manner in which it is applied and dried. As the thickness is increased, the absorptivity increases, particularly at low frequencies. However, for plasters of the type applied with a trowel, it is usually uneconomical to increase the thickness beyond ½ inch. If too much binder material is used, the plaster is not sufficiently porous. If an insufficient amount of binder is used, the plaster does not set hard and its tensile strength may be less than that required for adequate structural bond; under such circumstances, it may dust or pop off the wall. Likewise, if the undercoats of plaster are too wet ("green"), the binder material forms an impenetrable film at the surface; whereas, if the undercoats are too dry, the binder material is absorbed by the undercoats and the plaster will crumble. Since the absorption coefficients of acoustical plasters are dependent on such factors as the suction behind the plaster, the pressure applied to the trowel, and the manner of floating, texturing, or stippling, the journeyman should be instructed to exercise great care in applying and finishing these materials. A large measure of the success or failure which attends the application of acoustical plaster depends on the drying out of the plaster. The surface to which it is applied must provide the proper degree of suction. Accordingly, it is advisable to prepare scratch and brown coats which will draw the water from the acoustical plaster and thus prevent the formation of a non-porous film on the finished surface. It is also advisable to provide good drying conditions for the plaster, and to float or drag the surface of the plaster just before it takes its initial set. Unless these simple precautions are observed, the use of acoustical plaster may prove disappointing. On the other hand, if these precautions are carefully followed, a good brand of acoustical plaster will be found to be well adapted to many types of buildings where large surface areas are available for treatment, and where very high absorption coefficients are not needed. *In order to produce good results, acoustical materials for plastic application must be applied by competent journeymen in strict conformity with the specifications of responsible manufacturers.* In large buildings it is advisable to require the contractor to plaster a small room for test and approval before the material is applied to other parts of the building.

In selecting an acoustical plastic material it is desirable to consider its adhesive and cohesive properties, its resistance to fire and abrasion, its ease of application, its texture, and its maintenance (such as cleaning and decorating), as well as its coefficients of sound absorption. If it becomes necessary to use plastic material which will not withstand the wear and abrasion to which the walls near the floor will be subjected, it is a good plan to provide a wainscot of harder material, such as wood or hard plaster. The wainscot should extend up to a height of about 6 or 7 feet above the floor.

Acoustical materials for plastic application are classified into three groups in U. S. Federal Specification SS-A-118-a:

"Type I. Acoustic plaster. This shall be composed of a cementitious material such as gypsum, portland cement, or lime with or without an aggregate."

Included in this type are:

ATOZ	American Acoustics Inc.
HUSHKOTE ACOUSTIC PLASTER	Cleveland Gypsum Supply Co.
KALITE	Mission Lime Products Corp.
KILNOISE PLASTER	Kelley Island Lime and Transport Co.
MACOUSTIC PLASTER	National Gypsum Co.
PLASTACOUSTIC	R. Guastavino Co.
SABINITE	United States Gypsum Co.
SOFTONE	American Acoustics Inc.
ZONOLITE	Zonolite Insulation Co.

Fig. 6.4 Limpet being applied with a spray gun. (Courtesy The Stuart Co., Contractor.)

"Type II. Acoustic materials other than acoustic plaster which are applied with a trowel."

This type includes:

ACOUSTIPULP	Val-Porter Co.

"Type III. Fibrous materials combined with a binder agent and which are applied by being sprayed on with an air gun or blower."

Although the cost of these materials may be higher than that of type I, their absorptivities are usually somewhat higher (see Table A.2, Appendix 1). Figure 6.4 shows how this type of treatment is applied.

The following type III materials are applied with a spray gun or blower and can be applied to different required thicknesses:

LIMPET	Keasbey and Mattison Co.
SPRAY-ACOUSTIC	Sprayo-Flake Co.

Acoustical Blankets

The materials used most commonly in the fabrication of acoustical blankets are mineral wool, hair felt, wood fiber, and glass fiber. Although the thickness of these blankets is generally between ½ and 4 inches, blankets of greater thickness are sometimes used in special applications. These materials are more absorptive in the low-frequency range, principally because of their greater thickness, than are most other types, as will be seen from an inspection of the absorption coefficients at 128 cycles, given in the tables of Appendix 1. Hence blankets sometimes are useful for controlling the acoustical characteristics of studios and auditoriums that require "balanced" absorption, including a considerable amount at low frequencies.

The absorption coefficient of a blanket mounted against a wall depends on its density and thickness and on the frequency of the incident sound. Increasing the thickness of the blanket increases its absorptivity, principally at low frequencies, slightly at the "highs." This is illustrated by the curves of Fig. 6.5, which give the absorption coefficients of blankets made of type TW-F Fiberglas wool. Figures (*a*), (*b*), and (*c*) are for blankets with thicknesses of 1, 2, and 3 inches, respectively. The density of the wool is given along the side of each curve. A comparison of the absorption coefficients for blankets of the three thicknesses, having a density of 4 pounds per cubic foot, shows that at 128 cycles the absorptivity of the blanket increases as the thickness of

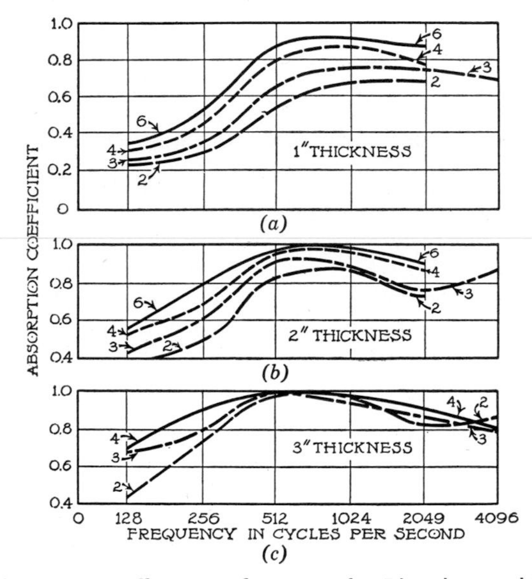

Fig. 6.5 Absorption coefficient *vs.* frequency for Fiberglas wool blanket type TW-F, having the densities in pounds per cubic foot indicated by the numerals. Figures (*a*), (*b*), and (*c*) are for blankets with thicknesses of 1 inch, 2 inches, and 3 inches, respectively. (Courtesy Owens-Corning Fiberglas Corp.)

the blanket increases. At 2048 cycles the variation of the absorptivity with the thickness of the blanket is not significant, at least for thicknesses of 1 to 3 inches. These data also illustrate that the absorption coefficient of a blanket made of a material of *relatively light density,* such as mineral wool, increases with increasing density. The effect of an air space behind a blanket is, in general, to increase its absorption at low frequencies. This is illustrated by Fig. 6.6, which gives the absorption coefficient for a 1-inch blanket of rock wool mounted 4 inches from a hard wall, and also when mounted directly against the wall. By a suitable choice of density, thickness, and manner of mounting, it is possible to obtain a wide variety of absorption characteristics with an acoustical blanket.

The fibers in certain types of blanket, especially some mineral-wool products, have a tendency to "settle," often as a result of

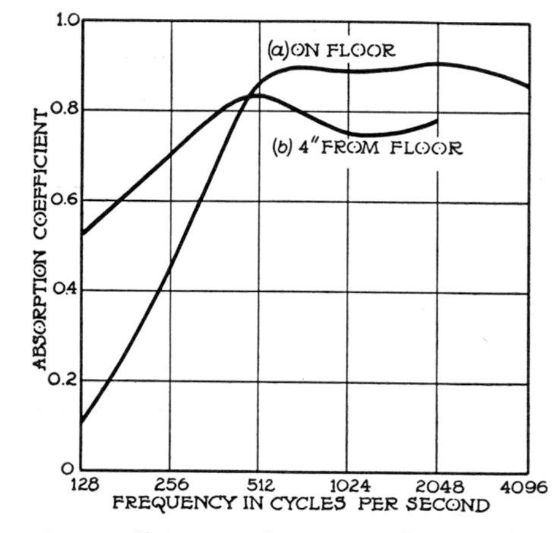

Fig. 6.6 Absorption coefficient *vs.* frequency of a 1-inch rock-wool blanket (*a*) mounted on floor, and (*b*) mounted on furring strips 4 inches from the floor. (L. J. Sivian and H. T. O'Neil.)

building vibration; this "settling" alters the acoustical characteristics of the blanket. For this reason, blankets fabricated of materials that tend to settle are frequently quilted at intervals of a few inches. In other cases, the materials are given additional structural strength by the addition of a binder material, or by a wire-mesh screen or hardware cloth on one or both sides of the blanket.

Tables of Absorption Coefficients

Tables of sound-absorption coefficients of acoustical materials are given in Appendix 1. Since the coefficients depend on the method of mounting the material, the type of mounting is given. The values determined by different observers, for the same material, usually are not identical. The discrepancies are due partly to differences in test procedures, specimen sizes, and nature of mounting. Since the authority for each set of measurements is listed, results for materials measured in the same laboratory can be compared. In some cases the average of the values obtained by a number of laboratories is given; in others, the coefficients are those believed by the authors to be most probable.

Included in Appendix 1 are tables of coefficients for the following types of materials:

Table A.1. Prefabricated Units—Tiles, Fiberboards, etc.
Table A.2. Acoustical Plaster and Other Materials for Plastic Application.
Table A.3. Acoustical Blankets, Felts, etc.
Table A.4. Hangings, Floor Coverings, Miscellaneous Materials.
Table A.5. Hard Plasters, Masonry, Wood, Other Standard Building Materials.
Table A.6. Audience, Individual Persons, Chairs, Other Objects.

In addition to the data concerning the acoustical properties of the materials, other information is tabulated: the light reflection coefficient (usually for the factory-painted tiles), the weight per square foot, and the unit size of the material.

Perforated Facings

A perforated facing such as plywood, metal, or fiberboard constitutes a very practical covering for an acoustical blanket. For example, Fig. 6.7 shows how Transite can be used for this purpose. Figure 6.8 is a photograph of a portion of an auditorium wall having patches of absorptive treatment in the form of a blanket covered with a perforated veneer plywood which serves as the wall surface. (Also see Fig. 15.4.) Except for the small holes, the appearance of the plywood covering the patches of absorptive material does not differ from other portions of the wall. Therefore in order to show the position of the patches the photograph has been retouched. This type of facing has the advantage that it can be easily cleaned and decorated, and repeated painting does not reduce its absorptivity if the holes are not bridged with paint. In this respect and also in the mechanism by which it absorbs sound, it is similar to type II-A prefabricated units.

Owing to diffraction, the facings are "acoustically transparent" over a wide range of frequencies. Thus, if a plane sound wave is normally incident on a wall containing a small aperture, the ratio of the flow of sound through this hole to the flow through an equal area of the incident wave front, may be quite large—

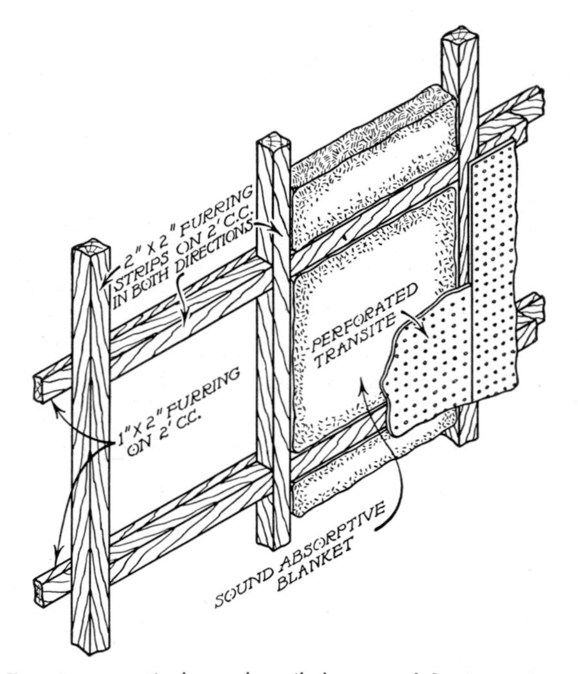

Fig. 6.7 Transite acoustical panels nailed to wood furring strips. (Courtesy Johns-Manville.)

even in excess of 10.[4] This ratio is largest at low frequencies, and smallest at high frequencies where the diffraction effects are negligible.[5] Therefore, the absorption coefficients of blankets

[4] F. Ingerslev and A. K. Nielson give the following ratios for a circular aperture 0.4 inch in diameter in a plate 0.04 inch thick: 50 at 100 cycles, 6 at 500 cycles, and 2 at 870 cycles. At very high frequencies the ratio would approach 1; *Ingeniøvidenskab. Skrifter,* No. 5 (1940).

[5] For a discussion of perforated facings, see L. W. Sepmeyer and R. W. Leonard, *J. Acoust. Soc. Am.* (forthcoming); J. Brillouin, *Cahiers du Centre Scientifique et technique du Bâtiment Cahier,* **31** (Jan. 1949); R. H. Bolt, *J. Acoust. Soc. Am.,* **19**, 918 (1947); V. L. Jordan, *J. Acoust. Soc. Am.,* **19**, 972 (1947).

covered with perforated facings in which only 5 to 10 per cent of the surface area is perforated are affected relatively little at low frequencies but decrease at the highs. This is also characteristic of type II-A materials. The greater the exposed area, the higher will be the frequency at which the absorption begins

Fig. 6.8 Portion of side wall of Arnold Auditorium, Bell Telephone Laboratories, Murray Hill, New Jersey, showing patches of absorptive treatment in the form of perforated plywood backed with an acoustical blanket. This photograph has been retouched to indicate the position of the patches. (Courtesy Bell Telephone Laboratories.)

to "drop off." For most applications, $\frac{3}{16}$-inch holes on $\frac{1}{2}$-inch centers, or $\frac{1}{8}$-inch holes on $\frac{3}{8}$-inch centers, will be found to be satisfactory. By spacing the holes farther apart, the absorption at high frequencies can be further diminished without appreciable loss of absorption at the low frequencies. If muslin or similar fabric is used to cover the blanket, it should be very porous. Any form of decoration or flameproofing should not clog the pores of the covering. Such a covering does not alter appreciably the absorption of the blanket.

The combination of a perforated panel like $\frac{1}{8}$-inch plywood with an air space and a blanket of rock wool or glass fiber provides a type of acoustical treatment that can be highly absorptive at low frequencies and progressively less absorptive at higher frequencies. This is the inverse of the action of many other available materials which are only slightly absorptive at low frequencies and progressively more absorptive at higher frequencies. As the percentage of open area in the panel is in-

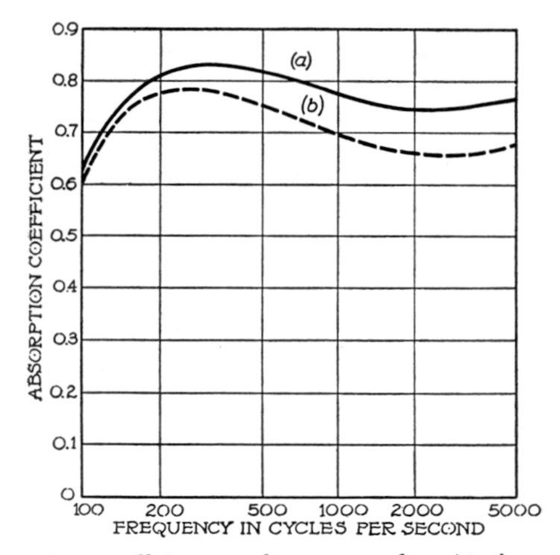

Fig. 6.9 Absorption coefficient *vs.* frequency of a 4-inch rock-wool blanket. Curve (*a*) is for the blanket alone; curve (*b*) is for the blanket covered by a 5⁄32-inch perforated panel with 10 per cent open area. (V. L. Jordan.)

creased, the characteristics of a combined perforated panel and absorptive blanket approach those of the blanket alone. Likewise, for a given percentage of open area, as the panel thickness decreases the absorption characteristic of the combination approaches that of the blanket alone. This is illustrated in Fig. 6.9. Curve (*a*) is for a 4-inch rock-wool blanket; curve (*b*) is for the same blanket covered by a 5⁄32-inch perforated panel having 10 per cent open area. Diffraction is not the only important factor contributing to the low-frequency absorption of an acoustical pad or blanket having a perforated facing. The facing itself is significant. Its lowest resonant frequency depends primarily on its mass and stiffness. In the frequency region of this resonance, vibration of the facing may be responsible for the greatest portion of absorbed energy. Motion of the mass of air within and adjacent to the holes of the facing may contribute absorption of the Helmholtz-resonator type; panel resonance and Helmholtz resonators are discussed in the next chapter.

Mounting Acoustical Materials

The manner of mounting acoustical materials can influence markedly their absorptive properties. Certain materials that are unsatisfactory when applied directly against a rigid wall may be satisfactory when they are mounted some other way; for example, with an air space behind them. Typical effects of two methods of mounting on absorption characteristics are illustrated in Fig. 6.1. The dashed lines give the absorption coefficients of the materials with solid backing—tiles cemented to a rigid wall, and also fiberboard rigidly fastened to the wall. The solid lines give the coefficients for the same materials when they are mounted on furring strips so that there is a 1-inch air space between the material and the wall. Since the absorptivity of various materials may differ widely, these curves are not necessarily typical of all acoustical materials. Nevertheless, they illustrate the principal effect of air-space backing—an increase in the absorption at low frequencies, due partly to the flexural vibration of panels of the material. The effect of varying the depth of the space behind the material will be discussed in Chapter 7 (see Fig. 7.1). Within certain limits, increasing the spacing from the wall increases the average absorption and alters the frequency at which maximum absorption occurs—the lowest resonant frequency of the panel. Since the size of a panel is a factor that determines this resonant frequency, the frequency region in which increased absorption takes place, owing to flexural vibration, depends on the separation between furring strips. For furred-out fiberboard or plaster on lath, the separation between strips is frequently determined by the mechanical properties of the material; and for prefabricated units the spacing is generally determined by the size and physical properties of the units. In other applications, 16 inches on centers is usually satisfactory. In music rooms, and in radio or sound-recording studios, random spacing of the furring strips may be used to distribute the resonant frequencies and thus provide a more uniform absorption throughout a wide range of frequencies. *Since the absorptivity of acoustical materials varies so widely with mounting conditions, a fair comparison of absorption data for*

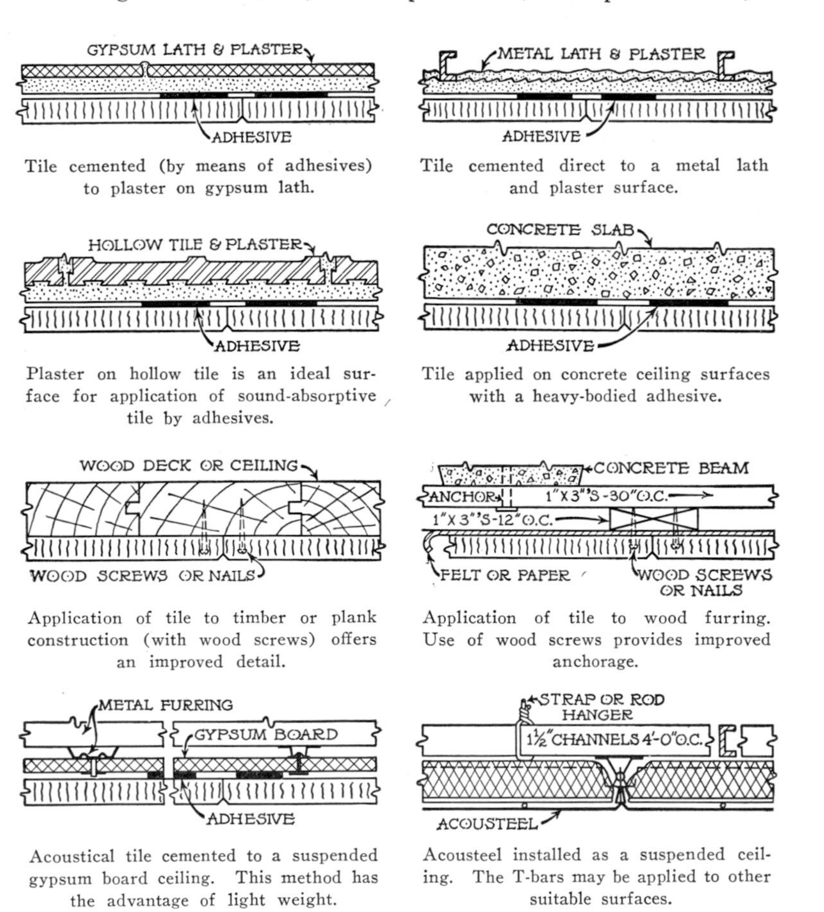

Tile cemented (by means of adhesives) to plaster on gypsum lath.

Tile cemented direct to a metal lath and plaster surface.

Plaster on hollow tile is an ideal surface for application of sound-absorptive tile by adhesives.

Tile applied on concrete ceiling surfaces with a heavy-bodied adhesive.

Application of tile to timber or plank construction (with wood screws) offers an improved detail.

Application of tile to wood furring. Use of wood screws provides improved anchorage.

Acoustical tile cemented to a suspended gypsum board ceiling. This method has the advantage of light weight.

Acousteel installed as a suspended ceiling. The T-bars may be applied to other suitable surfaces.

Fig. 6.10 Methods of mounting prefabricated acoustical units. (Courtesy Celotex Corp.)

two different materials can be made only if the data represent absorption coefficients for the same mounting conditions.

Acoustical tiles are most frequently mounted by adhesives, nails, or screws, or by a mechanical system such as "T-splines" which engage in a horizontal kerf along the side edges of the tiles. These types of mountings are illustrated in Figs. 6.10 and 6.11. In many instances a combination of two methods is used. No one type of mounting is best for all installations. The choice of the method of mounting will depend on the physical

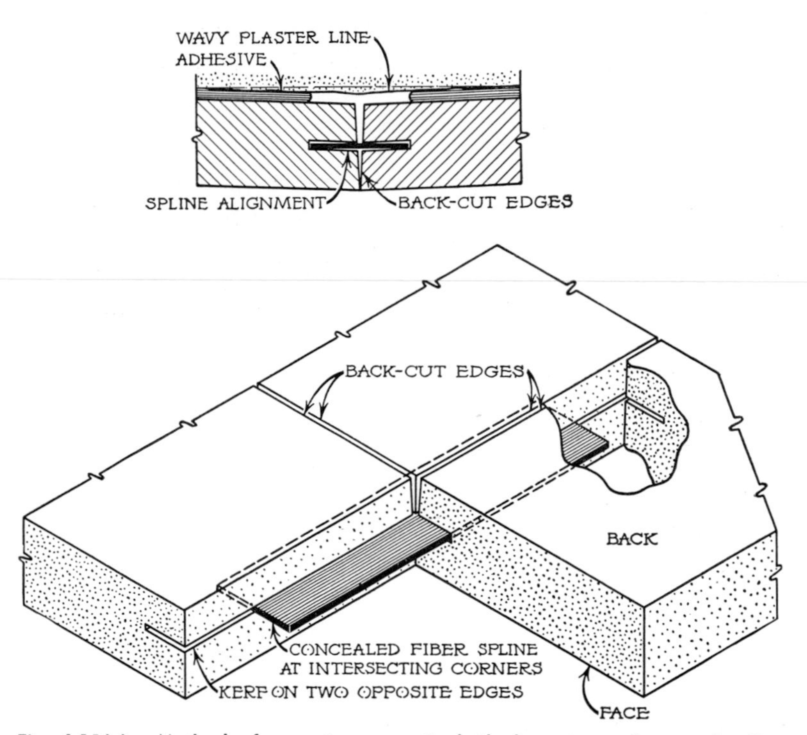

Fig. 6.11(a) Method of mounting acoustical tile by using splines and adhesive. (Courtesy U. S. Gypsum Company.)

properties of the acoustical material, the base to which it will be applied, the time required for installation, and labor costs. However, other factors are frequently the controlling ones. Thus, the adhesive method of mounting is particularly advantageous on a job where noise must be kept to a minimum during installation, as in a hospital. The adhesive method is also quick, economical, and clean. The adhesive should be bonded to both the material and the wall or ceiling. Since the failure to secure a good bond may result in the tile becoming loose, adhesive applications should be made only by individuals thoroughly skilled in the art. In general, tiles larger than 12 inches by 24 inches should not be applied with adhesives alone. Nails or screws

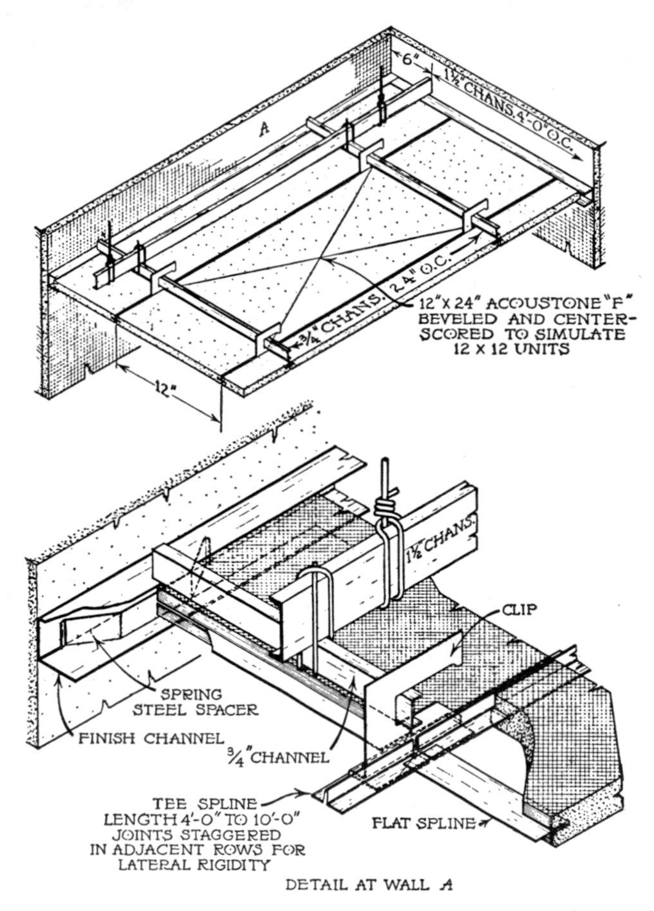

Fig. 6.11(*b*) Method of mounting acoustical tile by means of a mechanical system. (Courtesy U. S. Gypsum Company.)

may be necessary; they support the unit while the adhesive sets, thereby securing a stronger mechanical bond and offering a double protection against loosening. Cement-coated nails have about 40 per cent greater holding power in wood than do uncoated nails.

Tiles fastened by screws are usually held more securely than those fastened by nails. Also, they can be removed quickly. Facilities are now available for rapid application by electrically driven screws. Either nails or screws can be used to mount acoustical materials on wood furring strips. By this means a new ceiling can be furred down to any desired level, thereby concealing pipes, conduits, air-conditioning ducts, etc. The wood furring method of mounting also permits the tiles to vibrate flexurally; this vibration gives increased absorption in the low-frequency range. *Where the acoustical units are subject to breakage,* as they may be in a gymnasium, *a stiff backing for the tiles should be provided.* Gypsum board provides such a backing; it also is fire-resistant. It can be used to level off an existing irregular ceiling, or as an economical substitute for a conventional lath and plaster backing in new construction.

A number of manufacturers of acoustical materials have mechanical systems for the installation of prefabricated tiles which enable units to be removed relatively easily and replaced after the original installation has been made. Mechanical systems provide a convenient means of furring below ceiling obstructions with incombustible supporting members, as illustrated in Fig. 6.11(*b*).

Most acoustical materials are efficient thermal insulators. For this reason, care should be taken to prevent condensation on the underside of the slabs or decks on which the material is installed. An undesirable discoloration may result from air flow through the cracks between tiles or even through very porous tiles. This flow of air, called breathing, occurs most often in air-conditioned rooms. It can be minimized in installations by the application of a layer of building paper directly behind the tiles. This precaution may be advisable in some wood furring installations.

Although some acoustical materials, such as mineral tiles, are not affected by changes in humidity or moisture in a room, others—those made of cellulose products, such as wood or vegetable fibers—tend to expand upon absorption of water vapor and to contract upon drying. This possibility must be borne in mind when hygroscopic acoustical materials are being installed. The tiles should become adjusted to the moisture content of the room in which they are to be installed. Then, if the humidity is high, the individual units should be butted up against each other tightly, so that when dry conditions prevail noticeable gaps will not appear. On the other hand, if the tiles are installed in a very dry atmosphere, they should be fitted with a slight gap, about $1/64$ or $1/32$ inch. (Such gaps will not be noticeable if the tiles have beveled edges.) If this precaution is not taken, the tiles, by expanding, may warp or may exert enough force on one another to become loosened and thus become unsightly or even unsafe.

The Reflection of Light from Acoustical Materials

Acoustical materials in their natural state are, in general, very poor reflectors of light. For this reason they are usually prepainted. In addition to increasing the light reflection of a material, paint increases its moisture resistance and its washability. The light-reflection coefficients for prepainted acoustical materials are given in the Acoustical Materials Association Bulletin.[6] Perforated metal units such as Acoustimetal, Acousteel, or Sanacoustic Tile have somewhat higher light-reflection coefficients. The percentage of light reflected from such materials is approximately equal to the percentage of the metal facing that is unperforated (about 88 per cent for most tile of this type) multiplied by the optical reflection coefficient for the paint.

Effect of Paint on Absorption of Sound by Acoustical Materials

The sound-absorptive properties of materials may be greatly impaired by improper painting. Various types of materials are affected quite differently by paint, since their mechanisms of absorption are different. Thin panels that absorb by flexural vibration, and porous materials having large perforations, or covered with perforated facings, are practically unaffected by painting. On the other hand, fiberboards and acoustical plasters may be ruined acoustically by improper painting. Reduction in absorption takes place largely because the paint fills the sur-

[6] This bulletin can be obtained from Acoustical Materials Association, 205 West Monroe St., Chicago, Ill.

face pores; the reduction is most pronounced at frequencies above about 500 cycles. Materials which depend primarily on porosity for their absorption should not be decorated with oil or water paint, varnish, calcimine, distemper, or other viscous or heavy paints that bridge over or fill the surface pores.

Measurements by the United States Bureau of Standards [7] show that it is better to paint acoustical materials with a spray gun than with a brush; the sound absorptivity of the materials is affected to a lesser degree by this method than it is by brush painting. Other advantages are that the "hiding power" (ability of paint to obscure the surface to which it is applied) of paint sprayed on acoustical materials is higher than that of paint brushed on; and more coats can be applied before the paint bridges over the pore openings. If the coats are thin, so that a film is not formed across the pores, the principal factor that affects the decrease in absorption is the amount of pigment deposited on the surface. In such cases, the Bureau's measurements indicate that several coats of paint may be applied without seriously affecting the absorptive properties of the material, provided that the pores of the material are not clogged with dirt. A material that is known to be soiled, or that has been installed for several years, should be thoroughly cleaned with a suitable solvent before it is painted.

Acoustical plaster and fiberboard can be decorated with thin aniline dyes, gasoline or kerosene stains, thin lacquer sprays, or dry paint dusted on with a pounce bag, without impairing their sound-absorptive properties. On the other hand, perforated materials, or materials covered with a perforated facing, may be decorated with lead or oil and any other kind of paint, if the holes are not bridged. The effect of paint on a number of materials and plasters can be seen by comparing their coefficients, with and without paint, in the tables of Appendix 1.

Since the ability of many types of materials to absorb sound depends on the ease with which air can flow into the interstices of the material, an impervious membrane can reduce considerably their absorption coefficients. A simple device illustrated in Fig. 6.12 has proved very useful in making a quick comparison of

[7] Research Paper RP 1298, "The Effect of Paint on the Sound Absorption of Acoustic Materials," by V. L. Chrisler (1940); 10 cents a copy.

the effects produced by the application of different kinds of paint to acoustical plaster, fiberboard, or even certain types of acoustical tile. It consists of a five-gallon bottle, a source of air supply such as a tire pump, a check valve, a pressure gauge, and a 2-inch glass funnel with a collar for sealing (usually with putty) the funnel onto the specimen under test. A measure of the resistance of the materials to air flow is then obtained by determining the time required for a certain amount of air to be discharged

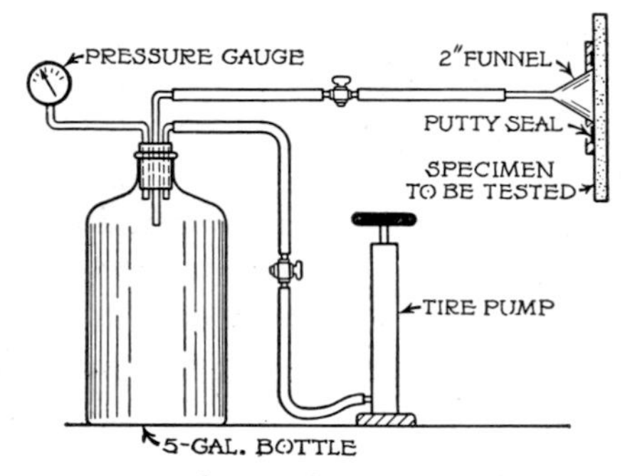

Fig. 6.12 Device consisting of a bottle, pump, and pressure gauge for comparing the resistance of acoustical materials to air flow.

through the specimen, that is, the time required for the pressure of the air in the bottle to be reduced a specified amount. If paint increases the resistance to air flow, then it will have reduced the sound-absorption coefficient of the material. A gauge pressure drop of from 2 pounds per square inch to 0.2 pound per square inch has proved satisfactory for practical measurements. Tests with this apparatus on different plaster surfaces show that there is practically no diffusion of air through hard gypsum or lime plaster, whereas the bottle will discharge its contents of air (from 2 pounds per square inch down to 0.2 pound per square inch gauge pressure) through acoustical plasters in a time ranging from about 1 minute to about 1 second. It is advisable to use this method, or a similar one, for checking a small area of painted material before large areas are decorated or redecorated. A simpler, but often satisfactory, method for checking resistance to air flow consists in pressing one's mouth against the material under test and blowing air into it. If the air can be blown quite easily into the porous material it is probable that it is absorptive; and *if the air flows into the plaster or fiberboard as readily after it has been painted as it does before it is painted, it can be safely concluded that the painting has not impaired the absorptivity of the acoustical material.* This simple test may provide the means for saving many acoustical plasters and fiberboards from almost complete loss of absorptivity as a result of indiscriminate painting.

Absorption by Patches of Material

The location and distribution of absorptive material in a room affects (1) the absorption due to the material and (2) the distribution of sound in the room. For example, twenty-five small areas—"patches"—of material, each 4 feet square, will absorb more sound than will one large patch having an area of 100 square feet. This dependence of absorption on the size of patch is frequently referred to as the "area effect."[8] Although the application of absorptive material in the form of small patches or narrow strips is more efficient than a uniform treatment, it is usually not the cheaper method of obtaining a specified amount of absorption with a given material since the cost of installation of patches on a per-square-foot basis is generally much higher than is the cost for uniform coverage.

The absorption by a patch of material is not independent of its position on the walls of a room. There is some experimental evidence which indicates that, from the standpoint of absorption efficiency, the most effective positions for the usual types of absorptive material are at (or close to) the corners, especially at frequencies at which the wavelengths are large compared with the dimensions of the patch.[9] The next most efficient positions are along the edges between two walls. It has been shown that the distribution of sound pressure in a room is a function of the distribution of the absorptive materials on the walls. Owing to the effects of diffraction, the application of acoustical material in patches, distributed more or less at random on the walls, provides a more diffuse sound field in the room than would be ob-

[8] R. M. Morris, G. M. Nixon, and J. S. Parkinson, *J. Acoust. Soc. Am.*, **9**, 234 (1938).

[9] C. M. Harris, *J. Acoust. Soc. Am.*, **17**, 242 (1946).

tained by uniform treatment of the walls. A certain amount of diffusion is generally considered to be a requirement for good acoustics. Since the distribution of an absorptive material in a room and its acoustical properties are so intimately connected, a detailed discussion of the subject will be given in Chapters 8 and 9.

7 · Special Sound-Absorptive Constructions

Many absorptive materials and constructions that are not described in the preceding chapter are useful or even indispensable for certain types of acoustical installations. Often these special treatments, when used with understanding and imagination, provide not only better acoustics than can be obtained by the use of the standard or "classified" materials, but also a more artistic appearance, sometimes at a considerable reduction in cost. Thus, in a recent design of a Lutheran church in which the architects wished to preserve a brick interior, which is normally highly reflective, a lattice of brick backed with a 2-inch layer of rock wool was used to provide the absorptive treatment for selected wall surfaces. This chapter is concerned with a number of these special acoustical constructions. Many others can be devised, but those considered are of sufficient variety to meet the practical requirements of nearly all types of buildings.

Panel Absorbers

The use of thin panels for increasing sound absorption at low frequencies already has been advocated. Panels of this type, if made of sufficiently durable and flexible materials like pressed wood fiber or paper boards, plywood, or plastic boards, can be employed for ceilings, wainscoting, or even for the entire walls of rooms where low-frequency absorption is required. Such materials, if used for walls or ceilings of small rooms, such as music studios, classrooms, and offices, reduce the amount of additional absorption required for optimum reverberation.

Maximum absorption by a panel occurs at frequencies in the regions of its resonant frequencies, which depend principally on its effective mass and stiffness. Hence, the nature and fabrication of the material, the size and thickness of the panel, and its method of mounting are important considerations of design. Since the relationship between these factors is quite complicated, it is difficult to predict the exact performance of such a structure. However, when a large number of panels is employed, each having slightly different resonant frequencies, the average absorption coefficient can be made fairly uniform over the lower frequency range and can be approximately predicted if measurements on individual panels are used as a guide.

The effective stiffness of a thin panel is influenced by the presence of an enclosed air space back of it, and therefore the air space affects the absorption characteristics of the panel. For example, Fig. 7.1 gives the absorption-frequency characteristics of birch plywood panels attached to frames 2 feet by 9 feet made from 1½-inch by 1¼-inch hard maple stock with two cross braces spaced 3 feet apart.[1] Sheets of plywood 2 feet by 9 feet were fastened to the frames and cross braces. The absorption-frequency characteristics with a 1¼-inch air space behind the panels is shown in Fig. 7.1(*a*). The effect of increasing the air space to 2¼ inches is illustrated in Fig. 7.1(*b*). The resonant frequencies are altered and the absorption is increased as a result of increasing the depth of the air space. If the panel is so thin that its stiffness is small in relation to the stiffness resulting from the enclosed air space, the resonant frequency of the panel plus air space is approximately equal to $170/\sqrt{md}$ cycles, where m is the mass of the panel in pounds per square foot of surface area and d is the depth of the air space in inches.

The absorption coefficient of a thin, wood panel can be increased by placing an absorptive material, such as a mineral-

[1] P. E. Sabine and L. G. Ramer, *J. Acoust. Soc. Am.*, 20, 267 (1948).

wool blanket, in an enclosed air space behind the panel, or by spot-cementing the absorptive material directly to the panel. This effect is shown in Fig. 7.2(*a*), which is for the same construction as Fig. 7.1(*a*) except that an absorptive material 1-inch thick is placed in the air space, leaving a ¼-inch unfilled space

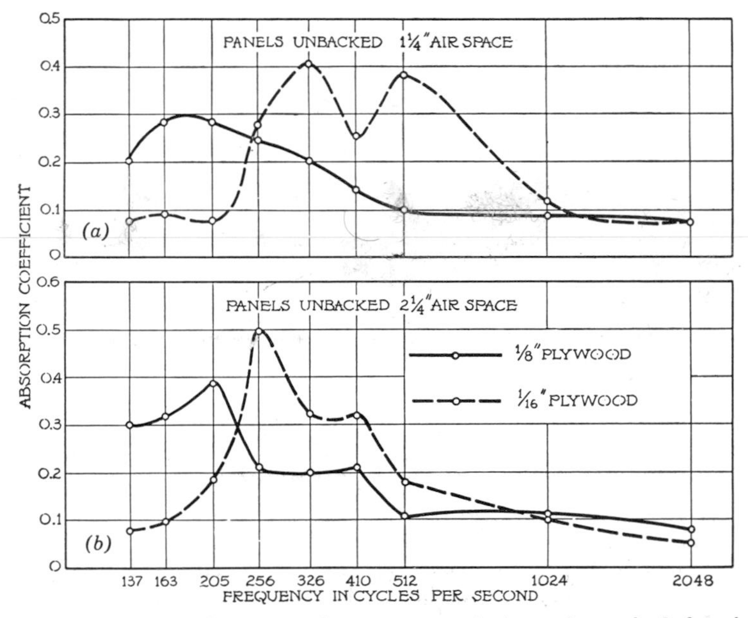

Fig. 7.1 Absorption-frequency characteristics of plywood panels 2 feet by 9 feet attached to frame with transverse cross braces 3 feet apart: (*a*) with 1¼-inch air space behind panels, (*b*) with 2¼-inch air space behind panels. (P. E. Sabine and L. G. Ramer.)

between the panels and the absorptive material. The addition of absorptive material has little effect on the coefficient at frequencies well above resonance. Its principal effect is to increase the absorption in the region of resonance and, usually, to broaden this region. Figure 7.2(*b*) gives the absorption coefficient for the same combination of materials as Fig. 7.2(*a*) but with the absorptive material spot-cemented directly to the panels. The results revealed by Fig. 7.2 suggest that, by altering the thickness and size of flexible panels and also the depth of the air space and the nature of the blanket behind the panels, the absorptive characteristics of this type of acoustical treatment can be varied over a rather wide range.

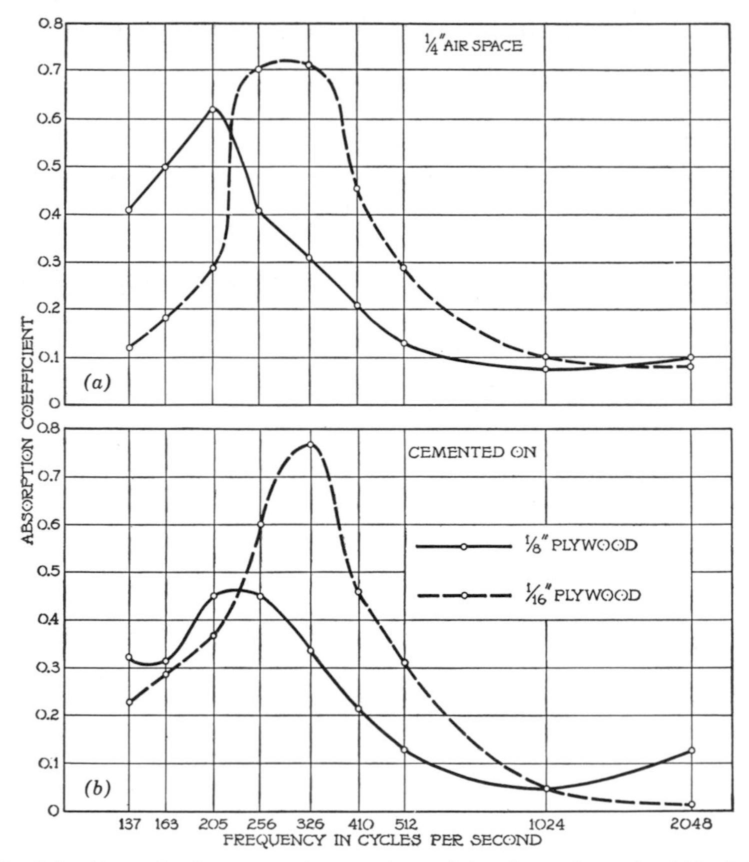

Fig. 7.2 Absorption-frequency characteristics of the plywood panels of Fig. 7.1 backed with mineral-wool blanket 1 inch thick: (*a*) with the absorptive blanket and panel, (*b*) with the absorptive blanket spot-cemented to the panels. (P. E. Sabine and L. G. Ramer.)

Helmholtz Resonator Absorbers

It is well known that it is possible to produce a tone by blowing across the mouth of a bottle or jug. If the volume of the jug (or chamber) is V, and the neck has a cross-sectional area A

and a volume v, the frequency f of this tone is given, approximately, by

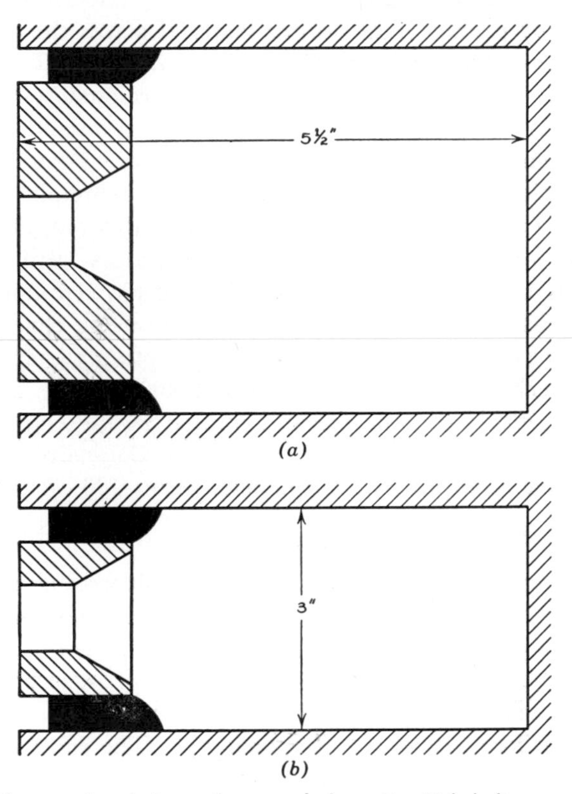

Fig. 7.3 Cross-sectional views of a sound-absorptive Helmholtz resonator built in as rectangular enclosure in a brick wall. This resonator has a resonant frequency of 199 cycles. (V. L. Jordan.)

$$f = 2160 \frac{A}{(vV)^{1/2}} \quad \text{cycles} \tag{7.1}$$

where all dimensions are expressed in inches. As the jug is emptied, the resonant frequency will decrease. The air in the neck of the resonator vibrates back and forth somewhat as a single mass, and the larger volume of air acts as a spring or restoring force. Systems of this kind are called *Helmholtz resonators* and have application in sound-absorptive treatments. In such resonators, frictional resistance is encountered by the alternating flow of air in and around the neck. Hence sound energy is absorbed, mainly in the region of the resonant frequency f. These losses can be increased by placing light porous material across the mouth or, to a lesser extent, by placing absorptive material in the chamber.

There is evidence which indicates that resonators were used in churches constructed in Sweden nearly a thousand years ago. They were made of clay in a wide range of sizes. Figure 7.3 shows a cross-sectional view of a sound-absorptive Helmholtz resonator used in the acoustical treatment of the assembly hall at the University of Aarhus (Denmark).[2] The resonators were made up of regular brick-enclosed chambers having circular openings in specially made tiles in the brick walls. By using chambers and openings of different sizes, the sound absorption was spread over the frequency range between 100 and 400 cycles. Helmholtz resonators, made of concrete and located in the ceiling of the concert studio of the Danish Broadcasting Company, have been used to give absorption at frequencies below 100 cycles.

Resonator-Panel Absorbers

More frequently used than the Helmholtz-resonator type of absorbers, however, are Helmholtz resonators combined with other absorptive structures, especially panels. For example, in some frequency regions this combined action is largely responsible for the absorption by prefabricated materials of type II-A materials (those with perforated facings).[2] The effect of varying the volume of air behind a perforated panel is to alter the resonant frequency of the vibrating system. This is illustrated in Fig. 7.4; in the lower part is shown a cross section of a perforated panel on hinges, backed with an air space of variable volume. The perforations are covered by a layer of porous cloth.[3] The volume of air space can be varied simply by turning the panel on its hinges. The curves in the upper part of Fig. 7.4 give the absorption coefficients for the structure, (*a*) when the panel is in the position shown, and (*b*) when it is entirely closed.

[2] V. L. Jordan, *J. Acoust. Soc. Am.*, **19**, 972 (1947). Also see papers by P. V. Brüel and A. J. King, *Acoust. Group Symposium, 1947,* Physical Society (London), 1949.

[3] The cloth has a flow resistance of about 10 acoustical ohms per square centimeter.

The maxima in these curves occur at a somewhat higher frequency than do the maxima for the unperforated panel absorbers whose absorptive characteristics are given in Fig. 7.1. The absorption of a resonator panel can be increased by means of light porous cloth mounted on either the inner or outer sur-

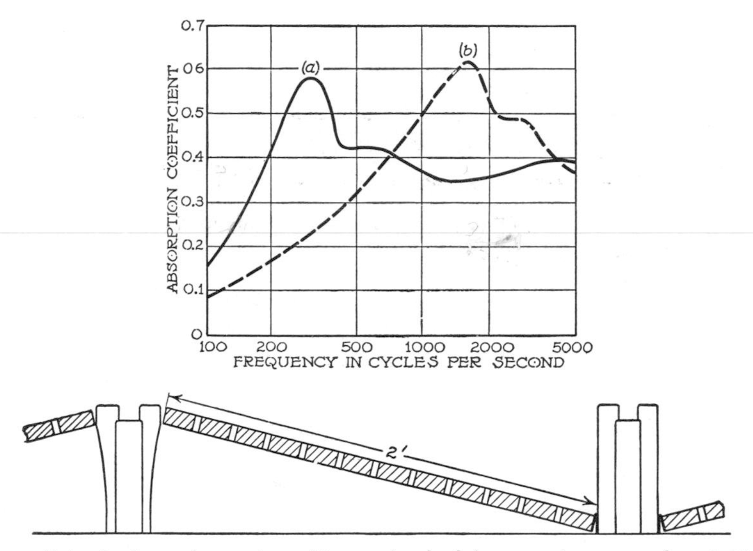

Fig. 7.4 Perforated panel on hinges, backed by an air space of variable volume. The curves give the absorption-frequency characteristic for the structure (*a*) when the panel is in the position shown, and (*b*) when the hinged panel is entirely closed. The wood panel is ⅝ inch thick and is perforated by circular holes 1 centimeter in diameter which are spaced 3.54 centimeters on centers. (V. L. Jordan.)

face of the perforated facing. Increased absorption may also be obtained by introducing absorptive material in the air-space backing.

Draperies

In general, draperies are not satisfactory for the absorptive treatment of an auditorium; although very absorptive at high frequencies, they are only slightly absorptive at low frequencies. An auditorium so treated may sound "boomy." Hence the use of draperies, unless especially designed, should be restricted to places such as doorways or prosceniums. For maximum absorption, they should be made of heavy, lined and interlined velours (or equivalent material) and should have a gather of 100 to 200 per cent. In order to increase their absorption at the lower frequencies, hangings used to cover highly reflective surfaces should be hung at least 6 inches from the wall and should be gathered into deep folds. The effect of spacing is illustrated in Fig. 7.5, which gives the measured values of the absorption coefficients of

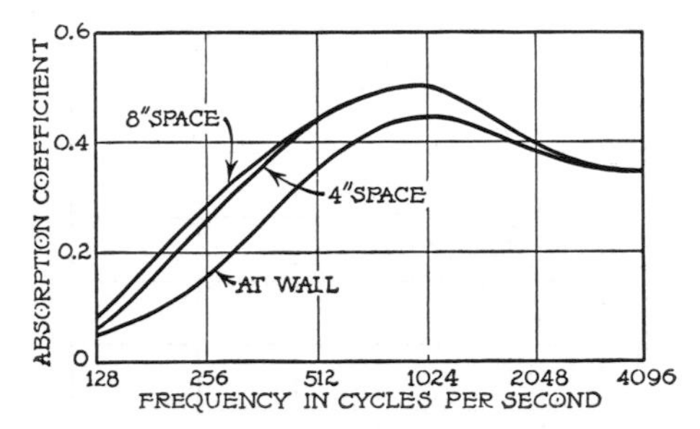

Fig. 7.5 Absorption coefficient *vs.* frequency of a curtain "straight hung," at different distances from a hard wall.

a curtain "straight hung" at several different distances from a very hard wall.

Movable draperies provide a convenient means of altering the total absorption in a room. Figure 20.3 is a photograph of one of the side walls in the National Broadcasting Company Studio 3A in New York City. Heavy, lined draperies hung 1 foot from the wall, with 100 per cent gather, are used as one of the devices for changing the acoustical qualities of the studio.

Variable Absorbers

Certain rooms, especially some broadcast studios and music rooms, make use of variable absorbers such as hinged panels, rotatable cylinders, or movable draperies. The hinged panels generally are absorptive on one side and reflective on the other; the cylinders incorporate various combinations of absorbers and reflectors. These special devices are utilized for varying and controlling the acoustical conditions in the room. In general, the control of the following three factors is desirable: (1) the magnitude of the average absorption over the greater part of the audible frequency range, (2) the shape of the absorption *vs.* frequency characteristic, and (3) the scattering or dispersion of sound in the room. Scattering influences the uniformity of the sound-pressure distribution within the room. This topic is discussed in Chapter 8.

A variable absorptive treatment should provide the possibility of varying the total absorption in the room over a wide range—in some instances by a ratio of at least 3 to 1. If hinged or rotatable panels are used which have an absorptive surface on one side and a thin reflective surface on the other, care must be exercised so that the reflective surface, such as a plywood facing,

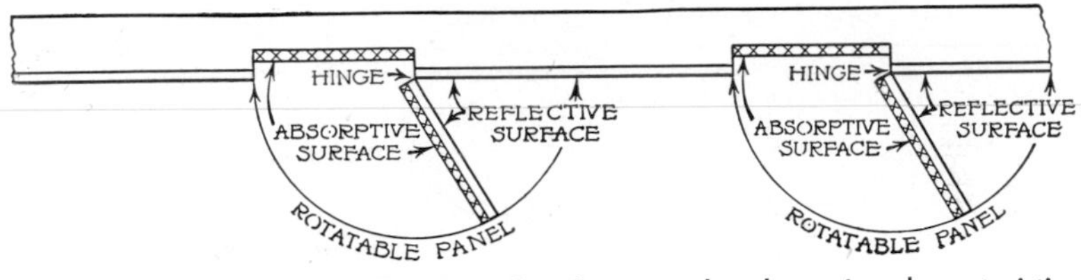

Fig. 7.6 Hinged-panel arrangement for changing the absorptive characteristics of a room.

does not act as a resonant panel backed by an absorptive material. If this precaution is not taken, the total variation in the absorption may not be so great as expected or desired.

In broadcast studios it is important that the operation of the variable treatment be silent, and not so complicated that it is frequently inoperative. As illustrations of practical designs, a few types of variable absorbers—hinged panels, rotatable cylinders, and rotatable panels—will be described briefly. Others equally satisfactory, or even better, for specific purposes, can be designed by competent engineers.

Hinged Panels. In rooms which require a variable absorptive treatment, where the expense or other design considerations do not justify the use of rotatable cylinders or panels, hinged panels may be found useful (for example, see Fig. 20.3). Their cost is not exorbitant, and they are easy to maintain. One side of the panels is highly reflective; the other side is treated with absorptive material. They can be used either on flat walls, as shown in Fig. 7.6, or on splayed walls. A hinged-panel construction incorporating resonators has been described in an earlier section.

ROTATABLE CYLINDERS. Figure 7.7 shows a section through rotatable cylinders designed to provide variable absorption for the ceiling of a music room. The convex surface of each cylinder is made up of three different materials, each extending the full length of the cylinder and 120° around it. The cylinders are fitted into openings of such size that 120° of each projects through a suspended plaster ceiling. The three materials differ greatly in their absorptive properties: material *a* is a 2-inch layer of Fiberglas, having a density of 6 pounds per cubic foot, covered with

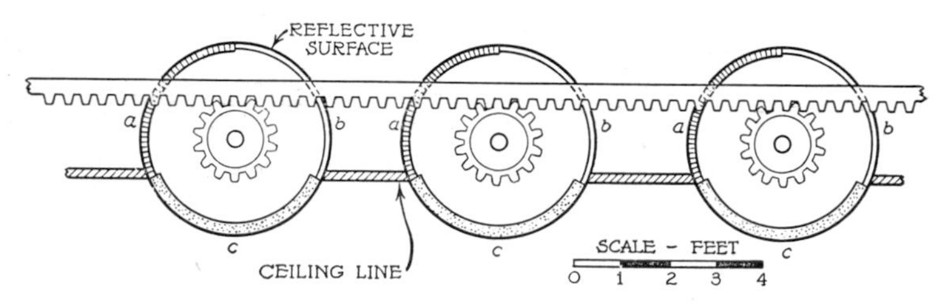

Fig. 7.7 Rotatable cylinders for varying the absorptive characteristics of a music room at the University of Washington.

¼-inch perforated plywood (there are 1024 circular holes per square foot; the holes are ⅛ inch in diameter and are spaced in a rectangular array ⅜ inch on centers both ways); material *b* is ⅛-inch unperforated plywood backed with a 2-inch layer of Fiberglas; and material *c* is ½-inch unperforated plywood. Material *a* is moderately absorptive at low frequencies and is increasingly absorptive at higher frequencies; on the other hand, *b* is most absorptive at low frequencies and is decreasingly absorptive at higher frequencies; and *c* is only slightly (but uniformly) absorptive at all frequencies. The cylinders are rotatable so that surfaces *a*, *b*, or *c*, or any desired combination of two of them, can be exposed to the room.

ROTATABLE PANELS. Rotatable panels are similar to the cylinders of the preceding section; both are used to form part of the boundaries of a room, and their rotation changes the total absorption in the room. They differ from the cylinders in that their rotation generally changes the shape of the walls or ceiling. Hence the rotation of panels has a pronounced influence on the diffusion as well as on the absorption of sound in the room.

A photograph of a wall in a broadcasting station in which such a variable absorptive treatment is used is shown in Fig. 20.8. One side of the panel is flat and is covered with acoustical tile; the other side of the panel is convex and is "treated hardboard." Sometimes the opposite sides of the panel may be made of the same material, differing only in shape. Such panels are used to control diffusion. They can be backed by an air space and an absorptive blanket. Various amounts of absorption can be obtained in the room by rotating the panels to intermediate positions.

Suspended Absorbers

In certain types of enclosed spaces (for example, in large machine shops having extremely high ceilings) it is difficult to apply the conventional type of acoustical treatment so that absorptive surfaces will be located near the sources of noise. In such cases, recourse may be had to relatively small prefabricated units of absorptive material hung from the ceiling. The use of such suspended absorbers is especially adaptable to locations where there are no extended surfaces on which to apply acoustical tile, or similar materials, and where it would be difficult or expensive to install a false ceiling because of pipes or other obstructions. Such treatment need not interfere with existing lighting or ventilating systems. Suspended units may be installed and removed relatively quickly but, unless well guyed or restricted to locations where they will not be set into motion by air currents, they may prove to be a possible source of distraction. Owing to diffraction, the effective absorption per unit area or per unit weight of small suspended absorbers can be very high. Sound waves impinge on "both sides" of the absorbers, thus enhancing their absorption. (In fact, such devices can very well have effective absorption coefficients greater than unity; such coefficients are possible also for small patches of highly absorptive acoustical materials having only one side of the material exposed to the sound.) This increase in absorption has been investigated by Sivian, Bedell, and O'Neil, who measured the coefficients of suspended panels of rock wool and also of cloth glued to heavy perforated metal. Figure 7.8 shows that a panel of muslin-covered[4] perforated metal 3 feet by 8 feet provides substantially greater absorption when suspended so that (*a*) waves impinge on it from both sides than when (*b*) it is mounted 2 inches from a hard wall. Their results also indicate that the increase in ab-

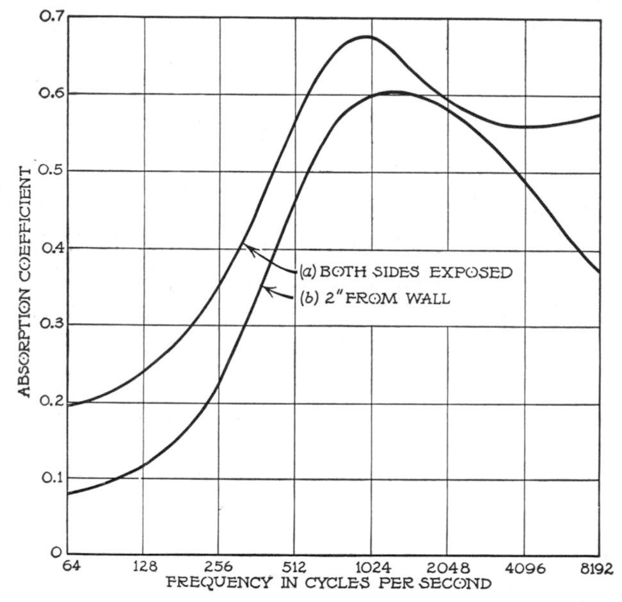

Fig. 7.8 Absorption coefficient *vs.* frequency for muslin-covered perforated sheet metal (*a*) when suspended so that sound has free access to both sides of it, and (*b*) when mounted 2 inches from a hard wall. (L. J. Sivian and H. T. O'Neil.)

sorption depends on the positions on the ceiling from which the panels are suspended.

One system of suspended absorbers has been described by Olson.[5] Two single conical shells, fabricated of a material made from shredded wood and a binder, were fastened together at the base of the hollow cone as shown in Fig. 7.9. The absorption coefficient (the absorption per conical absorber divided by the total surface area of the absorber) was measured for units

[4] Flow resistance of the muslin is about 60 acoustical ohms per square centimeter.

[5] H. F. Olson, *RCA Review,* 7, 503 (1946).

having different diameters at their base. These data are reproduced in Fig. 7.10. The total sound absorption per pound of material for this type of treatment is about twelve times that of conventional types of acoustical materials. The effect of in-

Fig. 7.9 Suspended absorber made up of two conical shells. (H. F. Olson.)

creasing the size of the unit is to increase the absorption at low frequencies and to decrease it at high frequencies. Figure 7.11 is a photograph of a drafting room treated with suspended absorbers of the conical type. These units were hung 10 feet

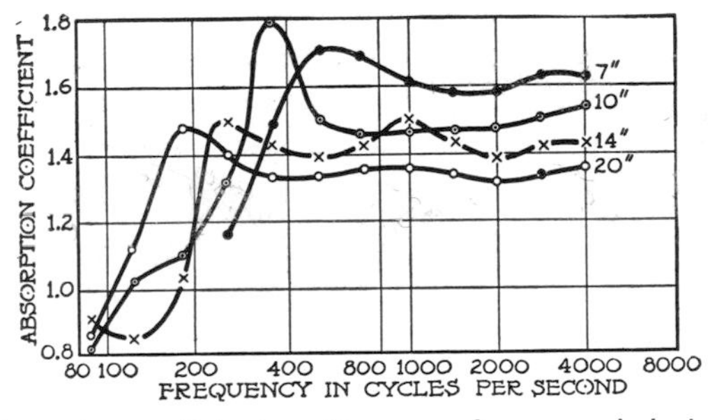

Fig. 7.10 Absorption coefficient vs. frequency for suspended absorbers of the type shown in Fig. 7.9 with the base diameters indicated. (H. F. Olson.)

above the floor level and were so spaced that there was one absorber for each 5 square feet of floor space. They had a 14-inch diameter and an absorptive "efficiency" of 7.4 square-foot-units (sabins) of absorption per pound of material. The room had a truss roof with a ceiling height ranging from 12 to 20 feet. Sus-

Fig. 7.11 Drafting room acoustically treated with the suspended absorbers shown in Fig. 7.9. (Courtesy RCA Laboratories.)

pended absorbers are therefore well adapted for an economical and effective treatment of such a room. Other absorbers of this type were fabricated from the same material in the shapes of cubes, spheres, and cylinders.[6] Measurements showed that units having equivalent volumes had approximately the same absorptivity, independent of shape, up to 4000 cycles.

[6] For a discussion of sound-absorbent spheres, see R. K. Cook and P. Chrzanowski, *J. Acoust. Soc. Am.*, **21**, 167 (1949).

8 · Principles of Room Acoustics

The principles of room acoustics include those properties of sound already considered in relation to the design of open-air theaters. But the acoustics of enclosed spaces involves certain additional principles concerned with the reflection, absorption, and scattering of sound at the boundaries of the enclosures. The effects of the boundaries give rise to reverberation and resonance, properties of paramount importance in controlling the acoustics of all rooms. Sound is prolonged in a room, even after the original source has stopped, because the sound continues as a succession of reflections from the walls, floor, and ceiling. Such persistence of sound is *reverberation.* The closely associated phenomenon of *room resonance* is not so well known to the layman, although its effects are generally familiar. A speaker at a considerable distance from a listener usually can be heard more distinctly in a room than he can in the open air. The reason is that *the boundaries of the room alter the distribution of sound emanating from the source;* they confine the energy that would, in the open, proceed outward into space. Thus, the boundaries, unless completely absorptive, increase the average sound level within the room.

The influences of the shape, the size, and the reflective and absorptive properties of a room on its acoustical characteristics are discussed in detail in this chapter. The above-mentioned phenomena, resonance and reverberation, and others such as *echoes, sound foci,* and *room flutter,* will be described. The principles that will be developed are applicable, in general, to all types of rooms. In Chapters 15 through 20 these principles will be applied to the acoustical design of particular kinds of rooms, such as auditoriums, lecture rooms, churches, theaters, sound-recording and broadcasting studios.

Room Resonance, Normal Modes

The use of hard plaster and tile for the interiors of rooms introduces many acoustical difficulties; it has, however, encouraged man's vocal aspirations. In such enclosures, the voice usually is enhanced by room resonance. It is not merely the voice of a singer that one then hears. It is the voice actuating the *normal modes of vibration* of the room. In a real sense, the singer is playing on an organ-like instrument capable of resonating at a large number of frequencies. Each time a note is sung, at least one, and usually more, of the normal modes of the room are strongly excited. (The normal modes of vibration are sometimes referred to as *natural modes of vibration.*)

Perhaps the reader has observed that many noises can set a lampshade or windowpane into vibration, and that such a vibrating object seems to produce a sound of definite pitch. Actually, the object vibrates at one of its *resonant frequencies* (sometimes called *normal frequencies* or *natural frequencies*). The closer the frequency of the exciting noise to a resonant frequency of the object, the greater will be the response of the resonant vibration. In the case of the organ pipe, it is the column of air that vibrates within the pipe. Its resonant frequencies are determined principally by the length of the air column.[1] Similarly, in a room, the air is set into vibration, and the resonant frequencies are determined approximately by the room dimensions.

Now let us consider one of the effects of the resonant frequencies of a room on its acoustical properties. First, suppose that at a given point in the open air a near-by source produces a sound

[1] The lowest resonant frequency of the air column of an open pipe has a wavelength which is approximately twice the length of the pipe. This normal mode of vibration is sometimes called the *fundamental* or *gravest mode.* The open pipe can also resonate at frequencies corresponding to integral multiples of the lowest resonant frequency.

pressure that has the same level at all frequencies. If this source is now placed in a room, the pressure at the same distance from the source will no longer be constant with frequency; instead, it will be much higher at some frequencies than at others—in particular, at the resonant frequencies of the room. This is illustrated by Fig. 8.1 which shows a record of sound-pressure

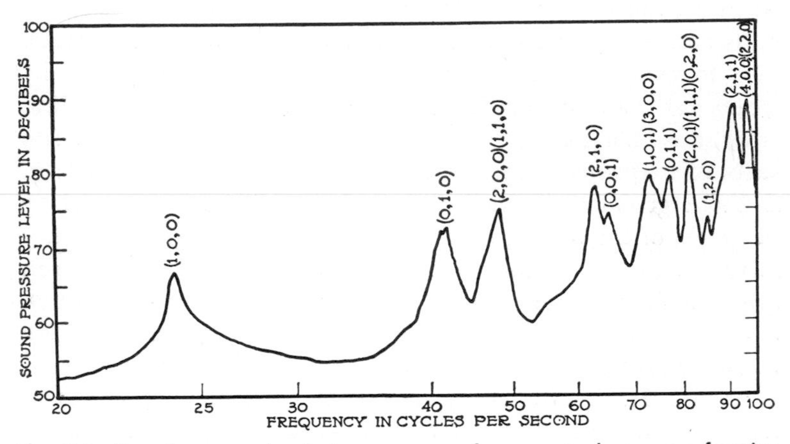

Fig. 8.1 Sound-pressure level at one corner of a room is shown as a function of the frequency of the sound generated by a loudspeaker in another corner. This frequency-response curve was taken in an empty room 23.0 feet by 13.4 feet by 8.4 feet. Note the increased level at the resonant frequencies. The normal modes of vibration corresponding to these frequencies are specified by the numbers within the parentheses above the peaks.

level *vs.* frequency for a source of sound in a room. The level was recorded at a fixed position in one corner of a room, as a function of the frequency of the sound generated by a loudspeaker in another corner. Records of this kind are called *transmission-frequency characteristics.*

The resonant frequencies of an enclosed space are usually difficult to calculate. For a rectangular room having smooth, hard walls, however, the computations are simple. Let L, W, and H be the length, width, and height, respectively, of a rectangular room, and let c be the velocity of sound. Then the resonant frequencies f of the room are obtained by substituting various integers for p, q, and r in the following equation:

$$f = \frac{c}{2}\left[\left(\frac{p}{L}\right)^2 + \left(\frac{q}{W}\right)^2 + \left(\frac{r}{H}\right)^2\right]^{1/2} \quad \text{cycles} \tag{8.1}$$

The values p, q, and r specify the mode of vibration. For example, the (1, 0, 2) mode is obtained by substituting $p = 1$, $q = 0$, and $r = 2$. The empty rectangular room in which the experimental data for the frequency-response curve of Fig. 8.1 were taken has the dimensions 23.0 feet by 13.4 feet by 8.4 feet. The walls and ceiling are of hard plaster and the floor is concrete. The lowest resonant frequency for this room, computed by means of the above equation, is 24.5 cycles. Others, in ascending order, are 42.5, 48, 49, $\cdots$ cycles, continuing in a triply infinite series of frequencies corresponding to increasing values of p, q, and r. Agreement between these values and those indicated by the peaks of the solid-line curve is very good. As the frequency increases, the peaks come closer and closer together and even overlap. Finally, the resonant frequencies of contiguous normal modes are so close together that the overlapping peaks coalesce and thus give a relatively uniform transmission-frequency characteristic. In this frequency region the total number of modes N, from the lowest frequency of response up to any frequency f_c, in a room of volume V, is given approximately by

$$N = 4V\left(\frac{f_c}{c}\right)^3 \tag{8.2}$$

This formula applies when f_c is greater than $4c/V^{1/3}$. For example, in a room having a volume of 110,000 cubic feet, there is a total of 440 modes up to 110 cycles, and up to 1100 cycles the total is 440,000. When the frequency is sufficiently high so that N is large, the number of modes per cycle (dN/df_c) at that frequency is given approximately by

$$\frac{dN}{df} = \frac{12Vf_c^2}{c^3} \tag{8.3}$$

Thus, for the above example, there are about 12 modes per cycle at 110 cycles and 1200 modes per cycle at 1100 cycles.

Equation (8.1) applies to a hard-walled rectangular room. A change in shape of the room from rectangularity shifts the frequencies of the normal modes, making them difficult (if not practically impossible) to compute. However, *altering the shape does not alter the total number of normal modes within the limits of accuracy of Eq. (8.2).* If the volume is maintained constant, the total number of modes up to a given frequency, as indicated by Eq. (8.2), remains approximately the same.

The resonant frequencies of a room usually are not distributed uniformly with respect to frequency (that is, the curve of dN/df *vs.* frequency is not smooth). In some frequency regions the modes "pile up." Sounds having a frequency within a region of concentration of modes are enhanced by the resonant action of the room. This effect has been particularly noticed in churches with masonry walls. In such rooms, a few notes of the organ seem to respond better than others. Many chants have been written in monotones which correspond to these resonant regions, and many a clergyman has adapted the pitch of his voice to conform to one of them. The distribution of the normal modes of vibration of a room is one of the considerations which determine the optimum shape of a broadcasting studio, as Fig. 20.9 indicates. It is appropriate to emphasize that irregularities in both the shape of the boundaries and the distribution of absorptive materials are frequently important in providing a uniform transmission-frequency characteristic for the room. However, these irregularities are effective only when their size is of the same order of magnitude as the wavelength of the resonant frequencies. Thus, ornamentation on the walls of a room, or pillars, etc., will not alter significantly the frequencies of the lower modes of vibration.

One effect of increasing the amount of absorptive material in a room is to smooth out its transmission frequency characteristic. Thus, if additional absorption were introduced in the room of Fig. 8.1, the peaks would be broadened and lowered so that they would overlap and coalesce.[2] Also, they would be shifted slightly in frequency. Other effects of the addition of absorptive material are considered in subsequent sections.

[2] E. C. Wente, *J. Acoust. Soc. Am.*, **7**, 123 (1935); F. V. Hunt, *J. Acoust. Soc. Am.*, **10**, 216 (1939).

Sound-Pressure Distribution

It is fortunate for those who occupy the rear seats in a balcony that an enclosure modifies the distribution of sound about a source; the average sound level [3] is generally somewhat higher than it would be at the same distance in the open air. The level would decrease at a rate of 6 db for each doubling of the distance away from a source of sound in the open air, away from

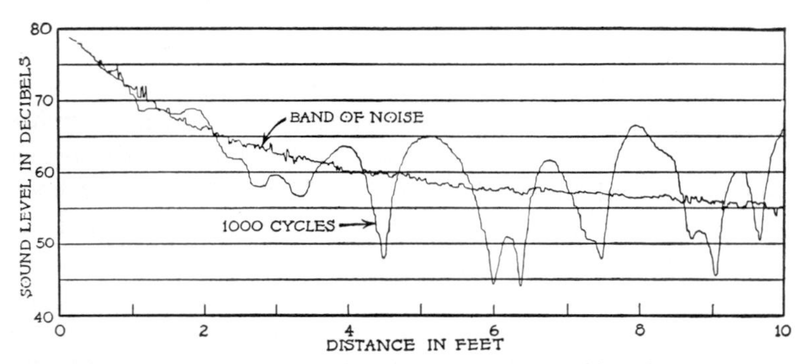

Fig. 8.2 Sound-pressure level *vs.* distance from a loudspeaker in a room, along the axis of the speaker. One curve shows the pressure distribution for a 1000-cycle tone. The other curve shows the pressure distribution for a sound containing many frequencies.

all reflective surfaces. Figure 8.2 shows that, when a source of sound is placed in a room, the average level does not drop off so rapidly. In this figure, two experimental records are given. One shows the sound level *vs.* distance from the loudspeaker when it emits a 1000-cycle tone; there are many dips caused by interference between the waves coming directly from the source and those that are reflected from the walls of the room. The other record shows similar data when the loudspeaker emits a band of thermal noise (600 to 1200 cycles). Here the sound level decreases relatively uniformly. The source of noise consists of many frequencies. If a number of minima, for different frequency components, occurred at the same position there would be a pronounced dip ("dead spot") in the sound pressure at that

[3] The term *sound level* is sometimes used loosely instead of the term *sound-pressure level;* at times, both terms are contracted to *level.*

point. However, the maxima for some frequencies are usually at the minima of others, so that the pressure distribution tends to be "smoothed out." As the frequencies present increase, the sound pressure in the room tends to become more and more uniform; that is, the sound field becomes more *diffuse*. The average level is about 3 db higher at a distance of 8 feet from the loudspeaker than it would be at the same distance in the open air.

Diffusion of Sound

Sound is said to be perfectly diffuse in a room if its pressure is everywhere the same and if, at all points in the room, it is equally probable that waves are traveling in every direction. Of course, it is quite impossible to obtain complete diffusion; in fact, this condition would not be desirable. In such a sound field, a listener would find it difficult to determine the direction from which a sound originated. But both performer and listener preference indicate that a certain amount of diffusion in a room is desirable. Diffusion promotes a uniform distribution of sound; it insures a relatively smooth growth and decay of sound; and it improves the "liveness" of the room. In short, it tends to enhance the natural qualities of speech and music. A certain amount of diffusion is especially advantageous in rooms in which microphones are used, as it greatly reduces the hazards of improper microphone placement.

Live and *dead* are qualitative terms applied to certain aspects of the acoustical properties of a room. At any point in a room, the "liveness" depends on the ratio of the amount of sound which is reflected to that point from the boundaries to the amount coming directly from the source.[4] The larger this ratio, the greater is the "liveness." Hence the farther one moves away from a source the greater will be the apparent *liveness*. Conversely, the nearer one approaches a source in a room the "deader" it will seem. In rooms containing a very large amount of acoustical absorption, there will be little reflected sound. Hence such rooms are referred to as "dead." Since the reflected sound depends on the boundary conditions, liveness is also a function of diffusion.

[4] J. P. Maxfield has considered certain aspects of room liveness; see, for example, *Western Electric Oscillator*, 6, 3 (1947).

Diffusion of sound in a room is increased by (1) objects within the room that scatter and thus randomize the directions of the sound waves, and by (2) the irregularities in the wall surfaces that similarly scatter the impinging waves by the reflective and absorptive properties of the walls. Thus, sound is more diffuse in a room after it is furnished than when it is empty. The chairs, tables, hangings, etc., are effective scattering agents. Irregularities in wall surfaces, such as splays, columns, coffers, or any form of relief or ornamentation, are especially helpful in increasing the diffusivity of the sound field. It should be stressed that these objects or irregularities are most effective in promoting diffusion when their size is of the same order of magnitude as the wavelength of sound. (Recall that a frequency of 512 cycles has a wavelength of about 2 feet.) They have very little diffusive effect when they are small in relation to the wavelength. In general, the diffusion of sound in a room increases with increasing frequency.

Another means of increasing the diffusion of the sound in a room is the use of an irregular distribution of absorptive material, preferably in patches.[5] In many rooms, where cost or other factors do not permit the use of splays or other types of surface irregularities, the use of distributed patches of absorptive material is the most feasible method for controlling diffusion. As an example, consider a room with 1000 square feet of absorptive material on one wall all in a single area. The diffusion would be increased if the one large area were separated into ten strips of 100 square feet each; there would be another increase in diffusion if these ten strips were distributed among the other walls and on the ceiling. The diffusion would be even greater if the strips were divided into small patches of unequal size. *Diffusion increases with increasing randomness of application of absorptive material.* As with splays and other scattering agents, patches of absorptive material are effective as diffusers only for wavelengths of sound that are of the same order of magnitude as their dimensions. A photograph of an auditorium wall in

[5] H. Feshbach and C. M. Harris, "The Effect of Non-Uniform Wall Distributions of Absorbing Material on the Acoustics of Rooms," *J. Acoust. Soc. Am.*, **18**, 472 (1946).

which both splays and absorptive patches are incorporated is shown in Fig. 15.4.

The effects of diffusion on room acoustics are quite prominent. Under steady-state conditions (while the source is producing sound continuously), diffusion increases the uniformity of the spatial distribution of sound pressure. Under transient conditions (while the sound is building up or decaying), diffusion increases the uniformity in the rate of growth and decay of sound in a room. The effect on decay is illustrated by the curves of Fig. 8.3. The upper curve is a record taken in a rectangular room having most of its absorptive treatment concentrated on the ceiling; there is little diffusion. There are large fluctuations in sound-pressure level as the sound dies out. The lower graph of Fig. 8.3 shows the decay of sound in the same room after the diffusion has been greatly increased. Now the waves are constantly being altered in direction, resulting in a more uniform rate of absorption. Note that the amplitude of the minor fluctuations is greatly decreased.

Physical and Geometrical Acoustics

The preceding sections were concerned with room resonance, with pressure distribution, and with diffusion of sound in a room. These phenomena were considered in terms of the normal modes of vibration of the enclosure which are determined by the size and shape of the room, by the absorptive properties [6] and shape of its boundaries, and by the objects within the room.

A rigorous analysis of the actual behavior of sound waves in an enclosed space must be based on such considerations. The influence of localized areas of absorptive material or the influence of non-uniformities in room geometry, such as splays, on both the transient and steady states of sound can be taken into account. These methods of determining the acoustical properties of a room are those of *physical acoustics.*[7] They can be used to calculate the behavior of sound in rectangular rooms (and a few other regular shapes) having special distributions of absorptive materials, so long as consideration is confined to only a few low-

[6] More exactly by the *acoustical impedance.*

[7] Sometimes referred to as *wave acoustics.*

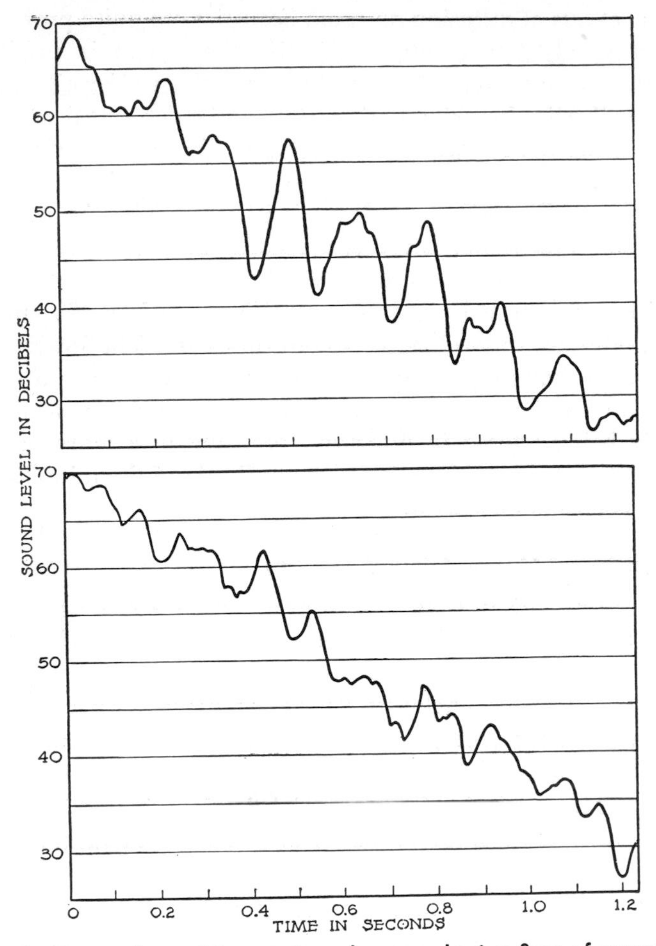

Fig. 8.3 Decay of sound in a rectangular room having floor of poured concrete and absorptive material concentrated on the ceiling. The upper curve is an actual decay record when there is relatively little diffusion. The lower curve is for the same room after the diffusion has been greatly increased by scattering obstacles. (R. W. Roop, courtesy Johns-Manville Research Laboratory.)

frequency modes. However, the formulas become so complicated in actual problems in which non-uniform boundaries and many modes are involved (the cases which interest the architect and engineer) that it is not yet feasible to present a complete set of *practical* design formulas based on physical acoustics.[8] Even those who are well versed in the mathematical complexities of physical acoustics are confronted with the most formidable tasks of computation, sometimes involving weeks of routine labor with computing machines, for even a relatively simple acoustical treatment of rectangular rooms.

The methods of *geometrical* (ray) *acoustics* avoid these complications. Although only approximate for dealing with most problems, and utterly misleading for some others, these methods are extremely useful, if not indispensable, when used with their limitations in mind. Geometrical acoustics assumes that sound travels as rays, in a manner similar to that of light rays. Upon reflection from a wall, part of the energy is absorbed and part reflected; the angle of the reflected ray is equal to the angle of the incident ray if the surface is large compared with the wavelength. As applied to rooms, geometrical acoustics generally involves the following assumptions: (1) that the sound in the room is perfectly diffuse after a large number of successive reflections of the rays which emanate from the source; (2) that only the frequency of the sound source is present during the growth and decay. These assumptions are an oversimplification of the conditions that prevail. Experimental studies show that, during the growth and decay of sounds in a room, frequencies other than the frequencies of the source may be present. They correspond to the resonant frequencies of the room.

The existence of normal modes of vibration is completely ignored in the geometrical acoustics. However, the example given in Fig. 8.1 shows that at low frequencies the presence of resonant frequencies of the room is very much in evidence. At these lower frequencies, a tone may strongly excite only a single mode. If the frequency of the tone corresponds to one of these resonant

[8] For the most up-to-date and comprehensive development of the pertinent theory and relevant formulas describing the behavior of sound in rooms in terms of physical acoustics, see Chapter VIII, P. M. Morse, *Vibration and Sound,* McGraw-Hill Book Co., 1948.

frequencies, the resulting sound level will be higher than it would be if the frequency of the tone were shifted "off resonance." This is in contrast to the conditions at the higher frequencies, where very many modes would be excited. Here, the resonant frequencies are so numerous and so close together that a slight shift in the frequency of the tone will not change the sound level; the transmission-frequency characteristic is relatively smooth in contrast with that of Fig. 8.1. Furthermore, Fig. 8.2 illustrates that the sound pressure in a room, under steady-state conditions, may not be diffuse but may have rather large deviations. In the preceding section it was indicated that these deviations are quite marked at frequencies where the wavelength of the sound is comparable to the dimensions of the room, and that they become less and less prominent at higher frequencies. In other words, when the wavelength of the sound becomes (1) very small compared to the dimensions of the room and (2) comparable with objects within it, the conditions of geometrical acoustics are approached.

Although only physical acoustics can give a complete description of the acoustical properties of rooms, it is fortunate that geometrical acoustics is, under appropriate conditions, an approximation to physical acoustics. The use of geometrical acoustics, *when employed with an understanding of the consequences and modifications that physical acoustics entails,* can lead to highly satisfactory results for application to the design of most rooms, such as auditoriums, theaters, offices, and classrooms. In short, it can be utilized wherever the wavelengths involved are small compared to the dimensions of the room, and are comparable to the size of objects within it and to the size of irregularities on the walls. Hence, for most rooms, geometrical acoustics is applicable for frequencies above about 250 cycles. In larger rooms or auditoriums, the lower frequency limit of applicability is somewhat below this. Broadcasting studios and music rooms must be designed for a wide frequency range involving relatively long wavelengths as well as intermediate and short ones. For such rooms physical acoustics must be used, at least qualitatively, to the fullest possible extent. In the present chapter we shall develop the approximate formulas of geometrical acoustics and supplement them wherever necessary throughout this text with relevant considerations of physical acoustics.

Growth of Sound in a Room

When a source of sound is "turned on" in a room, the sound pressure at any point in the enclosure increases, approaching some limiting value called the steady-state pressure. The growth generally does not take place uniformly; instead, the sound pressure fluctuates, sometimes rather violently, until the steady state is reached. This phenomenon is explained by physical acoustics in the following way. The source sets into vibration the normal modes of the room, exciting to the greatest extent those modes having resonant frequencies nearest the frequencies of the source. All frequencies except those of the source die out very rapidly, but while so doing they interfere with each other, producing pressure fluctuations in the growth. Finally, when the pressure reaches its "steady-state" value, only frequencies of the source are present.

When the number of excited modes is quite large, the results predicted by physical acoustics approach those given by geometrical acoustics. In the latter, it is assumed that the sound in the room is perfectly diffuse, that all directions of propagation of the sound waves are equally probable, and that the sound pressure is everywhere the same within the enclosure. Under these conditions, it is not difficult to show that the average growth of sound pressure at any point in a rectangular room is described by the equation

$$P = 1300\left(\frac{W}{a}\right)^{1/2}(1 - e^{-cat/4V})^{1/2} \tag{8.4}$$

where P is the rms pressure in dynes per square centimeter, W is the acoustical power of the source in watts, V is the volume of the room in cubic feet, c is the velocity of sound in feet per second, t is the time in seconds, and a is the total number of square-foot-units of absorption (given by $a = S_1\alpha_1 + S_2\alpha_2 + S_3\alpha_3 + \cdots$ where α_1 is the absorption coefficient of area S_1, etc.). Because of the assumptions on which its derivation is based, Eq. (8.4) represents the *average* process of growth of sound pressure; fluctuations due to interference phenomena are neglected.

As an example, consider a source having an acoustical power output of 0.0069 watt in an enclosure 40 feet by 70 feet by 100 feet. Here $V = 280{,}000$ cubic feet and the total surface area $S = 27{,}600$ square feet. Suppose that all the absorption takes place at the boundaries which have an average absorption coefficient of 0.05. Then $a = 27{,}600$ square feet times 0.05 or 1380 square-foot-units (sabins). Figure 8.4 is a plot of the pressure obtained by substituting these values in the above equation. A similar curve is given for an average absorption coefficient of 0.25. Note that the maximum value, in the latter case where the absorption is greater, is not so high and is reached sooner.

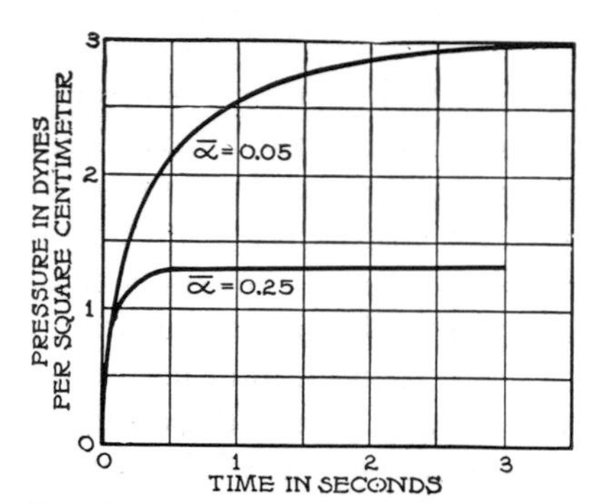

Fig. 8.4 Growth of sound pressure in a room computed for two different values of average absorption coefficient $\bar{\alpha}$ for the boundaries of the room.

Steady-State Value of Sound Pressure

According to Eq. (8.4), after a source of sound has been on an infinitely long time, the maximum value of the sound pressure is reached. Practically, the maximum value (called the "steady-state value") is reached in about 1 second in most rooms. When t becomes large in comparison with $4V/ca$, the exponential term becomes negligible and Eq. (8.4) reduces to

$$P = 1300 \left(\frac{W}{a} \right)^{1/2} \tag{8.5}$$

This is the approximate value of the steady-state sound pressure in a room at distances greater than about 5 feet from the source. The corresponding sound-pressure level is

$$L = 10 \log_{10} \frac{W}{a} + 136 \quad \text{db} \tag{8.6}$$

W is the acoustical power output of the source in watts, and a is the total absorption in sabins. According to Eq. (8.6), the average sound level in a room is independent of the volume and shape of the room. It depends only on the rate of emission W and the total absorption of the room a if the conditions for geometrical acoustics are fulfilled.

Equation (8.6) shows that decreasing the absorption in a room increases the steady-state sound level throughout the enclosure. The example in the preceding section where $W = 0.0069$ watt, illustrates this point. For an average coefficient of 0.05, $a = 1380$ sabins, so that the sound level is about 83 db; whereas, for an average coefficient of 0.25, $a = 6900$ sabins, so that the sound level is about 76 db. It might appear that for a large auditorium, used primarily for speech, the absorption should be low so that the sound level will be correspondingly high. However, it will be shown that the lower the total absorption in a room, the longer will be the persistence of each individual sound that is produced in the room. If unduly long, such persistence can be detrimental to the intelligibility of speech, and also to the quality of certain types of music. Hence a compromise is necessary. The amount and nature of the absorptive material in a room must be chosen carefully.

Decay of Sound in a Room

Sound which originates in, or enters, an enclosed space is repeatedly reflected by its boundaries. At each reflection, a fraction of the acoustical energy is absorbed. Nevertheless, the sound may persist for many seconds before it dies away (decays) to inaudibility. This prolongation of sound after the original source has stopped is reverberation. The greater the volume of a room, and the less absorption it contains, the longer will be the reverberation. A limited amount of it is desirable in most rooms. However, *excessive* reverberation is probably responsible for the majority of acoustical defects which have marred otherwise satisfactory buildings. For example, scores of famous cathedrals in Europe, such as St. Peter's in Rome are so very reverberant that a chord sounded by the organ may remain audible 10 or 15 seconds after the organ has stopped. In many places of worship, a chanted prayer is unrecognizable unless it has been memorized, and music is a confused discordant conglomeration of sound. One of the most striking examples of excessive reverberation known to the authors is a large reading room in a Los Angeles library. In the unfurnished room, a loud shout would remain audible for 25 seconds. This result could have been predicted, in advance of construction, by a ten-minute calculation utilizing equations given later in this chapter. It is difficult to carry on a conversation in such a room, for the duration of the average speech syllable is only about 0.3 second. Successive syllables overlap and commingle in an utterly unintelligible hodgepodge. The impact of a book dropped in the room sounds almost like thunder, and other sources of noise seem to be greatly magnified. *Excessive reverberation is one of the most damaging and annoying defects that can be inflicted upon an auditorium.*[9]

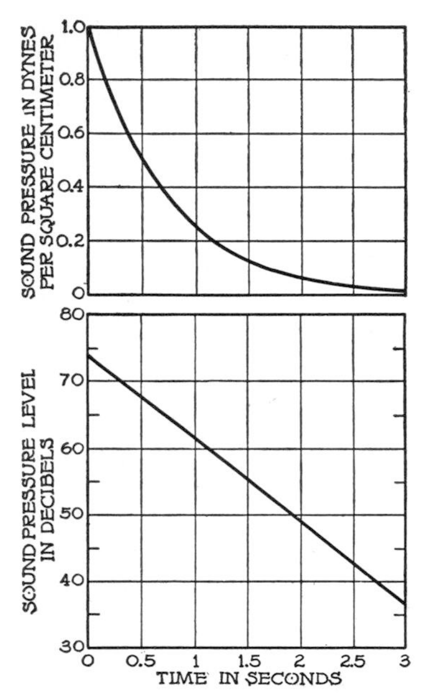

Fig. 8.5 Curves illustrating the decay of sound for a single mode of vibration of a room. The upper curve is for sound pressure vs. time and is exponential. The lower curve is for the same data plotted in terms of sound-pressure level in decibels vs. time. Note that this is a straight line.

Suppose that a source of sound in a room is "started" and that it emits a "pure" tone. (The frequency of the tone is called the *driving frequency.*) If the driving frequency corresponds to one of the resonant frequencies, and if there are no other resonant frequencies close by, then only the one isolated mode will be excited to any extent. When the source is stopped, energy will be absorbed by the walls and other absorptive agencies of the enclosure, and consequently the sound will die away. Figure 8.5 shows how the sound pressure diminishes with time when but a single mode has been excited. The pressure decreases at an exponential

[9] Not all persons object to excessive reverberation. A professor in an eastern university expressed his preference for an extremely reverberant lecture room. An anecdote attributed to this scholar relates that the room was so reverberant that he could come to class and catch the closing words of his previous day's lecture; thus he always knew where to begin.

rate, so that the sound-pressure *level* (in decibels) decreases at a constant rate. Analytically, the decaying sound pressure is proportional to e^{-kt}, where k is a number called the *damping constant* and t is the elapsed time in seconds.

Now suppose that there are two resonant frequencies near the driving frequency of the source. Then two modes of vibration

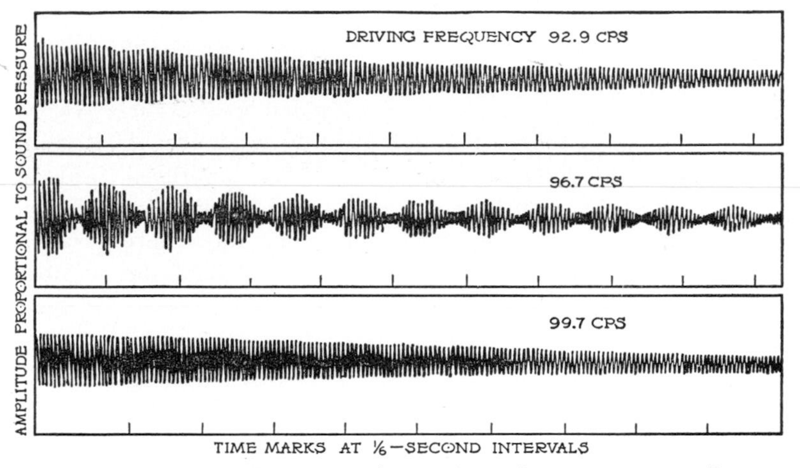

Fig. 8.6 Oscillograms showing the decay of sound pressure in a small room. In the upper and lower oscillograms single modes of vibration are excited. In the middle one two modes are excited, giving rise to beats.

will be excited when the source of sound is started. Since these resonant frequencies are nearly coincident, an interference between them, in the form of beats, will take place when the modes decay. This is illustrated by the three oscillograms shown in Fig. 8.6.[10] These records represent the decay of sound pressure in a room 8 feet by 8 feet by 9.5 feet for different driving frequencies. The driving frequencies for the upper and lower oscillograms, 92.9 and 99.7 cycles, respectively, are very nearly identical with two resonant frequencies of the room, 92.8 and 99.8 cycles; for these driving frequencies the sound pressure decays almost purely exponentially, and the frequency in each case is precisely one of the resonant frequencies of the room. At

[10] V. O. Knudsen, "Resonance in Small Rooms," *J. Acoust. Soc. Am.*, **4**, 20 (1932).

a driving frequency of 96.7 cycles, about midway between the two resonant frequencies, both modes of vibration are excited. In this case the record of decay of sound pressure *vs.* time shows beats of about 7 cycles—the difference between these two resonant frequencies. Each mode is excited and decays at its own characteristic rate, which is different, in general, for each mode. The resulting decay of the two modes is a pulsating diminution of sound pressure, heard by the ear as a prominent vibrato. When more than two modes are excited, the decay is more complicated but is made up of the interfering and beating frequencies of the excited modes. As the number of excited modes increases, so many beats are occurring simultaneously that irregularities in the decay curve tend to "smooth out" one another. This is illustrated by the records of decay of sound-pressure level shown in Fig. 8.7. In the upper curve only a few modes are strongly excited and the decay is highly irregular. When the number of excited modes is greatly increased, as for the lower record, the decay curve is relatively smooth. Here the average drop in sound level is approximately linear because the decay rates of most of the modes are approximately the same. When the rates are markedly different the decay curve is bent.

Average Decay Rate

A typical record representing the sound-pressure level at a given point in a room *vs.* time, after a sound source has been turned off, is given in the upper curve of Fig. 8.7. (This is called a *decay curve.*) Notice that the pressure does not decrease uniformly but fluctuates markedly. Just as in the build-up of sound in a room, these irregularities are caused by the interference of the excited normal modes of vibration, each of which decays at its own rate. The exact mathematical expression giving the decrease of pressure as a function of time is quite complex.[11] However, the *average* rate of decay can be obtained relatively easily.

Let us define the average absorption coefficient $\bar{\alpha}$ for the room by the equation

$$\bar{\alpha} = \frac{\alpha_1 S_1 + \alpha_2 S_2 + \alpha_3 S_3 + \cdots}{S_1 + S_2 + S_3 + \cdots} = \frac{a}{S} \qquad (8.7)$$

[11] See P. M. Morse, *Vibration and Sound,* McGraw-Hill Book Co., 1948, Eqs. (34.17) and (34.18).

where α_1 is the absorption coefficient of the area S_1, etc. Suppose, in accordance with the principles of geometrical acoustics, that the sound emanating from a source in a room is successively

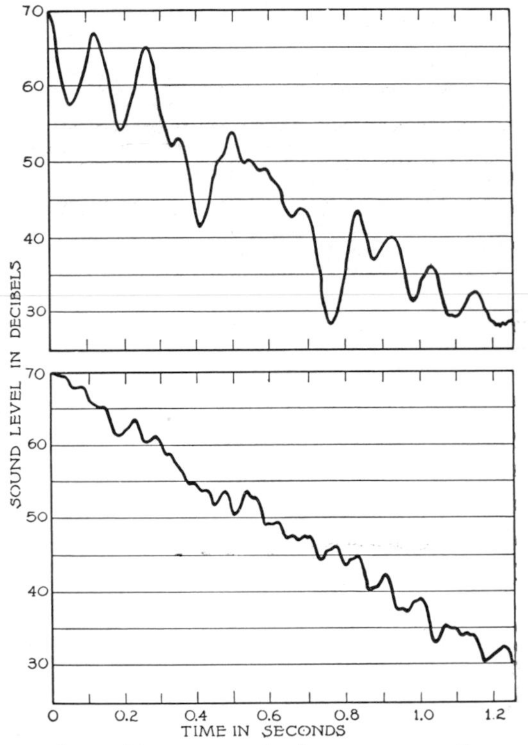

Fig. 8.7 Decay of sound in a room. In the upper curve the sound source is a narrow band of noise, and therefore relatively few natural modes of vibration are excited; the decay is highly irregular owing to interference among the modes. When the number of modes is greatly increased by increasing the band width of the noise, the decay is relatively smooth owing to the averaging-out effects of the many overlapping maxima and minima of the many excited modes. (R. W. Roop, courtesy of Johns-Manville Research Laboratory.)

reflected by boundaries having an average coefficient $\bar{\alpha}$. Each time a wave strikes one of the boundaries, a fraction $(\bar{\alpha})$ of the energy is absorbed, and a fraction $(1 - \bar{\alpha})$ is reflected. Since the pressure is proportional to the square root of the intensity, the ratio of the average reflected to incident pressure is given by

$$(1 - \bar{\alpha})^{1/2}$$

and the average decrease in the sound pressure level is

$$10 \log_{10} \left(\frac{1}{1 - \bar{\alpha}} \right) \quad \text{db/reflection} \tag{8.8}$$

The number of reflections per second is numerically equal to the distance sound will travel in 1 second divided by the average distance between reflections. Let us denote the distance per reflection, the *mean free path,* by m.f.p. Then the average decrease in sound-pressure level per second is

$$\text{Average decay rate} = \frac{c}{\text{m.f.p.}} 10 \log_{10} \frac{1}{1 - \bar{\alpha}} \quad \text{db/sec} \tag{8.9}$$

The mean free path of a sound ray in a room depends on the shape and size of the room, and to some extent on the distribution and nature of the absorptive material. However, in most cases, the mean free path[12] is approximately $4V/S$ feet. Substituting this value for m.f.p. and a value of 1130 feet per second for c into Eq. (8.9),

$$\text{Average decay rate} = 2825 \frac{S}{V} \log_{10} \frac{1}{1 - \bar{\alpha}} \quad \text{db/sec}$$

$$= 1230 \frac{S}{V} [-2.30 \log_{10} (1 - \bar{\alpha})] \quad \text{db/sec} \tag{8.10}$$

The quantity $-2.30 \log_{10} (1 - \bar{\alpha})$ is given as a function of $\bar{\alpha}$ in Fig. 8.8. *When $\bar{\alpha}$ is small compared to unity* (practically, when $\bar{\alpha}$ is less than 0.1), $-2.30 \log_{10} (1 - \bar{\alpha})$ approaches $\bar{\alpha}$. In this case, Eq. (8.10) becomes

[12] For an experimental investigation of mean free path, see V. O. Knudsen, *Architectural Acoustics,* John Wiley & Sons, Inc., 1932, p. 133; a theoretical treatment is given by A. E. Bates and M. E. Pillow, *Proc. Phys. Soc. London,* **59**, 535 (1947).

$$\text{Average decay rate} = 1230 \frac{S\bar{\alpha}}{V} \quad \text{db/sec} \tag{8.11}$$

As an example, consider a rectangular room 30 feet by 50 feet and 20 feet high. Suppose the average absorption of the ceiling is 0.50; of the walls, 0.30; and of the floor, 0.10. Then,

$$\bar{\alpha} = \frac{0.50(1500) + 0.30(3200) + 0.10(1500)}{6200} = 0.30$$

Substitution of these values into Eq. (8.10) gives for the average decay rate 91 decibels per second.

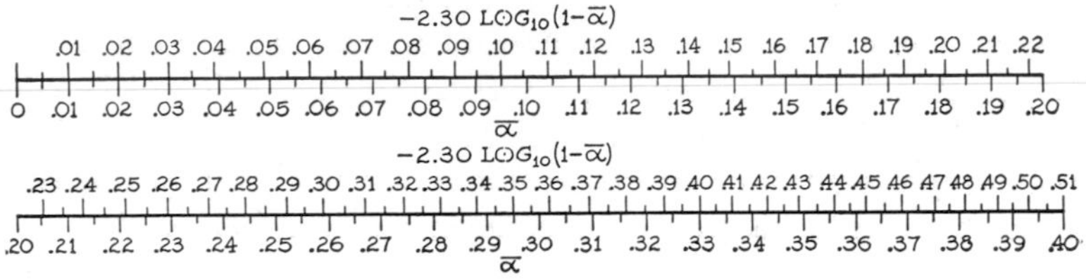

−2.30 LOG10(1−ᾱ)
.01 .02 .03 .04 .05 .06 .07 .08 .09 .10 .11 .12 .13 .14 .15 .16 .17 .18 .19 .20 .21 .22
0 .01 .02 .03 .04 .05 .06 .07 .08 .09 .10 .11 .12 .13 .14 .15 .16 .17 .18 .19 .20
ᾱ
−2.30 LOG10(1−ᾱ)
.23 .24 .25 .26 .27 .28 .29 .30 .31 .32 .33 .34 .35 .36 .37 .38 .39 .40 .41 .42 .43 .44 .45 .46 .47 .48 .49 .50 .51
.20 .21 .22 .23 .24 .25 .26 .27 .28 .29 .30 .31 .32 .33 .34 .35 .36 .37 .38 .39 .40
ᾱ

Fig. 8.8 Chart giving $[-2.30 \log_{10} (1 - \bar{\alpha})]$ as a function of $\bar{\alpha}$.

Reverberation Time

Because of the importance of the proper control of reverberation in rooms, a standard of measure called reverberation time (abbreviated t_{60}) has been established. This is the time required for a specified sound to die away to one thousandth of its initial pressure, which corresponds to a drop in sound-pressure level of 60 db.

From Eq. (8.10), which gives the rate of decay in decibels per second, it follows that the time for a decay of 60 db is

$$t_{60} = \frac{0.049V}{S[-2.30 \log_{10} (1 - \bar{\alpha})]} \tag{8.12}$$

When $\bar{\alpha}$ is small compared to unity, Eq. (8.11) applies; then,

$$t_{60} = \frac{0.049V}{S\bar{\alpha}} \tag{8.13}$$

If V and S are expressed in terms of meters, then in the metric system Eq. (8.13) becomes

$$t_{60} = \frac{0.161V}{S\bar{\alpha}} \tag{8.14}$$

It should be emphasized that when $\bar{\alpha}$, as given by Eq. (8.7), is not small compared to unity, Eq. (8.12) rather than Eq. (8.13) should be used. As an example, consider a room having a volume of 256,000 cubic feet, a total surface area of 25,600 square feet, and an average value of absorption coefficient $\bar{\alpha}$ equal to 0.4. Equation (8.13) will yield a reverberation time of 1.2 seconds, whereas the more exact value of 1.0 second is given by Eq. (8.12). It is apparent, therefore, that Eq. (8.12) is preferable to Eq. (8.13) for the acoustical designing of rooms; Eq. (8.12) is helpful for making estimates of t_{60} in very reverberant rooms. Figure 8.8 is useful, in the above calculations, for obtaining the value of $[-2.30 \log_{10} (1 - \bar{\alpha})]$ from the value of $\bar{\alpha}$. A more exact equation, including the effect of air absorption, is given in Eq. (8.17); this equation should be used for frequencies above 2000 cycles.

Limitations on Use of Reverberation Formulas

In using the reverberation formulas, Eqs. (8.12) and (8.13), it must be remembered that these equations apply only to rooms in which the sound is *diffuse,* a condition that was assumed in their derivations. When the sound is diffuse, the "smoothed-out decay curve" (one for which minor fluctuations have been averaged out) is a straight line. In contrast, when the sound is not diffuse, the smoothed-out decay curve is not straight but is curved or broken. Under the latter circumstances the reverberatory properties of the room cannot be described by the reverberation time or by any other single number. Thus large discrepancies may be noted between the observed and calculated values of reverberation time in a room whose shape or absorptive treatment or both do not promote diffusion. Examples of such rooms are: enclosures having one dimension extremely different from the other two, as in a long, narrow corridor, or in a large room with a very low ceiling; a church having the shape of a cross, or a shape with a very large, highly reflective dome; and rooms having almost all their absorption concentrated on a single surface, such as the ceiling. Suppose that a room has hard walls, a hard floor, and a ceiling which has been treated with acoustical tile of a type ordinarily used in buildings. Then, according to physical acoustics, we have the following picture of the decay of sound in the room. At first, all the modes of vibration begin to die out and the decay rate is rapid. During the latter stages of decay only those modes persist in which the wave motion grazes the absorptive ceiling. The reason for this is that the acoustical tile absorbs very much less sound at grazing incidence (that is, where the wave motion is tangential to the surface) than at other angles of incidence. The non-grazing waves die out most rapidly; therefore in the subsequent phases of decay the sound in the room is *not diffuse* and the decay rate becomes slower and slower.

The reverberation formulas based on geometrical acoustics do not apply to a room having decay characteristics such as the one here considered. But, more important, the room fails to have good acoustics. The faulty acoustics may be tolerated in a very noisy room; but it becomes increasingly conspicuous the more quiet the room is, and it is intolerable in rooms designed for speech or music. In a small office with the ceiling or floor covered with highly absorptive material, and with the walls smooth and highly reflective, speech or any other sounds will excite the slowly decaying modes of vibration, the long and monotonous persistence of which disturb the speaker and interfere with the hearing of speech. It is a source of annoyance in talking over the telephone, and it even reduces the intelligibility of speech at the other end of the line.

The above considerations are illustrated by some experimental data taken in a small room. At 512 cycles, the walls and floor of the room had an average absorption coefficient of 0.02; the ceiling, covered with an acoustical tile, had an absorption coefficient of 0.65. According to Eq. (8.12), the time of reverberation for the acoustically treated room is 0.58 second, but the measured time was 2.0 seconds. The calculated decay rate from Eq. (8.10) is 103 db per second and is indicated by the dashed straight line of Fig. 8.9. The actual decay curve is not a straight line, as shown by the "smoothed-out" solid-line curve of Fig. 8.9, and it has a decay rate of 90 db per second during the first 20 db of decay when the sound in the room is approximately diffuse. After an initial decline at this rate, the sound then decays more slowly and soon approaches a rate of 22 db per second. Now suppose that in this room speech has an average sound level of 70 db, and that the noise in the room has an average level of 40 db. Under the conditions shown in Fig. 8.9, the speech first decays for 20 db at a rate of 90 db per second, then dies away at the much slower rate of 22 db per second, until it is masked by

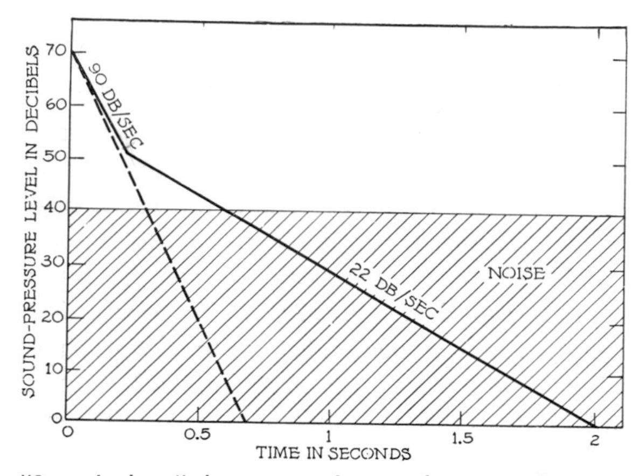

Fig. 8.9 "Smoothed-out" decay curve for sound in a small rectangular room in which only the ceiling is treated with absorptive material. Fluctuations in the actual decay curve are not shown. The noise level in the room is 40 db.

the noise. During the second and slower rate of decay, the persistence of the modes which graze the absorptive material is clearly audible; these prolonged modes interfere with the hearing of speech, and they are annoying to both speakers and listeners. If the noise level in the room is increased 20 db above that shown, the slow decay and long persistence of certain resonant frequencies in the room will not be heard. On the other hand, if the noise level in the room is less than that indicated, the resonant and "boomy" quality is accentuated. The acoustical defects just considered can be practically overcome by distributing the absorptive material in strips or patches over the walls and ceiling of the room, or by resorting to other means for diffusing the sound in the room (see p. 139).

Sound Decay and Reverberation Time at Different Frequencies

Custom has attached so much importance to the frequency of 512 cycles that, when the term "reverberation time" is used without the specification of frequency, it is generally understood to be the reverberation time at 512 cycles. If the absorptivity of the materials in a room were constant with frequency, the calculation of the decay rate at 512 cycles would be sufficient to represent the average rate at which sound dies away. However, a glance at the tables of absorption coefficients will indicate that *this is not so*. For example, an acoustical plaster, ¼ inch thick, may have coefficients of absorption of 0.06 at 128 cycles, 0.36 at 512 cycles, and 0.72 at 2048 cycles. Suppose that this plaster is applied to the entire inner surface of a room (a thin carpet on a concrete floor would have nearly the same absorption characteristic as this plaster). If the reverberation time is 1.2 seconds at 512 cycles in such a room, it will be about 7.2 seconds at 128 cycles and about 0.6 second at 2048 cycles. To describe the room as one having a reverberation time of 1.2 seconds would be misleading; it would ignore the intolerably long reverberation times at low frequencies. Such a room would have extremely poor acoustics; there would be complaints of "boominess," of excessive reverberation for the bass notes of music, and over-absorption (with consequent suppression) of the higher tones and harmonics of music. Even speech would be distorted, and speakers as well as listeners would be annoyed by its unnatural quality.

It is necessary, therefore, to specify and calculate the reverberation time for representative frequencies throughout the principal part of the audible-frequency range. If the materials used in the room do not have prominent irregularities in their absorption characteristics, calculations of reverberation time made at 128, 512, 2048, and 4096 cycles will meet the needs of design for nearly all practical purposes. But in the planning of rooms in which good acoustics is the prime requirement, such as radio, motion picture, recording, and television studios, consideration must be given to a wider range of frequencies. A large measure of success or failure in the acoustical design depends on the selection of materials that will give the proper reverberatory characteristics to the *entire* range of frequencies from 50 to 10,000 cycles. Therefore, the architect or engineer, in whose hands is placed the responsibility for acoustical design, should base his selection of interior materials upon calculations of reverberation not at the single frequency of 512 cycles but at representative frequencies, the number and range of which should be determined by the acoustical requirements of the room. The optimum times of reverberation at different frequencies for typical rooms will be considered in later chapters.

Effect of Air Absorption upon Decay Rate and Reverberation Time

In the earlier formulas of this chapter, the absorption of sound was considered to take place at the boundaries of the room—the absorption in the air was neglected. However, it has been pointed out in Chapter 5 (p. 68) that the absorption of sound in air may be very considerable, especially at the higher frequencies. As a result of the attenuation of sound in air, the pressure of even a plane sound wave decreases with the distance traveled. Thus, the pressure of a plane wave after traveling a distance x in a homogeneous medium is $P_0 e^{-mx/2}$, where P_0 is the *pressure amplitude* of the wave at the position $x = O$, and m is the *attenuation coefficient.*

If the effects of absorption in air are to be included in the derivation of the formula for the average decay rate of sound, Eq. (8.8) must be modified. The average ratio of reflected to incident pressure, when the loss in the air is taken into account, is

$$(1 - \bar{\alpha})^{1/2} e^{-mx/2} \tag{8.15}$$

In this case x is the distance that the sound wave travels per reflection, that is, the mean free path. Substituting $4V/S$ for the value of the m.f.p., and using Eq. (8.8), it follows that the total average decrease in the sound-pressure level is

$$10 \log_{10} \frac{1}{1 - \bar{\alpha}} + 4.34 \frac{4V}{S} m \quad \text{db/reflection}$$

Hence, the average decrease in sound-pressure level per second is given by

$$-2825 \frac{S}{V} \log_{10} (1 - \bar{\alpha}) + 4900m \quad \text{db/sec} \tag{8.16}$$

The reverberation-time formula, including the effect of air absorption, then becomes

$$t_{60} = \frac{0.049V}{-2.30S \log_{10} (1 - \bar{\alpha}) + 4mV} \tag{8.17}$$

The second term in the denominator, $4mV$, represents the effective absorption in the room contributed by the air. When m is negligibly small, as it is for frequencies below about 1000 cycles, Eq. (8.17) reduces to Eq. (8.12), which does not take account of the absorption in the air. For frequencies above about 1000 cycles, and for the usual humidities and temperatures in most rooms, m increases almost with the square of the frequency and consequently becomes very important at high frequencies. For example, at 4096 cycles the absorption in the air in a large auditorium may amount to as much as the absorption by the boundaries of the room.

The attenuation coefficient m at each frequency depends on the humidity and temperature of the air. The values of m for a temperature of 68° F are given in Fig. 8.10 as a function of relative humidity for frequencies of 1024, 2048, 4096, 6000, 8192, and 10,000 cycles.[13] (For temperatures between 60° F and 80° F, m increases about 8 per cent for each 5° F rise in temperature.) It will be seen that the attenuation coefficient m is very dependent on the humidity. Hence, reverberation of high-pitched sounds will be influenced by the humidity of the air. *Equation (8.12) can be used for calculating reverberation times for all frequencies below about 2000 cycles, but Eq. (8.17), with the appropriate value of m, should be used for all higher frequencies.* It is necessary to include air absorption for all frequencies above about 1000 cycles in large auditoriums, especially if the humidity is below 50 per cent.

[13] These values of m in Fig. 8.10 are about 10 per cent lower than those used for computing the curves of Fig. 5.7. The greater attenuation in the open air is due to the inhomogeneities in the atmosphere.

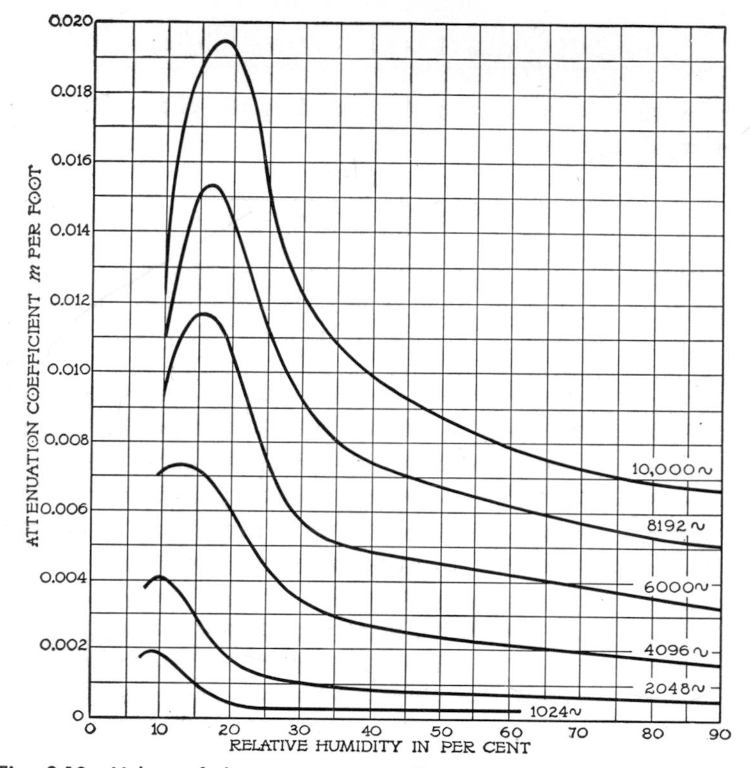

Fig. 8.10 Values of the attenuation coefficient m as a function of relative humidity for different frequencies.

Reverberation-Time Nomographs

Figure 8.11 gives two useful nomographs for making routine calculations of reverberation time based on Eqs. (8.12) and (8.17). The reverberation time of a room having a volume V is obtained from the nomograph in the following way:

(1) Calculate $\bar{\alpha}$ from Eq. (8.7).

(2) Determine the value of $-2.30 \log_{10} (1 - \bar{\alpha})$ by the use of Fig. 8.8.

(3) Multiply the value obtained in step (2) by the area S and denote this product as the quantity A. (If the effect of air absorption is to be included, the quantity $4mV$ should be added to $S[-2.30 \log_{10} (1 - \bar{\alpha})]$; that is, $A = [-S2.30 \log_{10} (1 - \bar{\alpha})] + 4mV$ where values of m are given by Fig. 8.10.

(4) Place a straightedge on the proper nomograph (the one applicable to the volume of the room) so that the straightedge intersects (1) the V line at the computed volume of the room and (2) the A line at the value computed in step (3). Then the intersection on the t_{60} line gives the reverberation time corresponding to Eq. (8.17).

As an example of the use of Fig. 8.11, suppose that $V = 200{,}000$ cubic feet and that the value of A for a frequency of 512 cycles is 8000 square-foot-units (sabins); then, by setting a straightedge as directed in step (4), it is seen that the reverberation time t_{60} at 512 cycles is 1.24 seconds.

Reverberation in Coupled Spaces

A great complexity of "coupled spaces" often is found in churches, including the nave, transepts, choir, sanctuary, chapels, balconies, and organ chamber. Many theaters and auditoriums are "coupled" (connected) by means of door openings to excessively reverberant rooms or corridors. In one large auditorium in which the authors have made measurements, the rate of decay of a tone of 512 cycles is approximately 40 db per second in the main seating area but only about 10 db per second in its huge, reverberant (concrete walls and ceiling, with scant hangings) stage recess. In such auditoriums, even though the reverberation in the audience space has been adjusted to the proper value, there will be a flow of sound from the reverberant rooms (or stage recess) into the main auditorium. Thus, auditors in a theater who are seated near an opening to a corridor, foyer, or anteroom will be disturbed by the reverberant sound from the adjacent space; and an auditorium with an excessively reverberant stage, as in the example cited, will have poor acoustics no matter how well-designed the audience space may be.

Most theater-type auditoriums are divided into at least three coupled spaces: the stage, the main portion of the auditorium, and the space under the balcony. Consider the following typical case: An auditorium has a conventional rectangular shape, a high-gabled ceiling, and a deep recess under the balcony. Portions of the side walls are treated with a very absorptive material to provide a fairly satisfactory condition of reverberation in

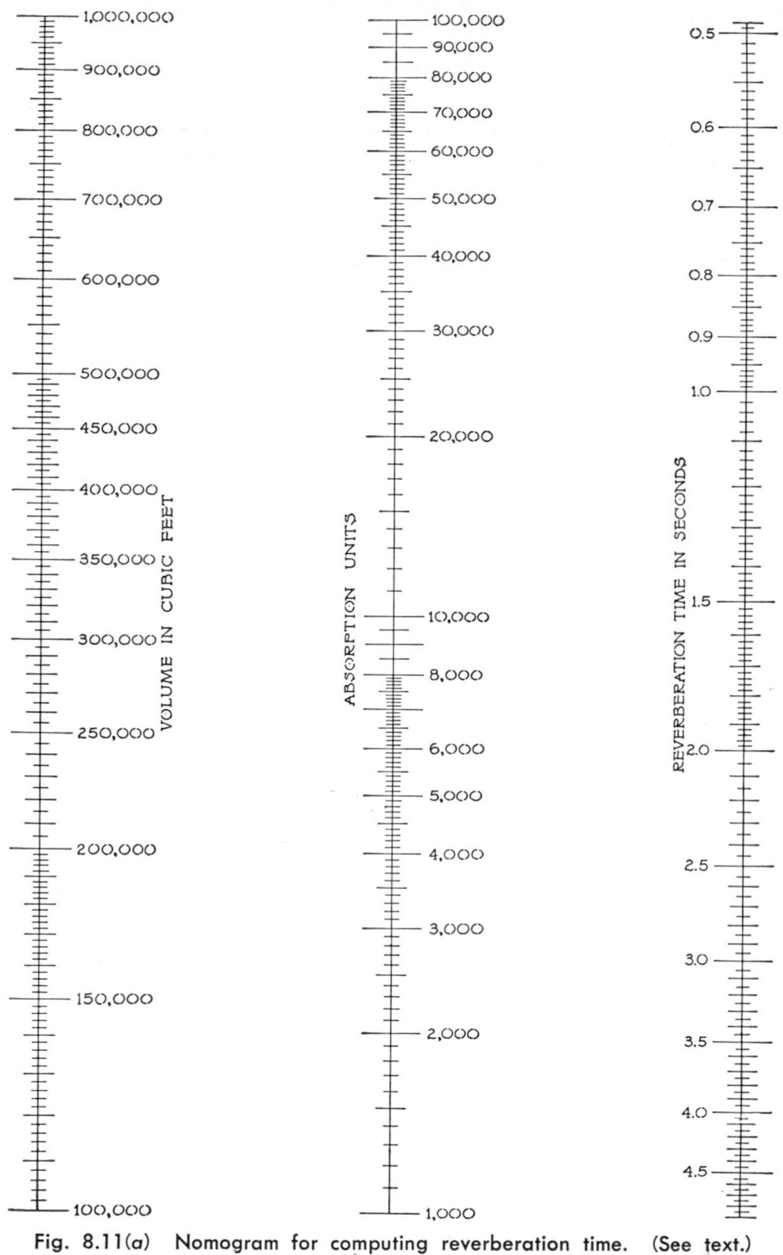

Fig. 8.11(*a*) Nomogram for computing reverberation time. (See text.)

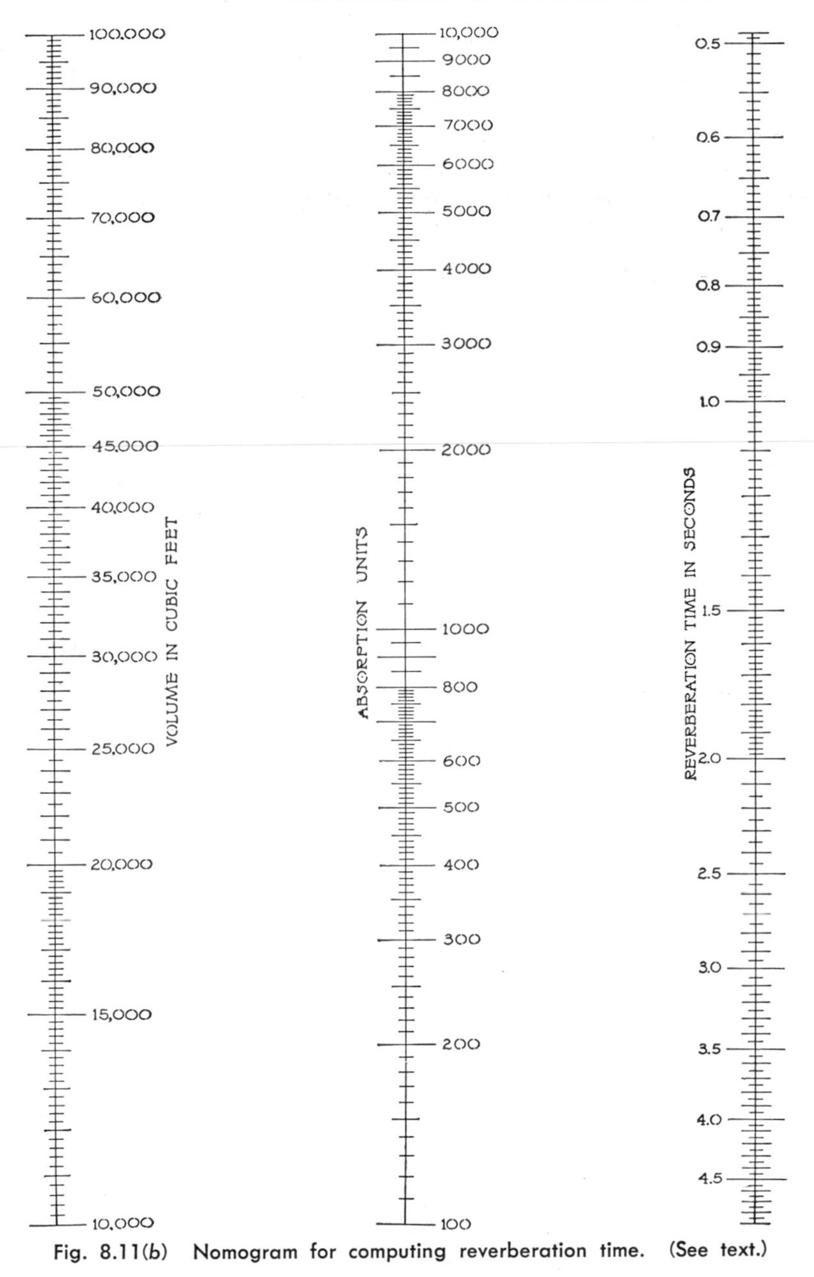

Fig. 8.11(*b*) Nomogram for computing reverberation time. (See text.)

the front part of the auditorium and in the balcony. The walls under the balcony, the soffit of the balcony, and the entire main floor are poured concrete. The wooden seats are not upholstered. As a result, the space under the balcony, especially when only a few persons are sitting in this section, is very reverberant. The rates of decay in the coupled spaces are not the same. During the decay of sound in such an auditorium, there is a transfer of energy between the two coupled spaces. During the steady state, the rate of transfer of sound from the main portion of the auditorium to the space under the balcony is equal to the rate of transfer in the opposite direction. Near the beginning of the decay these rates of transfer are nearly equal; but, since the sound decays much more rapidly in the main part of the auditorium than it does under the balcony, there soon will be established an excess rate of flow from the space under the balcony to the main part of the auditorium and the sound will be prolonged in both areas. In order to overcome this undesirable condition, it is necessary that the decay rates in both spaces be nearly equal.

The decay curve in Fig. 8.12(*a*) shows a record of the drop in sound-pressure level in a relatively "dead" room (having a low reverberation time) which is connected to a reverberant corridor. Both the sound source and the microphone were in the room. At first the sound level dropped quite rapidly, as indicated by the dashed line which represents the smoothed-out decay curve. This slope is practically that which would be obtained if the corridor were blocked off; it is characteristic of the room itself. During the process of decay some of the sound in the corridor returned to the room. At first this amount was negligible compared to that in the room. However, after the sound in the room died out, the sound "fed back" from the corridor to the room predominated. Thus the final slope is characteristic of the decay rate of the corridor. When the microphone was placed in the corridor, the results shown in Fig. 8.12(*b*) were obtained. The initial slope, that due to the decay of sound in and coming from the room, died out almost immediately. Then the sound died out slowly at a rate corresponding to that of the corridor alone.

The curves of Fig. 8.12 emphasize the difficulty that frequently is met in practice. Suppose that an office, connected to a highly reverberant corridor, is treated with sound-absorptive material. The acoustical results may be disappointing. In this case, the absorptive material may not be at fault. The trouble may be due to the untreated corridor, which acts as a reservoir of sound energy that continues to return the energy to the office, thereby prolonging the reverberation in the office.

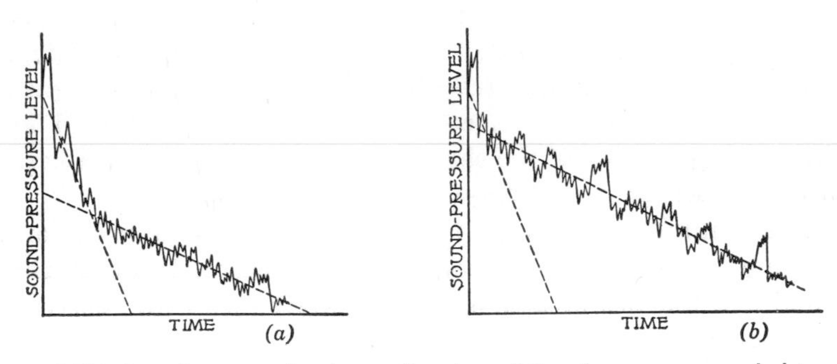

Fig. 8.12 Sound-pressure level as a function of time for a room coupled to a reverberant corridor. The source of sound is in the room; (*a*) represents the decay of sound in the room and (*b*) represents the decay in the corridor. (W. Tak.)

The reverberation formulas (8.12), (8.14), and (8.17) do not apply to coupled rooms unless the rates of decay in the two rooms are nearly identical, in which case the formulas are sufficiently valid for practical purposes. Since good design requires that the rates of decay in coupled spaces be approximately the same, these formulas are applicable to conditions that are fulfilled when the proper acoustical principles have been heeded. (See "Balcony Recess," p. 189, and "The Legitimate Theater," p. 314.)

Special Acoustical Phenomena Associated with the Shape of Rooms

Architectural acoustics, until the beginning of the present century, was based almost entirely on considerations of the shape and size of rooms. It was then but a small segment of geometrical acoustics. When regarded at all—this was the exception rather than the rule—acoustical design was determined principally by simple analyses of the paths taken by rays of sound (on typical sections of the proposed plans) which progress from the source to the listeners. Such an analysis can be quite useful. Straight lines representing sound rays are drawn outward from the position of the source in a room (for example, see Fig. 15.1). When a ray strikes a surface, it is assumed that the angle at which it is reflected is equal to the angle of incidence. The angle of the surface can then be adjusted to direct the reflected sound where it is most needed, toward the rear of the room. In former days rooms that gave beneficial reflections were occasionally designed by this means. Moreover, echoes were sometimes avoided by the proper planning of the shape and size of rooms on the basis of the simple law of reflection.

The control of sound by reflective surfaces is so important that much of the following chapter will be devoted to the principles and practice of designing the correct forms for reflective surfaces in rooms. Sound reinforcement for the more remote seats and the diffusion of sound throughout a room are helpful effects that can be obtained from properly designed reflective surfaces. In addition, as pointed out in the discussion of shells for open-air theaters, a certain amount of reflected sound is desired not only by the listeners but also by the performers. However, when not properly controlled, reflected sounds can be responsible for such acoustical defects as *echoes, sound foci, dead spots,* and *room flutter.* These defects, their causes, and suggested remedies will be discussed in this section.

Echoes. Sound that reaches a listener in a room by a path involving reflections from its boundaries always travels a greater distance than does the sound that comes by the direct path. If the difference in these two path lengths is as great as 65 feet, which corresponds to a time difference of about 0.06 second, the delay in the arrival of the reflected sound is sufficient to enable a listener to hear it as a separate sound; that is, the delayed reflection produces an *echo.* Even when the difference in the two path lengths is somewhat less than 65 feet, but greater than about 50 feet, the delayed reflection may have a damaging action. It tends to blur or even mask the direct sound. Delayed reflections are most detrimental when they are concentrated or focused by a highly reflective concave surface. (Figure 4.2 illustrates how sound is reflected from a concave surface.) In contrast, they are least damaging when diffused by convex surfaces. (See Fig. 4.3.) In other words, if these delayed reflections are sufficiently attenuated by means of absorption or divergence, their blurring effect becomes negligible.

Sound Foci. Figure 8.13 shows a number of examples of poor acoustical planning which are typical of rooms found in almost every community. These rooms suffer from the focusing effects of concave surfaces. The points S and S' in each of the drawings are called *conjugate foci.* In general, there are many pairs of conjugate foci in rooms having shapes such as these; for example, in (*d*), for every position such as S on the platform within the hemicylindrical wall, there is a conjugate focus at S' in the seating area of the auditorium. The normal location for the speaker in (*d*) is at the center of curvature of the cylindrical wall. Sound originating at this point is reflected back to C by the concave wall. In this position the speaker has the illusion that he is speaking very much louder than he actually is. Since he overestimates the loudness of his voice, he tends to speak with inadequate loudness for the audience. At the same time, he is likely to be annoyed by the incessant bombardment by his own voice. If the distance from S to S' by the reflected path exceeds the direct path distance by 65 feet or more, the reflected sound will be heard as an echo, and a very prominent one because so much reflected sound has been concentrated in a small region.

Whispering Galleries. A phenomenon closely associated with the reflections from curved surfaces is the tendency for sound, especially in the high frequencies, to travel or "creep" around a large concave surface. This phenomenon has become famous in connection with St. Paul's Cathedral in London, and it can be observed in many other concave structures. A whisper directed along such a concave surface may be heard distinctly at least 200 feet away. Experiments in St. Paul's show that the whispered sound, for the most part, is concentrated in a narrow band skirting the circular base of the dome, and that the thickness of this band decreases with diminishing wavelength. Impulsive sound, such as a hand clap, will travel around the gallery several times. The orchestra shell of the Hollywood Bowl, which is made up of one half of a truncated right circular cone,

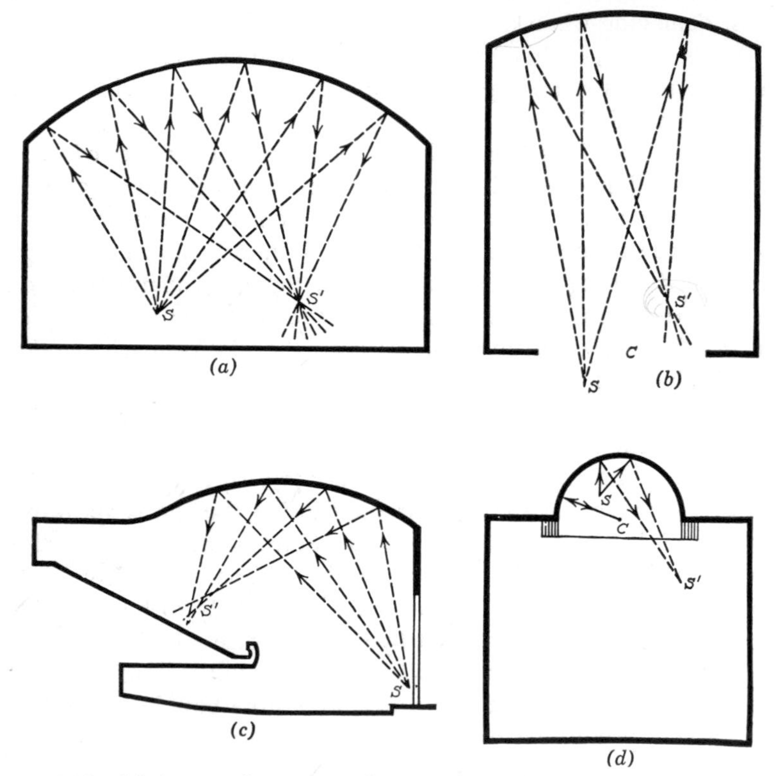

Fig. 8.13 (*a*) Section illustrating reflection of sound from a barreled ceiling which has a radius of curvature equal to the height of the room. (*b*) Plan showing reflection of sound from a concave rear wall having its center of curvature at C. (*c*) Reflection of sound from a domed ceiling. (*d*) Plan showing reflection of sound from cylindrical wall behind a speaker.

has a series of large triangular grooves which extend along the entire semicircular span of the shell. Two persons standing at the opposite ends of one of these grooves, although 90 feet apart (measured across the stage), can carry on a whispered conversation even when there is loud conversation or music on other parts of the stage. Similar effects can be observed in nearly every circular or elliptical structure where the curved surfaces are continuous for long distances. Although the phenomenon is often quite harmless from the standpoint of good acoustics, it may under certain circumstances introduce difficulties. Therefore, it is good practice to avoid shapes that are likely to become whispering galleries.

DEAD SPOTS. It has been shown that, owing to the focusing effects of concave surfaces, rooms having shapes such as those illustrated in Fig. 8.13 suffer from localized concentrations of sound. But these excessive concentrations are formed at the expense of deficiencies of reflected sound at other localities in the room. The regions of deficiency are sometimes called "dead spots." In these localities, especially in large rooms, the sound level usually is inadequate for satisfactory hearing. The concave surfaces of the types considered here are largely responsible for non-uniform distributions of sound throughout a room. *Except in special instances* where it is necessary to concentrate reflected sound in restricted regions, *it is not advisable to incorporate large concave surfaces of the type illustrated in the design of a room.* Another type of "dead spot," already mentioned, results from destructive interference among two or more combining sound waves. This occurs when the condensation from one wave (such as the direct sound wave) unites with the rarefaction of another (as the reflected wave). Interference of this type is not so deleterious to the acoustics of a room as is popularly supposed.

All forms of "dead spots" are less prevalent and less important than they are generally thought to be. For example, the lay observer often reports that he can hear very well in all parts of a certain auditorium except in the "thirteenth and fourteenth rows, center," or in some other specific location where, he avers, there is a "dead spot." At times, such localities do receive insufficient sound, because most of the reflected sound is concentrated elsewhere or because there may be some destructive interference among the sound waves at these localities. But such reports should be carefully scrutinized. They often mean that there is a real defect in the acoustics of the room, but the defect is not so localized as it is claimed to be, and it may be the result of excessive reverberation, inadequate sound level, too much noise, interference from echoes or delayed reflections, or improper diffusion. The layman's diagnosis of "dead spots" as the principal defect in the acoustics of a room too frequently covers a multitude of unsuspected acoustical ills. Sometimes such a diagnosis is made even when exhaustive tests fail to reveal any acoustical defects in the room; it is advisable under these circumstances to test the hearing of the "diagnostician"; he may have defective hearing and he may not know it, or will not admit it.

ROOM FLUTTER. Among the undesirable acoustical phenomena associated with the shape of a room is *flutter echo*. It usually occurs between a pair of parallel (opposite) walls in a room. It is most noticeable in a rectangular room when one pair of opposite walls is smooth and highly reflective and the other two opposite walls are treated with absorptive materials.[14] A single impulsive sound, such as a sharp hand clap, produces a multiple echo as the impulse is reflected back and forth between the pair of reflective walls. If the distance between these walls is large (50 feet or more) the flutter is slow, and a single hand clap is heard as a series of "puts" (put, put, put, put, . . .), gradually dying away to inaudibility. If the walls are somewhat nearer each other, the successively reflected impulses recur more frequently and the series is heard as a prominent flutter or even as a "dry rattle." When the parallel walls are only 8 to 10 feet apart, the sound of a single pulse, like a hand clap or a snap of the fingers, is heard as a "buzz" which dies away quite rapidly.

Room flutter frequently occurs in uncarpeted rooms where the ceiling and floor are highly reflective and the walls are broken with windows, doors, hangings, pictures, etc. It can be especially troublesome in broadcasting or sound-recording studios and is annoying in all speech, music, living, or work rooms. Flutter echoes can be eliminated by avoiding the use of parallel pairs of walls, or by "breaking up" the uniformity

[14] See D. Y. Maa, *J. Acoust. Soc. Am.*, **13**, 170 (1941). This paper is recommended to the reader because it treats the phenomenon of flutter by the methods of both geometrical and physical acoustics, and it shows how the former approaches the latter where diffraction and interference effects are not important.

of such walls with doors, windows, book shelves, hangings, paintings, splays, or patches of absorptive materials. A very small departure from parallelism between the pair of flutter-producing walls, such as a change in the direction or slope of one wall by as little as one part in twenty, will suffice.

9 · Acoustical Design of Rooms

Planning for good acoustics in a building begins with the selection of the building site and continues through all stages of designing. The architect will avoid inexcusable errors in design if he sets up a check list of the necessary and sufficient measures to be taken for obtaining good acoustics. These steps, approximately in chronological order, are as follows:

(1) The selection of the site in the quietest surroundings consistent with other requirements.

(2) The making of a noise survey to determine how much sound insulation must be incorporated in a building to meet specified requirements of quietness.

(3) The arrangement of the rooms within the building.

(4) The selection of the proper sound-insulation constructions.

(5) The control of the noise within the building, including solid-borne as well as air-borne noise.

(6) The design of the shape and size of each room that will insure the most advantageous flow of properly diffused sound to all auditors, and that will enhance the aesthetic qualities of speech and music.

(7) The selection and distribution of the absorptive and reflective materials and constructions that will provide the optimum conditions for the growth, the decay, and the steady-state distribution of sound in each room.

(8) The supervision of the installation of acoustical plaster, plastic absorbents, or other materials whose absorptivity is dependent on the manner of application.

(9) The installation of sound-amplification equipment under the supervision of a competent engineer, wherever such equipment is necessary.

(10) The inspection of the finished building, including tests to determine whether the required sound insulation, sound absorption, and the other acoustical properties have been satisfactorily attained.

(11) Maintenance instructions, in writing, to be left with the building manager, indicating (*a*) how the acoustical materials can be cleaned or redecorated, (*b*) which furnishings in the building must be retained to maintain good acoustics, (*c*) how, in large speech and music rooms where high-quality reproduction is desired, the humidity should be maintained in order to avoid excessive absorption of high-pitched sounds, and (*d*) how the sound-amplification system should be maintained.

The foregoing steps or their equivalents, if carefully executed, will lead to good acoustics. Steps (6) and (7) are the primary subjects of the present chapter; the others are treated in subsequent chapters.

Requirements for Good Acoustics

In the design of rooms intended for speaking purposes the prime objective is the realization of conditions that will provide good *intelligibility of speech*. This phrase, as used by telephone engineers, signifies how well speech is recognized and understood. In the design of music rooms the prime objective is the most favorable enrichment of the tonal quality and tonal blending of the sounds. It is necessary to provide not only the optimum conditions for *listening* to music but also the best possible conditions for the rendition of music by skilled artists. When a radio, violin, or any musical instrument is played in an enclosure, the enclosure is, in effect, a part of the instrument; that is, the instrument is *coupled* to the room, and the instrument excites the resonant frequencies of the room. (See "Room Resonance," p. 134.) A high-quality radio or a world-famous Stradivarius cannot produce high-quality music in a room that has poor acoustics.

The above check list is a practical aid to the fulfillment of the requisites for good acoustics. These requirements, which are applicable to all rooms used for speech and music, may be stated as follows:

(1) All noises, whether of outside or inside origin, should be reduced to levels that will not interfere with the hearing of speech or music.

(2) The room's shape and size should be designed to (*a*) give proper diffusion to the sound, (*b*) reinforce the sound reaching the audience, especially toward the rear seating area, and (*c*) contribute to the attainment of a favorable ratio of direct to reflected sound for all auditors. Although these desirable conditions are also affected by the distribution of the absorptive materials and by the sound-amplification system, they are largely controlled by the shape and size of the room. It is often necessary to design special wall and ceiling surfaces to act as reflectors for the reinforcement of sound at the rear of the room, and it is sometimes essential to introduce splays or other surface irregularities to provide proper diffusion of sound.

(3) The reverberation time *vs.* frequency curve should approach the optimum characteristics, which are determined by the volume and type of room. The fluctuations in the growth and decay curves should be such as to yield optimum reverberation conditions.

(4) Provision should be made for reinforcing the speech and music in a room so the sound level will be adequate in all parts of the room. In a small room, this requirement can be met by the proper design of reflective surfaces (walls, floor, and ceiling); in a large room, in addition to the proper design of the reflective surfaces, a high-quality sound-amplification system is indispensable.

In general, the above four requirements are necessary and sufficient for providing satisfactory acoustics in all rooms. In view of the differences between speech and music, the requirements stated above are not identical for speech rooms and for music rooms. There are, however, certain broad features that apply to both: freedom from disturbing noise, proper shape (a room shape that is good for music usually will be satisfactory also for speech), and a sufficient sound level for all auditors.

The requirements concerned with the reduction of noises and with adequate loudness of speech or music have been considered in Chapter 5 as they apply to open-air theaters. For the average listener, a sound level of about 65 db is adequate for good intelligibility of speech in reasonably quiet surroundings (noise levels of about 40 db), and this level is the optimum average level based on listener preference for both speech and music. In a noisier environment higher levels of speech and music are preferred and required. In all good music rooms, as in all other rooms in which listening is a required function, the noise level should be low—at least 2 or 3 db lower than the unavoidable noise level of an attentive audience during the *silent* pauses in music. The noise level of an attentive audience is of course a variable quantity, depending on the size, age, and other aspects of the individuals, but the average level is about 40 db.[1] The noise level in the room when no audience is present should not exceed about 35 db. In the case of speech, the unvoiced consonants, which are relatively feeble, must be loud enough to emerge above the masking effect of the noise. Although the need for amplification of sound is less for music than it is for speech (since its acoustical power is generally much greater than that of speech), both require amplification in large rooms.

The intelligibility of speech declines rapidly with the distance from a speaker in the open air, owing to the inverse relationship between sound pressure and distance. In a room the distribution of sound radiated by a source is greatly affected by the boundaries of the room: the distribution is altered; the sound levels are generally raised (see Fig. 8.2); and other phenomena such as room resonance, reverberation, and diffusion are introduced. The room can be so designed that its effect on the distribution of sound throughout the room can be very advantageous for good

[1] The noise levels referred to in this book are the levels that would be measured with a sound-level meter incorporating a 40-db frequency-weighting network.

listening. In fact, one of the chief purposes of this chapter is to present the most favorable shapes of auditoriums, shapes that will provide the best possible distribution of sound to all listeners.

Individuals in a large ensemble, like a chorus or orchestra, are dependent on useful reflections to hear each other adequately. This is an indispensable requirement. Consider a large orchestra seated on a stage. A player in the second-violin section may be 50 feet from some of the woodwinds. If this orchestra is in the open air without benefit of an orchestra shell or other reflective surfaces, the second-violin player may hear the near-by violins at a sound level 20 db higher than that of the woodwinds, which are ten times as far away. Much of the time, under such conditions, he will not hear the woodwinds at all and may get "out-of-step" with them. *The separate players must hear each other if they are to play in perfect synchronism.*

Speech ordinarily requires somewhat less reverberation than music, and it is chiefly in this respect that the acoustical properties of music and speech rooms differ. In order to provide the best possible acoustical environment, a music room must also have optimum reverberation characteristics for both performers and listeners. This requirement is not easy to fulfill, for the performers consistently want more reverberation than do the listeners. Reverberation supports and enhances the tones produced by a musical instrument. Incidentally, it also helps to mask imperfections. Therefore mediocre musicians should give it their thanks! The desire of the musical performer to have his instrument or voice supported by reflective surroundings, and his need of such reflecting surfaces in order to play or sing in precise time and proper balance with others, call for carefully designed wall, floor, and ceiling reflectors. In general, the part of a room (stage, platform, or one end of the room) occupied by the performers should have surroundings that are somewhat more reflective than the part occupied by the audience.

The foregoing conditions and requirements for good acoustics should be taken into account in the design of speech and music rooms. Since the intelligibility of speech can be rated by techniques that have been developed for testing the performance of telephones, it is possible to devise means for the quantitative rating of speech rooms. The rating of music rooms is more elusive.

One cannot measure the *intelligibility* or any comparable aspect of music that would furnish a quantitative basis for such a rating. But it is not impossible to find suitable criteria for obtaining the preferred listening conditions in music rooms. Experiments have demonstrated, for example, that different listeners are in fairly good agreement in their choice of the optimum reverberation-time characteristic for a specific type of music.

Rating of Speech Intelligibility—Articulation Testing

Articulation testing is a useful method for determining quantitatively how well speech can be heard and understood. Since it is a standard means for rating speech intelligibility, it can be used for establishing figures of merit for the various conditions relevant to listening to speech in rooms. In these tests of speech intelligibility, one person calls out a list of meaningless monosyllabic speech sounds. An observer writes down, as accurately as possible, the speech sounds he hears. The percentage of syllables heard correctly is called the *percentage syllable articulation,* or simply the *percentage articulation.* The test itself is referred to as an *articulation test.* For example, if a speaker calls out 1000 speech syllables, and the observer hears 850 correctly, the percentage articulation is 85 per cent. If the syllable articulation is 85 per cent, only approximately 3 discrete words out of 100 will be misunderstood; the conditions are very good. This is the minimum acceptable value if sound amplification is used. If the syllable articulation is 75 per cent, approximately 6 words out of 100 will be incorrectly understood; the conditions are satisfactory, but attentive listening is required. If the syllable articulation is 65 per cent, approximately 10 discrete words out of 100 will be misunderstood; the hearing conditions are barely acceptable and are fatiguing to the listener. For values of syllable articulation below 65 per cent, speech is usually not heard satisfactorily.

RELATION BETWEEN ARTICULATION, SPEECH LEVEL, AND NOISE. A knowledge of the relationship between percentage articulation and the sound level of speech is useful in the acoustical design of rooms. Figure 9.1 gives the percentage syllable articulation as a function of speech level for two conditions of noise: (1) for a noise-free environment having a noise level of 0 db, and (2) for a noise level of 43 db having the frequency spectrum shown in Fig. 2.8, which corresponds to "average room noise." These data assume that reverberation is absent and that the threshold of hearing of the observer corresponds to the "minimum audible threshold." The curves show that the articulation increases rapidly with speech level. For the 0-db noise condition, according to Fig. 9.1 the articulation would be about 88 per cent for a speech level of 40 db; it would be about 96 per cent for a speech level of 70 db. Raising the speech level above this value produces no further increase in articulation. For typical room noise having a level of 43 db, the articulation would be about 17 per cent for a speech level of 40 db, and about 93 per cent for a speech level of 70 db. These curves are for undistorted speech.[2] If a low-quality sound-amplification system which distorts the speaker's voice is used, the increase in articulation may not be so great as is indicated.

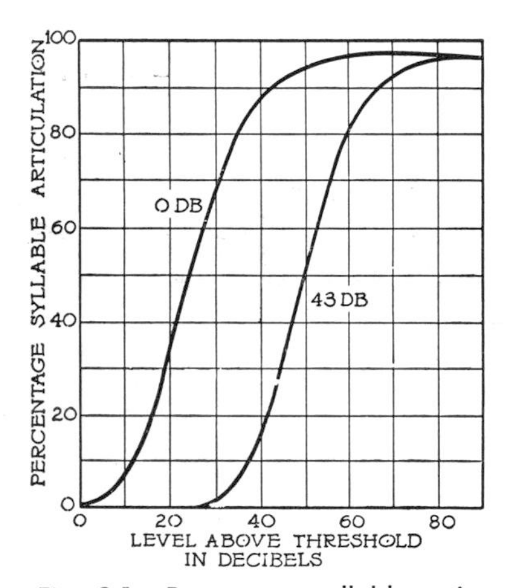

Fig. 9.1 Percentage syllable articulation *vs.* speech level above the threshold of hearing for two conditions of noise: (1) for a very quiet environment having a noise level of 0 db, and (2) for a noise level of 43 db having the frequency spectrum of typical room noise.

Relation between Articulation and Reverberation. Figure 9.1 shows the necessity of an adequate sound level of speech in a room if speech is to be understood by the listener without undue strain. These data apply to conditions where there is freedom from reverberation, as in an open-air theater. Reverberation affects the hearing of speech in two principal ways, one beneficial and the other detrimental. Increasing the reverberation time in a room increases the sound level of a source of sound in that room; and, as Fig. 9.1 indicates, this increase can be quite helpful to the intelligibility of speech when the level is low. On the other hand, reverberation causes one sound to persist or "hang over" while the next one is spoken, and therefore acts essentially as noise. This is illustrated by the sound spectrograms in Fig. 3.8. From this standpoint, reverberation is detrimental. If the reverberation time is sufficiently small, the prolongation of an individual sound is so "short-lived" that its disturbance of the following sound is almost negligible. Then, the reverberation is beneficial be-

[2] For other conditions, the percentage articulation can be calculated by a method given by N. R. French and J. C. Steinberg, *J. Acoust. Soc. Am.*, **19**, 90 (1947).

cause of the concomitant increase in sound level. But, as the reverberation time increases, the detrimental effects of the reverberant sound soon become so great that they offset the advantage of the increased sound level. If the reverberation time is increased further, the speech articulation progressively decreases. This is illustrated in Fig. 9.2, which gives the probable percentage syllable articulation *vs.* reverberation time for an average speaker in rooms having the volumes indicated. These curves were computed from the results of speech-articulation

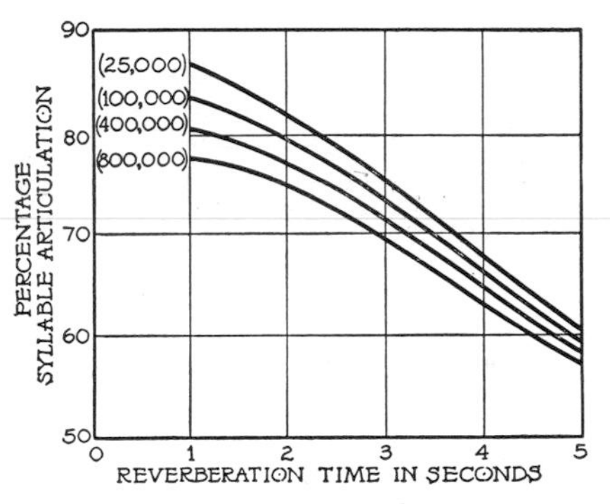

Fig. 9.2 Average percentage syllable articulation *vs.* reverberation time in auditoriums having volumes of 25,000, 100,000, 400,000, and 800,000 cubic feet.

data obtained in a small room in which the reverberation time was controlled, and in a series of large auditoriums having different times of reverberation. The curves are based on the assumptions that the rooms have no outstanding acoustical defects such as sound foci, echoes, or improper diffusion, and that the noise level has been reduced to 30 db. The positions of their maxima indicate the optimum times of reverberation for the best hearing of *unamplified* speech. The curves also show the maximum reverberation times that can be tolerated in order to maintain the articulation above a specified value, as, for example, 75 per cent. Under less favorable conditions, such as a greater background-noise level, the upper limits of reverberation time that can be tolerated are even lower than those indicated in Fig. 9.2. If an articulation of 75 per cent is regarded as the minimum required for satisfactory listening conditions, it is apparent that in an auditorium of 1,000,000 cubic feet the *unamplified* voice of an average speaker will not be heard satisfactorily for any condition of reverberation. In an auditorium of this size a sound-amplification system is indispensable.

These results apply to the average speaker; a speaker with a weak voice may not be heard as well as the curves would indicate. Figure 9.3 gives the calculated percentage syllable articulation for different speakers in an auditorium having a volume of 400,000 cubic feet, a noise level of 30 db, and a reverberation time of 1.0 second at 512 cycles. Curve *A*

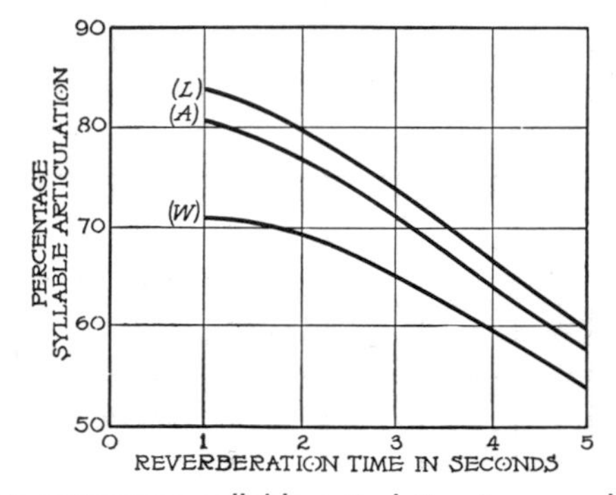

Fig. 9.3 Average percentage syllable articulation in an auditorium having a volume of 400,000 cubic feet. The different curves show how the loudness of a speaker's voice affects the hearing of speech in an auditorium. (See text.)

applies to the average of a group of fourteen professors in the University of California at Los Angeles. Curve *L* is for the speaker with the loudest voice in the group, and curve *W* is for the speaker having the weakest voice. These results quantitatively indicate one reason why some speakers are heard much more satisfactorily than others.

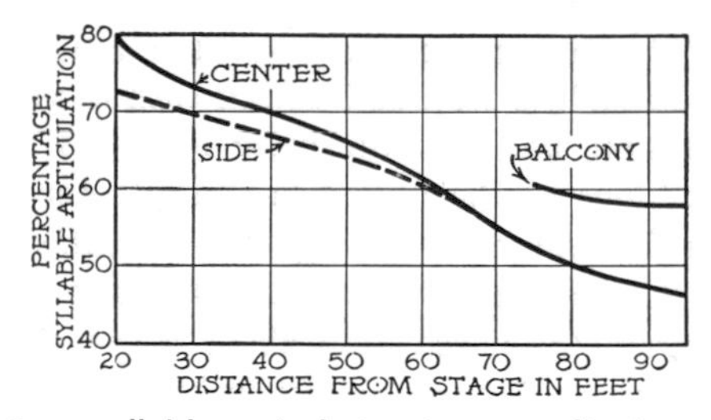

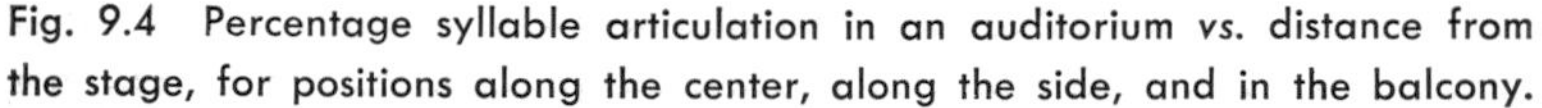
Fig. 9.4 Percentage syllable articulation in an auditorium *vs.* distance from the stage, for positions along the center, along the side, and in the balcony.

Figure 9.4 gives the percentage syllable articulation for different locations in an auditorium having a reverberation time of 2.1 seconds at 512 cycles with an audience present. The three curves show the articulation of unamplified speech as a function of the distance from the stage, for positions along the center, along the side, and in the balcony. Notice that the percentage articulation is greater at the front of the balcony, where there is a free flow of sound, than it is toward the rear of the orchestra, where the sound reaches the listeners after it has grazed the audience for a considerable distance.

Design of Room Shape

The epochal researches of W. C. Sabine demonstrated, but probably overemphasized, the large role that reverberation time plays in the determination of the quality of the acoustics of rooms. Manufacturers and distributors of building materials responded eagerly to the need for sound-absorptive materials in the control of reverberation and developed sales and service organizations to assist architects and builders in the "acoustical treatment" of rooms. Too frequently this assistance, which is sought by the architect usually after the design of the building is practically completed, consists primarily or even entirely in recommending which surfaces in the rooms should be treated with absorptive materials. As a result, there has developed a tendency to regard the control of reverberation time as the dominant and almost determining element in room acoustics, and at the same time to give too little regard to other important aspects of design, such as shape and size. As emphasized at the beginning of this chapter, good acoustical planning is based upon many significant factors that affect the insulation, generation, transmission, absorption, reflection, diffusion and hearing of sound. Each element is important; the neglect of any one may mar or ruin an otherwise good design.

The shape of a room is one of the important factors affecting its acoustical properties. Hence the determination of the most desirable shape is a problem that the architect should know how to solve. In the present section, we shall discuss the pertinent principles and procedures that will give a solution to this problem.

FLOOR PLAN. The design of an auditorium or a lecture room usually begins with the layout of the floor plan. The seating should be arranged so that the audience is as near the stage as is consistent with the requirements set by the distribution of sound from the source and with those for good visibility. Thus, although an audience can be brought nearer the speaker in a room having a square floor plan than in one in which the length is greater than the width, the latter is preferable. One of the reasons for this preference can be visualized easily by referring to Fig. 1.8, which shows how sound is distributed around the head of a person who is speaking. The sound level, especially in the higher frequency range which is responsible for a large percentage of the intelligibility, drops off rapidly at right angles to the direction the speaker faces Hence, the front seats along the sides are not very satisfactory for hearing of speech in a large, square room, and these seats are usually outside of the "beam" of the loudspeakers as normally used in sound-amplification systems. In a small room the sound level is sufficiently high for good hearing for a wide range of the ratio of length to width. It is apparent, then, that the optimum ratio of length to width for a room is not a fixed number, but varies with the size and shape of the seating area; it also depends on whether a sound-amplification system is used. For most rooms, ratios of length to width of between 2:1 and 1.2:1 have been found satisfactory.

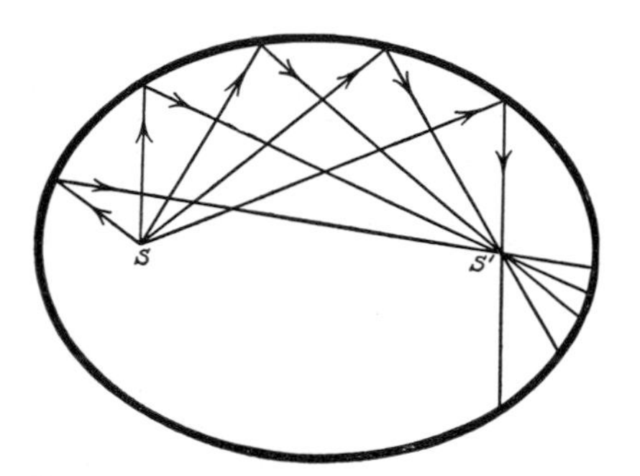

Fig. 9.5 Elliptical floor plan illustrating an acoustical defect, focusing by the walls of the room.

Circular and elliptically shaped floor plans nearly always give rise to focusing effects, non-uniform distribution of sound, and echoes. Two prominent defects are illustrated in Figs. 9.5 and 9.6. In the circular plan (Fig. 9.6) sound originating at *S* and directed at nearly grazing incidence to the walls, as in the direction *SA*, tends to creep along the side of the wall.[3] Sound reflected from the rear portion of the cylindrical walls, as rays *SM*, *SN*, *SP*, and *SQ*, are brought to a focus at approximately *S′*. This focusing defect is even more pronounced in the elliptical

[3] See "Whispering Galleries," p. 167.

plan (Fig. 9.5), especially when the source is at one focus, as indicated. Here, the concentration of reflected rays in a small region could be partially overcome if the source were moved from a focus; nevertheless, the distribution of sound would remain very non-uniform. In both elliptical and circular plans, the acoustical conditions can be greatly improved by the addition of cylindrical diffusing surfaces, as in Fig. 9.7. Figure 9.8 is a photograph of a room having such wall surfaces.

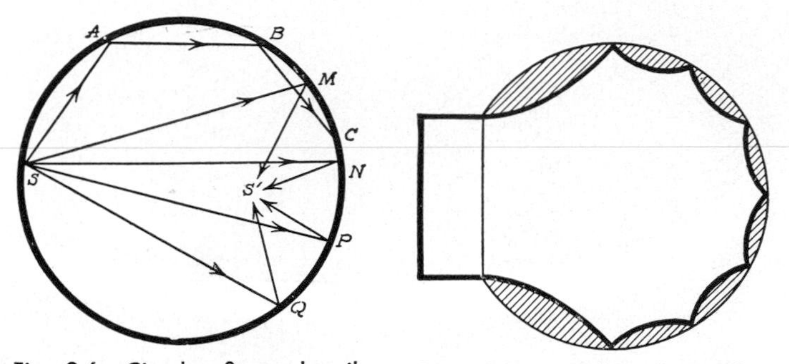

Fig. 9.6 Circular floor plan illustrating two outstanding defects: (1) the tendency for sound to creep around the walls (shown by ray *SABC*); (2) the focusing of sound due to reflection from the rear part of the circular wall.

Fig. 9.7 Circular floor plan modified by convex diffusing surfaces which greatly reduce the focusing and creeping effects illustrated in Fig. 9.6.

In order to bring a large audience as close as possible to the stage of an auditorium, it is advantageous to design a floor plan with diverging side walls. Reflections from these walls can aid in the establishment of a higher sound level at the rear of the auditorium, but these reflections must be carefully controlled. Path-length differences of 65 feet or more between direct and reflected sound give rise to echoes. Path-length differences from about 50 to 65 feet produce a blurring quality which may result in a lack of "intimacy," especially for auditors in the front seating area. Intimacy is a qualitative term used to describe the extent to which sound appears to come from the screen in a motion picture theater. If the included angle of the sound received by an auditor is small, he will judge the auditorium to have intimacy. In this respect, reflections from the side walls are more significant than those from the ceiling, for one's ability to localize sounds in the horizontal direction is somewhat greater than it is in the vertical direction.[4]

It is good design to utilize the floor area which has the best

Fig. 9.8 Photograph of the Voder Room showing a circular floor plan with convex diffusing wall surfaces. The walls of the room are tilted to aid in diffusion. (Courtesy American Telephone and Telegraph Co.)

acoustical environment for seating and to use the poorest areas for non-listening purposes. Thus, wherever possible, the area directly in front of a speaker should be used for seating rather than for an aisle.

Elevation of Seats. Since an audience constitutes a highly absorptive surface, sound waves which graze it are greatly attenuated. Hence, it is good design in an auditorium, from the standpoint of hearing as well as of seeing, to elevate the seats in order to provide a free flow of direct sound from the source to

[4] Some interesting oscillograms showing the arrival of reflected sound in a theater reputed to have poor intimacy are given by C. A. Mason and J. Moir, *J. Inst. Elec. Engrs. London,* **88**, Part III, 175 (1941).

the listeners. A good line of sight will do this. The first few rows can be level since they will have a good line for both sight and sound. The higher the source is elevated, the farther back the level area can be extended. Let us denote by d the distance which should not be exceeded between the source and the last row of the level seating area. A useful formula for computing this distance is

$$d = r(2.5h - 1) \tag{9.1}$$

where r is the distance between rows and h is the height of the source. For example, suppose that the rows are 3 feet apart, and that the lips of a speaker are 5 feet above the floor level. Then, according to Eq. (9.1), the level area should not extend more than 35 feet from the speaker. Of course, the floor can begin to slope up at any convenient distance which is nearer the speaker. The angle of elevation of the inclined floor depends on the nature of the room; in an auditorium, it should not be less than about 8°; in a demonstration-lecture hall it should be about 15°. Steeper elevation is desirable if it can be provided at a reasonable cost and without making the aisles too steep for reasons of safety. It is advantageous not only to elevate the seating area but also to stagger the seats.

CEILINGS. The ceiling and walls should provide favorable reflections of sound, especially for the seats far removed from the stage. In some instances, the ceiling also should aid in the diffusion of sound. However, if adequate means of diffusion are furnished by the floor and wall surfaces, no additional diffusion is needed for the ceiling; hence it may be utilized to the utmost for the advantageous reflection of sound. Lecture rooms, chamber music rooms, council chambers, and Christian Science auditoriums are types of rooms in which a low, smooth, highly reflective ceiling may be used to good advantage.

There is no formula, either simple or complicated, for calculating the optimum ceiling height of a room. Consideration must be given to the optimum volume, which is the subject of a subsequent section. In general, the ceiling height of a room to be used for speech and music should be about one third to two thirds of the width of the room—the lower ratio for very large rooms, and the higher ratio for small rooms. Thus for an auditorium 100 feet wide and 150 feet long, a ceiling height of about 30 to 35 feet is indicated; for a room 18 feet by 24 feet, a ceiling height of 10 to 12 feet is about optimum. Experience in the design and use of radiobroadcast studios has led to certain recommended ratios of length to width to height. Some of these ratios lead to excessively high ceilings for theaters and other large rooms and should not be used except for studio design. If the ceiling of an auditorium is too high, not only will the volume per seat be excessive, but also long-delayed reflections from this surface will be a source of echoes.

Ceiling splays in the front of a room, or appropriately tilted portions of the ceiling, can be devised to reinforce the sound reaching the rear parts of an auditorium (see Figs. 15.1 and 16.2). They serve the same purpose as do the front splays of the side walls. The law of reflection (angle of reflection equals angle of incidence) can be used to determine the most propitious angle of inclination. Similarly, a splay between the ceiling and the rear wall can be designed to reinforce the sound in the rear of the room, and at the same time to prevent echoes from the rear wall.

Concave surfaces such as domes, cylindrical arches, and barreled ceilings should be avoided wherever possible. If they are required by the architectural style, the radius of curvature should be either at least twice the ceiling height, or less than one-half the ceiling height. If coves, bays, or other small concave surfaces are employed, their radii of curvature should be quite small compared to the ceiling height. The most serious defects (sound foci or echoes) occur when the radius of curvature of a ceiling surface is about equal to the ceiling height; see Fig. 8.13(*a*).

In order to avoid flutter echoes, a smooth ceiling should not be strictly parallel to the floor. If the floor and ceiling are both smooth, level, and highly reflective, the flutter between the floor and ceiling will be very prominent.

Side Walls. The side walls should reinforce the sound that reaches the rear parts of a large room. This is especially desirable for auditoriums in which a sound-amplification system is not utilized for all spoken and musical programs. The location of the walls is, of course, determined principally by the general contour of the floor plans. The angle that any portion of the wall surfaces, such as a splay, makes with the wall contour line should be such as to reflect sound beneficially to those seats where the sound level is not adequate. The law of reflection can be used to determine this angle. The side walls should be designed so that the sounds they reflect to the audience will not be too long delayed. Some parts of the side walls may be suspected of causing probable echoes or unduly delayed reflections; this may happen in very large auditoriums. In such instances the suspected surfaces should not be smooth and reflective. Instead they should either be made "acoustically rough" to diffuse the sound, or they should be covered with highly absorptive material. Examples of side walls based on good acoustical designing for different types of rooms are given in Chapter 15.

Flutter echoes frequently occur between the side walls. They can be avoided by a number of means: by diverging, non-parallel, or tilted walls; by splayed or vee'd walls. Splays not only serve to prevent flutter, but they also can contribute both to desirably directed reflections and to the diffusion of sound within the room. As little as $\frac{5}{8}$-inch splay to the running foot will prevent flutter. A slight inclination of walls can have the additional advantage of directing useful reflections upon the audience.

REAR WALL. In the design of all rooms, large concave rear walls should be avoided. Unfortunately, they are of common occurrence because it seems so simple and economical to most architects to have the rear wall follow the curvature of the last row of seats. Walls with this shape are responsible for troublesome echoes and delayed reflections in many theaters and auditoriums. This is illustrated in the upper part of Fig. 9.9, which is a longitudinal section showing a vertical rear wall. Sound rays reflected from the ceiling near the rear wall at P are next reflected from the rear wall at Q to seats in the vicinity of R; there results an echo at R if the path difference at R exceeds 65 feet. Sound rays striking the rear wall at M are reflected to the ceiling at N, and then to O at the back part of the stage. Often these reflections from concave rear walls are concentrated in regions near the microphones of the sound-amplification system; then feedback trouble is induced. These detrimental reflections can be converted into beneficial ones by introducing a ceiling splay between the ceiling and the rear wall, as shown in the lower sectional drawing of Fig. 9.9. Here the rays *SP′* and *SM′* are reflected to the rear seats; thus 1 or 2 db are added to the sound level in that area. Absorptive material on the rear wall elimi-

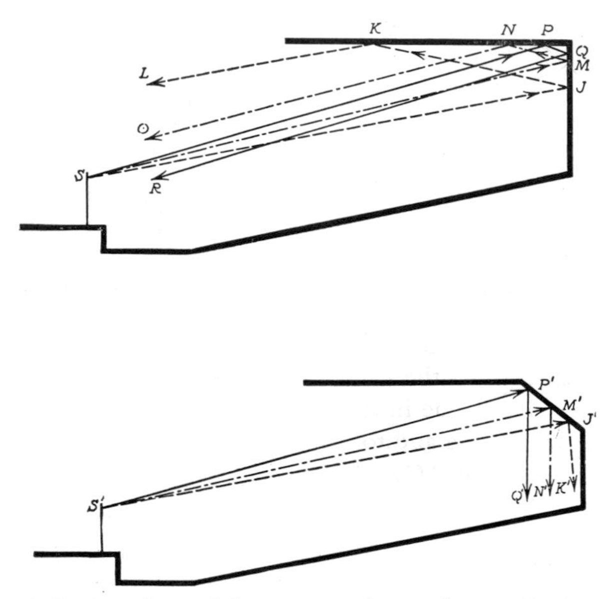

Fig. 9.9 Reflection of sound from rear surfaces of an auditorium, showing how a suitable ceiling splay can be utilized to prevent echoes from reaching the front seating area, and at the same time to reinforce the sound for the last few rows of seats.

nates the echo at *R*. Concave surfaces in certain situations can be made as effective as splays, and they are sometimes better adapted than splays to the general appearance of the room. However, unless properly designed, they can lead to focusing effects. In some designs, splays between the ceiling and side walls are useful in preventing long-delayed reflections and in directing advantageous reflections to the audiences.

If reflections from either a vertical or a tilted wall are capable of producing echoes, the offending surface should be treated with absorptive material. There will still be some reflection from this surface, but the sound level is thus reduced so greatly that its detrimental effects are negligible.

In some large rooms, reflections from a portion of the rear wall can be utilized effectively by tilting the wall; for example, see "Balcony Recess" below. Proper rear wall design can increase the sound level in an auditorium where the increase is most needed. Caution must be observed, however, to avoid the concentration of reflections in small areas, especially for excessive path-length differences between the direct and reflected sounds. In rooms where the rear wall is relatively high or where the seating area rises rapidly, it is not advisable to tilt the entire rear wall. To do so might reflect the direct sound toward the front of the room so that echoes could be produced.

BALCONY RECESS. Good design of a balcony recess usually requires a shallow depth and a high opening. For an auditorium or legitimate theater, the depth should not exceed twice the height of the opening. This plan permits sound to flow readily into the space under the balcony. Good design also requires that the reverberation time in the balcony recess approximate that of the main part of the auditorium (see "Reverberation in Coupled Spaces," p. 161, and "Control of Reverberation Characteristics," p. 195).

By applying the above rules, it is possible to design the recess so that the sound level in this space is about the same as it is in other equally distant parts of the auditorium. However, if the opening is low and the recess relatively deep, the sound level will be considerably lower in this area, especially at the rear of the recess. For example, if the depth is equal to four times the height of the balcony opening, the level may be 8 db lower at the rear wall than it is at the opening. In large auditoriums and theaters it is advisable to "break up" the rear wall in order to provide proper diffusion of sound throughout the balcony recess. *A large, unbroken concave rear wall always should be avoided,* since it invariably gives rise to a non-uniform distribution of sound. Trouble of this kind also may arise from large vertical surfaces of glass in front of the standee rail.

The balcony rail (front) should not be overlooked when the acoustical design of an auditorium is being worked out. Since it is frequently a large, concave surface having a width that is large compared to the shorter wavelengths of speech and music, the balcony front can give rise to an echo or "slap-back." By tilting this surface downward and making it convex it is sometimes possible to utilize the resulting reflections to increase the sound level at the rear of the auditorium. Otherwise, the front should be highly absorptive or should have a contour such that reflections from it will be diffused and not concentrated in small areas.

The balcony soffit and rear wall should be designed so that a large portion of the sound coming directly from the source will be reflected to the auditors under the balcony, and the remainder absorbed by the rear wall. An example of one such plan is shown in Fig. 9.10(*a*), which is a section of the balcony recess of

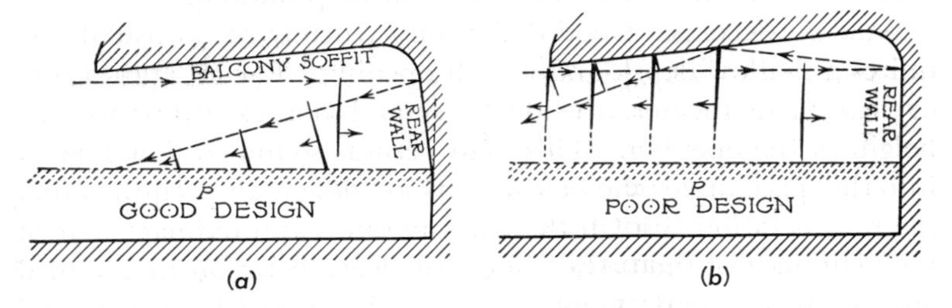

Fig. 9.10 (*a*) Section of the balcony recess of the Philips Theater, Eindhoven. A sound ray is reflected from the inclined rear wall to the area under the soffit. (*b*) Section of a balcony recess having an untreated vertical rear wall. Sound waves striking the rear wall are reflected to the main part of the auditorium without reaching the audience at *P*. (R. Vermeulen and J. de Boer.)

the Philips Theater in Eindhoven. Measurements made in a scale model indicate that this design leads to a distribution of sound on the floor of the auditorium that is fairly uniform; see Fig. 9.14(*b*). If the time lag between the direct sound and the reflections from the rear wall is short, the auditor will not be aware of the direction from which these reflected sounds come. He will have the illusion that all the sound comes directly from the stage, for auditory localization is poor in the vertical direction. Furthermore, it is much more difficult to discriminate between sounds coming from directly ahead or behind than between sounds coming from one side or the other. Hence, these reflected contributions may be utilized effectively. In contrast to the section of Fig. 9.10(*a*) is the section shown in Fig. 9.10(*b*).

Here the rear wall reflects sound to the front part of the auditorium. The results of scale model tests using this section are shown in Fig. 9.14 (*c*); they indicate that the sound level under the balcony would be low in comparison to that in other locations, and that the design of Fig. 9.10(*a*) is superior.

Volume per Seat

The most desirable volume for a room is closely correlated with the design of the ceiling. There is no fixed optimum ratio between ceiling height and width and length. The optimum height, and therefore the optimum volume per seat, is dependent on both the seating capacity of the room and the purposes the room is to serve.

The optimum volume per seat for a room is the lowest value consistent with the visual and aesthetic requirements, with the comfort of the audience, and with the general appearance. Thus, although it is desirable to have a low value of volume per seat, it should not be attained by seating the auditors so close to each other that they do not have sufficient leg room or by sacrificing other functional features. In motion picture theaters seating 1000 people, the optimum volume per seat may be as small as 125; for theaters with a seating capacity of 2000, the volume should not exceed about 175 cubic feet per seat. In music rooms seating more than 1500, a volume per seat of 200 cubic feet has been found to give satisfactory results.

There are many advantages in keeping the volume per seat at a low value. The building cost is greatly reduced. Maintenance costs for lighting, cleaning, redecorating, air conditioning, etc., are correspondingly lowered. There are also important acoustical advantages. Thus, suppose that a given reverberation time is sought. Then, from Eq. (8.12), the smaller the volume of the room the fewer will be the units of absorption required to obtain this reverberation time. In an auditorium with a low volume per seat, if the furnishings (seats, carpets, draperies, etc.) have been carefully chosen, there may be no need for additional acoustical materials to control the reverberation. Then the architect has greater freedom in his choice of materials for finishing the interior and for decoration. Also, it follows from Eq. (8.6) that, the lower the volume per seat, the higher will be the sound level in the room for a source of a given power. In speech and music rooms where sound-amplification systems are not employed, this increased level is most beneficial. Even if a "sound system" is used, the smaller the room, the smaller will be the required power rating of the sound system.

For a given seating capacity, an auditorium design which incorporates a balcony usually has a lower volume than one that does not. Other factors which affect the volume per seat are the arrangement of the aisles, the inclination of the floor, the distance between seats, whether or not the seats are staggered, and the floor plans.[5]

Optimum Reverberation in Rooms

In this section we shall discuss the factors that determine the most propitious conditions of reverberation in rooms. The term *optimum reverberation* includes not only the reverberation time *vs.* frequency characteristic throughout the audible range of frequencies, but also the optimum nature of the growth and decay of sound in rooms and the optimum ratio of reverberant to direct sound reaching the auditors. Each individual sound should build up and die away with a certain degree of smoothness and uniformity. The transient acoustical properties should be such that the sound-pressure level decays at a rate which has a uniform trend, but which is characterized by minor fluctuations. These minor fluctuations are similar to a vibrato in music; they should be prominent enough to avoid monotony and to add desirable liveness, but not so prominent as to introduce sudden surges of sound that mar smooth growth and decay. The optimum degree of smoothness will be closely approximated if steps have been taken to provide proper diffusion in the room. (See "Diffusion of Sound," p. 139.) In determining the optimum reverberation time *vs.* frequency characteristic for a room, we shall assume that the room's shape, and the distribution of absorptive and reflective surfaces in it, are such as to give proper diffusion of sound, and thus proper uniformity and smoothness of the decay.

The question arises, then, what reverberation-time character-

[5] For a volume per seat analysis of several typical motion picture theaters, see C. C. Potwin and B. Schlanger, *J. Soc. Motion Picture Engrs.*, 32, 156 (1939).

istic is most desirable? Even a casual examination of this question indicates that large variations with respect to frequency are not desirable, since those frequency components for which the reverberation time is long would be prolonged and overemphasized, whereas those components for which the reverberation time is short would die away too fast and be underemphasized. The optimum reverberation time *vs.* frequency characteristic is one that will allow all frequency components of speech and music to grow and decay at such rates during their transient states, and to be maintained at such levels during their steady states, as will provide high intelligibility of speech and excellent conditions for the rendition of and the listening to music. The curves of Fig. 9.2 indicate that high speech articulation requires a low reverberation time—less than 1 second (at a frequency of 512 cycles) in small rooms and somewhat longer times in larger rooms. Certain aspects of speech and music, and of hearing, indicate that the reverberation time at low frequencies may be longer than the optimum time at 512 cycles.[6] However, listener preference tests conducted in radiobroadcast studios in which the reverberation characteristic can be varied lead to a somewhat flatter reverberation characteristic than those based on these theoretical considerations. The most favorable characteristic for a room is represented by a chart which gives the optimum reverberation time as a function of frequency throughout the relevant audible-frequency range; for most rooms that are used for both speech and music, it is sufficient to specify the frequency range from 128 to 4096 cycles; for music rooms, and rooms in which good acoustics is the prime requirement, the relevant frequency range extends downward to about 50 cycles and upward to about 10,000 cycles.

A careful consideration of the available data on the preferred reverberation time *vs.* frequency characteristic for rooms is summarized in Figs. 9.11 and 9.12, which give the authors' recommendations. Figure 9.11 shows the optimum time of reverberation, at 512 cycles for different types of rooms, as a function

[6] For a discussion of criteria for determining the most favorable reverberation time *vs.* frequency for speech rooms and for music rooms, see V. O. Knudsen, *Architectural Acoustics*, pp. 382–388 and 406–414, John Wiley & Sons, Inc., 1932.

of room volume. The optimum times for speech rooms, motion picture theaters, and school auditoriums are given by single lines; the optimum time for music by a broad band. In music rooms, the optimum reverberation time is not the same for all kinds of music; the best choices for certain types of music are indicated. For example, slow organ and choral church music require more

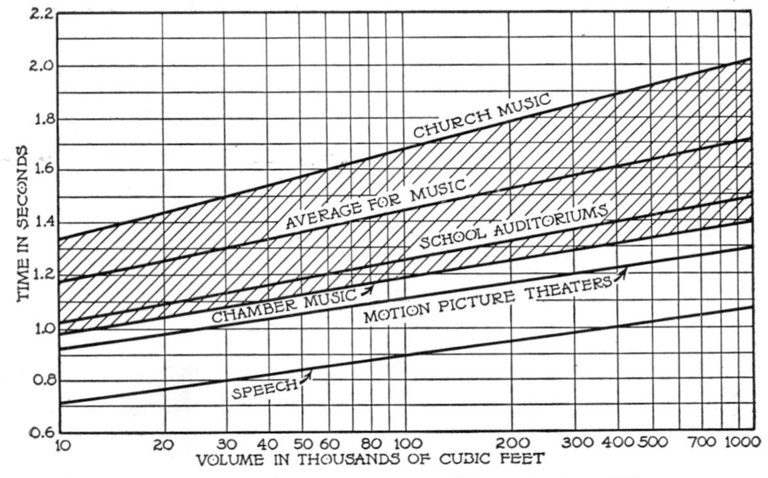

Fig. 9.11 Optimum reverberation time at 512 cycles for different types of rooms as a function of room volume.

reverberation than does a brilliant *allegro* composition played on the woodwinds, piano, or harpsichord.

The optimum reverberation time *vs.* frequency characteristic for a room can be obtained by the use of Figs. 9.11 and 9.12 in the following manner: First, knowing the volume and purpose of the room, determine from Fig. 9.11 the optimum reverberation time at 512 cycles. Then, to obtain the optimum reverberation time at any other frequency, multiply the 512-cycle value by the appropriate ratio for that frequency, which is given by Fig. 9.12. If R is the value of this ratio at frequency f, the reverberation time t_f at that frequency is given by

$$t_f = t_{512}R$$

where t_{512} is the reverberation time at 512 cycles given in Fig. 9.11. R is unity for frequencies above 500 cycles, and it is given by a band below 500 cycles. The ratio R for large rooms may have any value within the indicated band; preferred ratios for small rooms are given by the lower part of the band. The exact value of R to be used for frequencies below 500 cycles is not

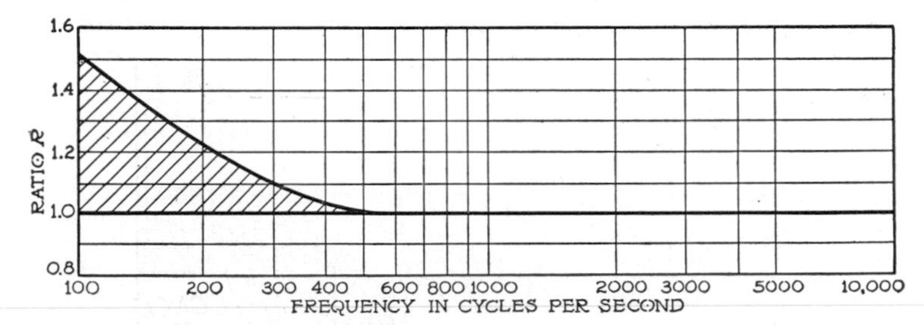

Fig. 9.12 Chart for computing optimum reverberation time as a function of frequency. The time at frequency f is given in terms of a ratio R which should be multiplied by the optimum time at 512 cycles, as given by Fig. 9.11, to obtain the optimum time at frequency f.

critical, but in general it should not fall outside of the indicated band. Deviations of less than 10 per cent from the specified optimum reverberation time will be acceptable. In general it is much easier to fulfill design specifications if there is a rise at low frequencies in the reverberation time *vs.* frequency curve.

Control of Reverberation Characteristics

The reverberation characteristics of a room can be controlled by the amount and placement of absorptive material within it. The total amount of absorption in a properly designed room determines the rate at which sound will decay in it. Proper distribution of the absorption aids in controlling the diffusion of sound and also the nature of the time fluctuations of the sound during its decay.

The first step in planning the acoustical treatment of a room is to determine the optimum reverberation time, from Figs. 9.11 and 9.12, and to find the total number of square-foot-units (sabins) of absorption required to give this time. A large part of this absorption will be furnished by agents other than acoustical materials, for example, by the chairs, rugs, audience, walls, ceiling. It is customary to assume that the size of the audience in an auditorium will be equal to two thirds of the seating capacity. Then the amount of absorption that must be added is the difference between the total required units and the number of units furnished by the above-named agents.[7] (A specific example is worked out in Table 15.1.) These questions remain:

"THE ACOUSTICS WERE SPLENDID – I COULD HEAR EVERY WORD THE WOMEN BEHIND ME WERE SAYING!"

Fig. 9.13 People are not in complete agreement as to what constitutes ideal room acoustics! (Courtesy Register and Tribune Syndicate.)

Where should the material be placed? What materials should be used?

As a general rule, the surfaces surrounding the stage should reinforce, by useful reflections, the "voices" of the performers.[8] On the other hand, the rear wall must be designed so that long-

[7] In making these computations, it is of course necessary to take account of the loss of absorptivity of a given surface that has been covered by the acoustical material; the net gain in absorption per unit area of such a surface is the coefficient of the acoustical material minus the coefficient of the surface it covers.

[8] This is not true for motion picture theaters, where the space immediately behind the loudspeakers should be highly absorptive. See "Motion Picture Theaters," p. 317.

delayed reflections from it are prevented from reaching the audience. This requirement usually necessitates the use of a highly absorptive rear wall; the portion of the wall above the wainscot (the wainscot should extend not more than about 1 foot above the heads of the audience) should have an average absorption coefficient in excess of 0.75. In auditoriums where the acoustical design indicates the desirability of tilting forward the rear wall (see "Rear Wall," p. 187) so that reflections from this surface may be beneficially utilized, very much less absorptive material is needed and it should be applied in patches or panels. After allowances have been made for the rear wall treatment, the remainder of the required additional absorption should be distributed on the side walls, preferably in patches, strips, or panels having dimensions of the order of 3 to 5 feet. (For example, note the distribution shown in Fig. 16.3.) The application of the absorptive material in the form of patches not only promotes diffusion but also helps to suppress flutter echoes.

The problem of non-uniform application can be worked out in many ways. Some examples follow:

(1) Finish the entire wall with a material such as plywood applied to furring strips (taking care to brace the plywood at irregular intervals so that the resonant frequencies of the resulting panels are distributed in frequency). Then perforate some of the panels with small holes and back these perforated panels with an absorptive blanket, and thus obtain patches of absorption in such numbers, sizes, and locations as will give the desired reverberation and diffusion.

(2) Finish the entire wall or large portions of the wall with perforated board and install absorptive material behind selected portions of the perforated covering. Thus, although the appearance of the surface is uniform, a non-uniform absorptive treatment is obtained.

(3) Treat selected panels, strips (horizontal or vertical), or splays with absorptive material as required to give the optimum reverberation time and good diffusion. If the side walls are treated with absorptive material, the wainscot should have a height of about 4 or 5 feet. The wainscot, if made from a thin, flexible material like plywood or Flexwood, backed with an ab-

sorptive pad or blanket, can furnish low-frequency absorption, which usually is needed to provide the optimum reverberation characteristics.

The reverberation time of a balcony-type auditorium can be calculated in the following manner:

(1) When the depth of the balcony recess is not greater than twice the height of the opening, the volume under the soffit is added to the volume of the main part of the auditorium, and the reverberation time is computed as usual, from Eq. (8.17).

(2) When the depth of the balcony recess is greater than twice the height of the opening, it is necessary to make separate calculations for each of these spaces. For these computations the coefficients of absorption shown in Table 9.1 are applicable to the balcony opening.

TABLE 9.1

ABSORPTION COEFFICIENTS FOR BALCONY OPENINGS

Ratio of Balcony Depth to Height	*Frequency*		
	128 cycles	*512 cycles*	*2048 cycles*
2½	0.30	0.50	0.60
3	0.40	0.65	0.75

If the chairs are highly absorptive, as they should be, it usually will not be necessary to add any absorptive material to the balcony recess other than the absorptive material on the rear wall. If the chairs are not absorptive, it may be necessary to add some absorptive material to the soffit or side walls of the recess in order to provide the optimum reverberation in this space. When this is done it is desirable to distribute this material in panels, strips, or patches.

If the design considerations of this chapter are followed, absorptive material will not be required on the ceiling unless the seats and floor both have a low coefficient of absorption. But absorptive material should be applied to all surfaces in a room which are responsible for, or are likely to cause, echoes.

The acoustical materials selected to provide the required absorption in a room must satisfy the above requirements. However, certain additional qualities must be considered, such as fire resistance, decorative possibilities, and the other properties discussed in Chapter 6. It is important to choose materials which will provide the optimum reverberation characteristics throughout the entire relevant range of frequencies, *not* at just one frequency. In order to do this, it is sometimes necessary to use a combination of materials having different absorption *vs.* frequency characteristics. Obviously, it is an advantage to have the acoustical characteristics of an auditorium as nearly independent of the size of the audience as possible. Table A.6, Appendix 1, shows that plain wooden chairs have a relatively low absorption—approximately 0.3 square-foot-unit (sabin) of absorption. If a person is seated in such a chair, the combined absorption of the person plus chair is approximately 5 sabins. Thus, the acoustical conditions in a room containing unupholstered seats usually are dependent on the size of the audience. Heavily upholstered chairs will overcome this difficulty; they will also provide increased absorption, and thus they will reduce the amount of acoustical material required for the optimum reverberation time. Such chairs have an absorption almost equivalent to that of a person, and most of this absorption is on the portion that is covered when the chair is occupied.

Scale Model Tests

It is highly desirable to check the shape of the proposed design of a large auditorium in order to determine possible improvements before construction and to avoid potential acoustical defects. Scale models offer a simple means of conducting such an investigation. The models are relatively easy to build, and changes in size or form can be made readily.

Two kinds of model tests have been used with good results. In the first, generally involving optical methods, wavelengths are used that are very small compared to the dimensions of the model; these are the conditions assumed in geometrical acoustics. In the second type, wavelengths are used that are scaled down in the same proportions as the dimensions; these methods may be utilized to investigate diffraction and other phenomena of physical acoustics.

Optical Methods. A scale model is constructed so that its interior has an optically reflective surface. A source of light is then placed in the normal position of the speaker or the other sources of sound. Since the wavelength of light is extremely small compared to dimensions of the model, these methods of testing only yield information regarding the behavior of high-frequency sounds in rooms.

(*a*) *Light Distribution Method.* This method [9] utilizes a three-dimensional model of the auditorium, fabricated of sheet aluminum. The metal surface has an optical reflection coefficient of about 50 per cent so that, after about three reflections, light rays emanating from a source placed at the normal position of the speaker are reduced to about 12 per cent of their initial value. An opal glass surface is used to represent the audience because it has approximately the same reflection coefficient for light as the audience has for sound. The brightness of the opal glass indicates the steady-state sound-pressure distribution over the seating area for the first few reflections of the sound waves. Figure 9.14(*a*) is a photograph of the model of the Philips Theater in Eindhoven which was used in obtaining the photographs shown in Figs. 9.14(*b*) and 9.14(*c*). In Fig. 9.14(*a*) the opal glass has been removed in order to expose the interior. The light source on the stage can be seen on the right; the balcony soffit is on the left.

(*b*) *Ray Method.* In this method a scale model is used which has walls and ceiling of highly optically reflective material, for example, polished metal or mirror surfaces. The light source is enclosed by a cylindrical screen containing narrow parallel slits, which limit the emergent light to a large number of radial beams. If the model is filled with smoke, the rays will become visible and thus indicate their paths for their first one or two reflections. Figure 9.15 shows the course of the rays in a model of the Assembly Hall in the United Nations Building at Geneva. The importance of the reflector behind the speaker in distributing the sound throughout the room is indicated. Since the surface *C* is acoustically absorptive, it was made optically absorptive in the model so that rays are not reflected from it. A large path

[9] R. Vermeulen and J. De Boer, *Philips Tech. Rev.*, 1, 46 (1936).

(*a*)

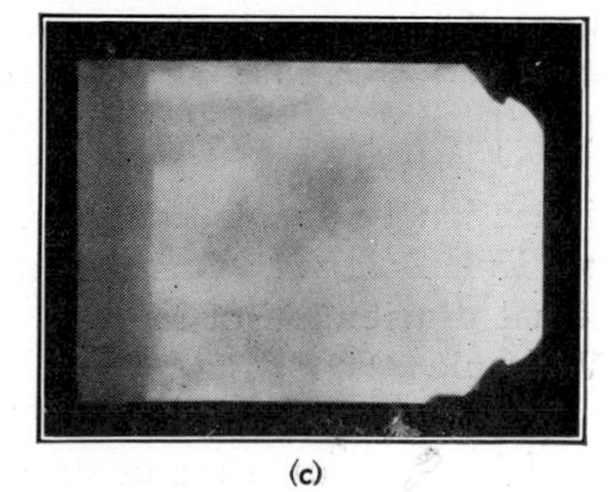

(*b*) (*c*)

Fig. 9.14 (*a*) Optical scale model of the Philips Theater—see Fig. 9.10(*a*) for a section of its balcony—showing the source of light on the stage at the right of the photograph. The balcony soffit is shown on the left. (*b*) Photograph of opal glass in the optical model. The stage is at the right. The relative uniformity of illumination indicates that the distribution of sound in the theater similarly would be uniform. (*c*) Photograph of opal glass in model used when modified so that the balcony recess has the cross section shown in Fig. 9.10(*b*). The distribution of light is very non-uniform, indicating that the sound level would be non-uniform and quite low in the balcony recess. (R. Vermeulen and J. de Boer.)

difference between the direct rays and those reflected from the high ceiling D is indicated. The long-delayed reflections from the rear wall give rise to echoes which converge to a region near the speaker's platform.

It is not always easy to ascertain the paths of sound rays in an auditorium from a purely geometric consideration of the two-

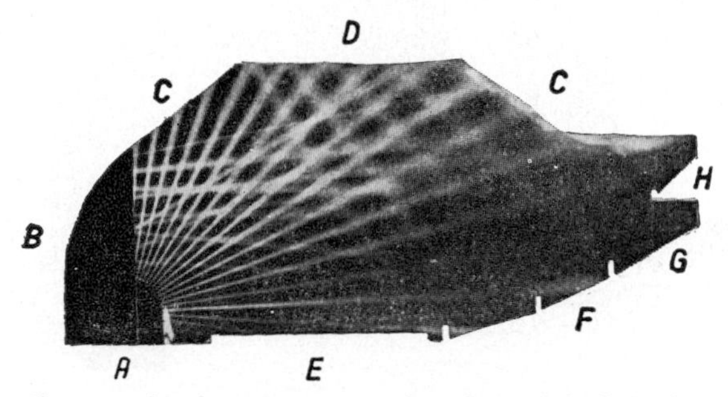

Fig. 9.15 Photograph taken in an optical scale model of the large assembly hall in the United Nations Building in Geneva (sectional view). The path of the sound rays is indicated by light beams in the smoke-filled model. The source of light is placed at the speaker's rostrum *A*, which is surrounded by a sound reflector *B*. The audience is seated in the areas *E*, *F*, *G*, and *H*. (R. Vermeulen.)

dimensional plans and sections. In these cases, model tests are particularly useful. One such method of investigating the directional distribution of sound in a room has been described by Vermeulen.[10]

Wave Methods. If the lengths of the waves used in the scale model investigation are of such size that the ratio of the wavelengths to the dimensions of the model are approximately the same as the ratio of the wavelengths of sounds in speech or music to the dimensions of the auditorium, it is possible to study diffraction effects, as well as sound distribution, in the model. From the standpoint of cost, it is desirable to work on a small scale. Hence it is necessary to use waves of correspondingly small wavelength. Among the more convenient means of producing such waves are: (1) an ultrasonic generator, (2) an intense sound-pulse generator (such as the discharge of a highly charged electrical

[10] R. Vermeulen, *Philips Tech. Rev.*, **5**, 321 (1940).

condenser), and (3) ripples on a shallow tank of water. The first type of waves can be used in a three-dimensional study of auditoriums, studios, or open-air theaters. The two latter methods are usually limited to two-dimensional studies of wave propagation. However, it is frequently possible to ascertain certain aspects of the nature of the propagation of sound in the auditorium from a composite study of the action of waves in two-dimensional models representing the plan, the longitudinal section, and the transverse section.

(*a*) *Ultrasonic Method.* Sound waves above the audible range of frequencies (that is, ultrasonic waves) are useful in scale model investigations. The wavelength of the sound generated by the source should be such that it is scaled down in the same proportion as the linear dimensions of the model. Thus, if the model of a room is built on a scale of 1/4 inch to 1 foot, a frequency of 24,000 cycles would correspond to a sound of 500 cycles in the room. The physical size of the ultrasonic source should also be scaled down so that its directional characteristics approximate the actual source.

In general, steady-state measurements are made in this method; that is, the source emits sound continually, and the resulting pressure distribution in the model is determined by a tiny microphone.[11] It is important that the pressure detector be small in size so that it does not alter the distribution of sound in the model.

By a careful choice of absorptive properties of the boundaries of the room, it would be possible to investigate some effects of diffraction due to non-uniform distribution of absorption. This is in contrast to the methods that follow, which generally employ scale models having hard boundaries, and which display diffraction phenomena due only to the geometry of the boundaries.

(*b*) *Sound-Pulse Method.* The sudden electrical discharge of a condenser produces a sound not unlike the report of a pistol. Although a wide frequency band of sound is generated, the energy is concentrated in the region of very short wavelengths so that

[11] The method can be extended to the generation of sound pulses, but this would involve considerably more experimental equipment.

these waves may be used in scale model studies. The models are usually constructed of wood, or of plaster of Paris, to a scale of about 1/16 inch to the foot. The sound pulse, produced by the discharge, consists principally of a single pressure condensation followed by a rarefaction which spreads out from the source with the velocity of sound. Because the advancing wave front is much denser than the surrounding air, light, when passed through such a region, is refracted. Hence, by illuminating the model stroboscopically, the sound pulse can be photographed or viewed. Figure 4.6, a photograph made by this method, clearly shows the effects of diffraction.

(*c*) *Ripple-Tank Method.* One of the oldest methods for investigating the paths of sound waves in an auditorium makes use of water wavelets. Ripples are generated in a tray which has as its boundaries a small-scale section of the auditorium. Some device, such as an electromagnetic plunger, is used to start wavelets at a point which corresponds to the one normally occupied by a speaker. If the depth of the water is greater than the wavelength of the ripples, the analogy between the ripples and sound waves can be useful in studying the behavior of sound in the enclosure.

A wooden sectional model, having a scale of the order of 1/4 inch to 1 foot, is placed on a glass plate to form a miniature tank, sometimes referred to as a *ripple tank.* The model is filled to a depth of about 1 inch. According to the scale, if the ripples have a wavelength of 1/2 inch, they are comparable to sound waves having a wavelength of 2 feet. The progress of the ripples can be followed visually or photographically on a frosted glass plate if light is passed upward through the clear glass bottom of the tank onto the plate.

Checking the Completed Room

Certain checks should be made on a room when it is completed, if good acoustics is a prime requirement. Before a building is accepted, its acoustical properties should be determined. The architect should consult the check list recommended at the beginning of this chapter. Then if a step has been neglected, or slighted, immediate measures should be taken to correct any resulting defects.

All large rooms should be tested for echoes. A simple but effective test consists in producing a sharp hand clap and listening for an echo from suspected surfaces. For example, if the rear wall is suspected of producing an echo in the front part of the auditorium, the person conducting the test should stand on the stage, or the usual position of the sound source, and clap his hands. After each sharp clap, an assistant should listen from positions in the front of the auditorium. If the rear wall has been properly designed and treated, there should be no recognizable echo. If none is heard in this test it can be safely assumed that the rear wall is not a source of echoes. Other suspected surfaces, as a high ceiling or large reflective surfaces on the side walls, should be tested similarly. If echoes are heard in these exploratory tests, more elaborate tests should be made, preferably by an acoustical engineer. A suitable piece of apparatus for making these tests consists of a high-frequency, directive, sound source. By directing the beam of sound from this reflector at all possible surfaces that might give echoes, and listening along the reflected beam, the offending surfaces can be located.

A hand clap is an effective means for detecting "room flutter." A single clap is made by the observer while he stands between opposite surfaces that are parallel, or nearly parallel, to each other. If no flutter is heard (and none should be in a well-designed room) this possible defect has been avoided. If a flutter is heard, and especially if it is prominent, recourse should be made to the control measures previously discussed.

Next, and most important of all in the majority of large auditoriums, is the checking of the reverberation characteristics. Although a well-trained observer with keen auditory perception may rely on ear tests for this important check-up, it is wise to have a quantitative measure of the reverberation time at 128, 512, and 2048 cycles. This measure will suffice to describe the reverberation characteristics if the design and construction of the room have been based on the principles and procedures described in this book. If there is any uncertainty about these matters, it is advisable to obtain graphic records of the decay of sounds at many frequencies. The test sounds, which may be either narrow bands of thermal noise or warble tones, should be produced at the normal location of the sound source, and their decays recorded with the microphone located at typical listening positions throughout the seating area of the room. These tests should be made with an audience present if the room is not equipped with heavily upholstered chairs; otherwise, appropriate corrections must be made for the effect of an audience. The decay should conform to the requirements for optimum reverberation described earlier in this chapter. These records can reveal improper diffusion that mars the acoustics of the room.

Defects in the reverberation characteristics of a room are likely to result from faulty application of acoustical plaster or other plastic materials applied by workmen on the job, or from improper painting of absorptive materials so that their surface pores are filled or bridged over with heavy paint. Plastic materials should be tested for specified thickness as well as for porosity.

The building should be examined with regard to noise levels and sound insulation. If noise from outdoors or from the ventilation or other equipment within the building is objectionable, the noise-reduction procedures described in Chapters 11, 12, and 13 should be followed.

Articulation Testing

If poor hearing conditions (for speech) are suspected in certain parts of a room, articulation tests should be conducted; especially those seating areas that are alleged or reputed to be defective should be checked. If no sound amplification is used, the poorest seats are likely to be at the rear or in an area most disturbed by noise from outside, an adjacent corridor, or a duct opening. If sound amplification is used, there may be certain areas that are not adequately "covered" by the high-frequency loudspeakers. The level of the amplified speech should not be unduly high, and the quality should be so natural that the listeners are not conscious of artificial amplification. Typical speakers should be heard distinctly and without undue effort in all seats in the room.

The average percentage syllable articulation for an auditorium can be obtained from the results of tests with listeners stationed in all representative locations in the auditorium. The speaker should be located on the front portion of the stage or platform, or at the usual speaker's position. It is desirable to place observers in the following locations: front center, front side, middle center, middle side, rear center, rear side, balcony center, and balcony side. If each of the persons in the group acts as a caller for a list of syllables, and the listeners change positions progressively after each list is called, the average score on all the recorded lists will give a figure of merit for the speech-hearing quality of the room. In addition, the hearing quality of different parts of the auditorium also can be determined and compared.

Unless a list of syllables is selected with some care, considerable training must be given the observers participating in the speech-articulation tests. However, a method of articulation testing which uses special word lists has been devised by Fletcher and Steinberg which reduces the training time to a minimum; less than an hour is required to become familiar with this technique. In this procedure, the fractions of the vowels and indicated consonants that are understood correctly are determined separately, and the percentage syllable articulation is calculated from

$$PA = 100[1 - (1 - V_w C_w^2)^{0.9}] \tag{9.2}$$

where PA = percentage articulation (for syllables)
V_w = fraction of vowels correctly heard
C_w = fraction of designated consonants correctly heard.

Equation (9.2) is expressed graphically by the chart in Fig. 9.16. Suppose the fraction of vowels correctly understood is 0.97, and the corresponding fraction for consonants is 0.91. Then $V_w C_w^2 = 0.80$; hence, according to Fig. 9.16, the percentage articulation is equal to

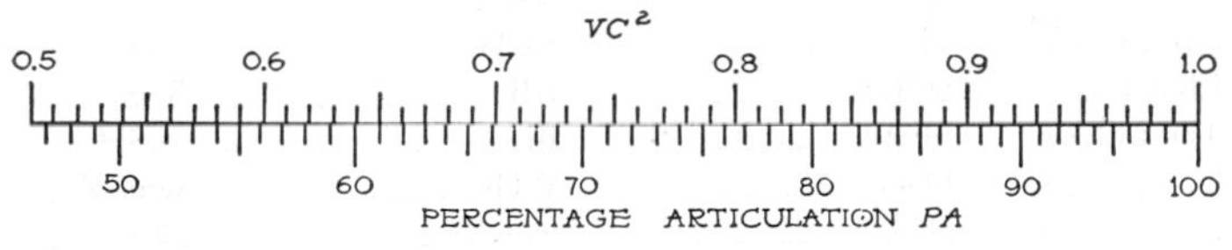

Fig. 9.16 Chart giving the relationship between percentage syllable articulation (*PA*) and the fraction of vowels and consonants correctly heard. See Eq. (9.2).

77 per cent. A sample word list is given in Fig. 9.17. The words have been chosen so that the vowels and consonants appear with about the same frequency as they do in ordinary conversational speech. Each group is preceded by an announcement in order to introduce the conditions of reverberation that would normally exist in the room. Only the errors in vowels are counted in the vowel list, and only the errors

in the consonants are counted in the consonant lists. The vowels and consonants tested in each word are indicated by capital letters.

It is obvious that the same list cannot be used repeatedly because the

Articulation Test Record

Test No._____ Observer_________ Date____________________

List No._____ Caller___________ Articulation: *V*____ *C*____ *S*____

Conditions of Test____________________________________

No.		Called		Called		Called
1. The first is	bAIt	______	Set	______	Ret	______
2. Now try	bUt	______	riP	______	Ben	______
3. Next comes	bOAt	______	STy	______	riFF	______
4. Group 4 is	bAke	______	Guy	______	wiN	______
5. The next is	bAlk	______	Jen	______	THigh	______
6. Now comes	bAt	______	Pen	______	Yen	______
7. Listen to	bIte	______	riM	______	Let	______
8. Try to hear	bEt	______	Men	______	riNG	______
9. Next comes	bOUght	______	Tie	______	THy	______
10. Group 10 is	bOOk	______	riCK	______	wiNG	______
11. Please try	bIt	______	SHy	______	wiCK	______
12. Can you hear	bOUt	______	WRy	______	Jet	______
13. Now try	bEAt	______	High	______	Yet	______
14. The next is	bOOt	______	Net	______	Wen	______
15. Listen to	bEck	______	Vet	______	wiLL	______
16. Try to hear	bOOT	______	Hen	______	Sigh	______
17. The next is	bOOk	______	Fen	______	riG	______
18. Now comes	bIke	______	whiZ	______	Pet	______
19. Now try	bAck	______	THen	______	riD	______
20. Group 20 is	bIt	______	riCH	______	wiTH	______
21. Listen to	bOUt	______	riB	______	Ten	______
22. Try to hear	bEAk	______	Fie	______	wiSH	______
23. Can you hear	bUck	______	WHiz	______	whiST	______
24. Next comes	bOAt	______	whiCH	______	Den	______
25. The last is	bArk	______	Vie	______	Wet	______

Fig. 9.17 Sample word list for syllable articulation test.

observers soon memorize groups of words. It is necessary, therefore, to prepare a number of similar word lists. These can be made by writing the monosyllables from each column on separate packs of cards. Each pack is shuffled separately. The words from the packs are then recorded in columns to form a new word list. For example, another list prepared in this manner might start as follows:

1. The first is	bEt ______	riM ______	wiN______
2. Now try	bOOt______	Vet ______	Yen______
3. Next comes	bEAk______	Set ______	riG ______
4. Group 4 is	bIt ______	THen______	Jet ______

Under ideal conditions for speech intelligibility—a level of about 65 to 70 db, no noise, and no reverberation—tests conducted in accordance with the above procedure will give an articulation of about 96 per cent. It may seem that the articulation should be more nearly 100 per cent under these ideal conditions, but a few errors are unavoidable, such as the confusion of *th* as in *thin,* and *f* as in *fin,* or possible errors of pronunciation or of perception or remembering.

When the acoustics of a room has been designed and checked in accordance with the procedures described in this chapter, and the results of the tests have been good, the architect can be assured that the room is free from acoustical defects with regard to its use for both speech and music. He will have obtained from the speech-articulation tests a quantitative rating of the acoustics of the room as a speech room. There is no precise method of rating the acoustics of a music room; the artists who perform in the room, and their audiences, will render a verdict concerning its degree of excellence. However, if the room has successfully passed the tests decribed in this section, its acoustics will be acceptable for music, and the artists and listeners who use it may be expected to give it a rating that will vary from good to excellent.

10 · Noise Control

Freedom from the harassing effects of noise is one of the finest qualities a building can possess. The architect is obliged to seek, by every possible means, those features of design and construction that will impart to his building the utility and charm of quiet surroundings. An intelligent approach to the problem of constructing quiet buildings must be based upon a knowledge of (1) the magnitude, nature, and distribution of noise in buildings and out-of-doors; (2) acceptable noise levels in various types of buildings; (3) the propagation, and especially the attenuation, of sound through the free air, through openings and ducts, and through or around obstacles, embankments, and landscaping; (4) the reduction of sound, and the suppression of vibration by varied types of partitions and flexible connectors; (5) the reduction of machinery noise at its source by appropriate selection of equipment from a noise-producing standpoint; and (6) the reduction of noise by the proper use of sound-absorptive treatment. For a room or building in which quietness is a prime requirement, it frequently is desirable to make measurements of the noise at the site as a means of determining the types of structure that are needed to insulate adequately against the prevailing noises. For routine design, however, it is sufficient to make use of the data from noise surveys in buildings and outdoors.

Too frequently the architect overlooks noise control or depends on luck and a few "sound-insulation blankets." The *noise level,* that is, the sound-pressure level of the noise, in speech and music rooms should be low enough so that it will not interfere with the hearing or with the production of speech or music. In offices, factories, and other work rooms, noise should be reduced to levels that will not impair the health, contentment, or efficiency of the workers in these rooms. In restaurants, residences, and hospitals quietness is especially desirable.

In the development of the mechanization of industry, machines of greater power and higher speed, often with correspondingly augmented noise output, replace smaller ones. Undoubtedly, the growth of mechanization has been accompanied by an increase in noise. Although commendable efforts are being made to reduce the noise of machines and appliances, there has been no marked reversal in the upward trend in city and industrial noise.[1] On the other hand, the public is becoming increasingly conscious of the ill effects of noise. The architect and builder, therefore, face a growing need for the reduction and control of noise in buildings.

The harmful effects of noise are well known. Even quite feeble noises interfere with the hearing of speech and music; moderately loud noises produce auditory fatigue; and very loud noises, if long endured, induce permanent losses of hearing. Although the influence of noise on the working efficiency and general health of human beings is generally recognized as harmful, those who have scientifically investigated these effects are not in complete agreement about their nature and extent. There is evidence from one carefully conducted investigation that both the working efficiency and the total output of weavers increased when they wore ear plugs which reduced the noise level from 96 to 81 db.[2] The detrimental effects of the noise were observed to be greatest at the beginning and near the end of work periods, possibly indicating that persons go through a process of adaptation to noise, endure it without noticeable effects for a time, but finally suffer

[1] The efforts of the National Noise Abatement Commission to reduce noise have been beneficial. Improvements made recently in automobiles, street cars, and home appliances have minimized their noise. But we yet live in a noisy world.

[2] H. C. Weston and S. Adams, *Ind. Health Research Bd.* (*England*), Report 70 (1935).

from its incessant attack. The bulk of other evidence indicates that the reduction of noise and reverberation, following the usual acoustical treatment of offices and factories, results in increases in output of labor and in human well-being that more than offset the cost of the acoustical treatment. Although it is difficult to measure fatigue, most observers agree that excessive noise exacts a heavy toll in frayed nerves and physical exhaustion.

No one has determined the price we pay in loss of sleep resulting from avoidable noises. Several years ago, one of the authors kept a record of the number of times he was awakened each night. Approximately three fourths of all awakenings could be attributed to noise. Among the most frequent "offenses" were the honking of automobile horns, barking of dogs, the screaming of ambulance sirens, the late arrival of some members of the family, and the chirping of birds. The wearing of ear plugs, which attenuated these noises about 30 db, reduced the total number of awakenings to less than one half. Important factors in determining the disturbing effects of noise are its over-all level, frequency distribution, and time pattern.

The Addition of Noise Levels

When two sounds occur simultaneously, the resultant sound has a higher average level than either of the two. In this section we are concerned with the determination of the total level of a number of sound sources. The total is not the sum of the individual levels; the decibel scale is logarithmic. The mathematical equation giving this information is quite complex for the general case.[3] Fortunately, the results are simple for the case of primary concern here—the addition of "random noises." This type of noise contains sounds of all frequencies and random phases and has a frequency distribution similar to the curve of Fig. 2.8. It is similar in nature to the background noise in buildings or on busy streets. Suppose that L_1 is the average sound-pressure level of one such source of noise, that L_2 is the average level of a second source, and that L_2 is greater than L_1. Let their difference be denoted by D. Then the total noise level

[3] For a discussion of the computation of composite noise, see E. Dietze and W. D. Goodale, Jr., *Bell System Tech. J.*, **18**, 605 (1939).

in decibels is equal to $L_2 + N$, where N is a number determined from the chart in Fig. 10.1, which corresponds to the difference D. Thus, suppose that $L_1 = 40$ db and $L_2 = 50$ db. Here, $D = 10$ db. From Fig. 10.1, the corresponding value of N is about 0.4. Therefore, the total noise level is $50 + 0.4$, or 50.4 db. Likewise, if $L_1 = 30$ db and $L_2 = 35$ db, N is 1.2, or the total noise level is 36.2 db. For the special case where two noise sources are equal in level, D equals zero, and the total noise level is 3 db higher than the level of either source. As a practical example, suppose that the average background sound level in a legitimate

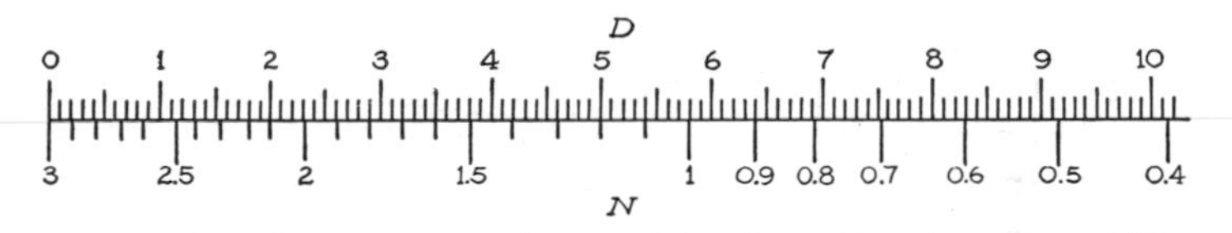

Fig. 10.1 Chart for computing the sound level resulting from the addition of two combining random noises. If D is their difference in decibels, N is added to the higher level to obtain the total level.

theater due to audience noise is 35 db. Assume that the noise transmitted through the walls of the building from outside has a level of 25 db. Then the total level inside the theater resulting from both sources would be about 35.4 db, and the transmitted noise would probably not be detected by the average person. In general, the addition of noise which raises the total background level by less than 1 db will not be objectionable unless the weaker noise is quite different in character from the louder one.

Noise in Buildings

A comprehensive survey of the noise in several thousand locations has been conducted by the Bell Telephone Laboratories in order to determine typical noise conditions indoors and outdoors in business and residential areas, and in many different types of buildings both in summer and in winter.[4] The summary of these data provides information useful in predicting noise reduction requirements.

The principal sources of room noise may be grouped into three

[4] D. F. Seacord, *J. Acoust. Soc. Am.*, **12**, 183 (1940).

broad classifications: people, machinery, and outdoor sources. The relative noise contributions from these three types of sources depend, to a large extent, on the use of the room in which the noises commingle. For example, in about 45 per cent of the business locations at which measurements were made, the predominant source of noise was people; in 25 per cent of these locations, machinery was the predominant noise source; and in 30 per cent of them, outside noise sources were predominant. In factories, the relative percentages were of the order of 10, 80, and 10 per cent, respectively. The greater the number of

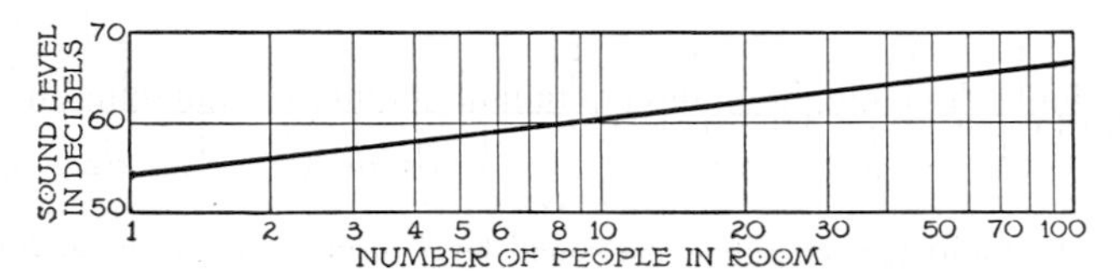

Fig. 10.2 Relation between room noise at business locations and the number of people in the room. (D. F. Seacord.)

people in a room, the noisier it is; and, the higher the noise level, the louder an individual must speak in order to be heard above the noise. This suggests that the effect of acoustical materials in reducing the noise level in a room may be greater than one might expect; since the absorptive treatment lowers the average noise level, individuals can speak in a lower voice and be heard. Hence, at least part of the noise level (that due to speaking or conversation) is reduced "at its source" as well as by absorption. An approximate relationship between the noise in a business office and the number of people in the room is given in Fig. 10.2. These data show that the noise level, in decibels, increases directly as the logarithm of the number of people in the room.

A summary of the average sound-pressure levels of room noise [5] at telephone locations (obtained in the above survey) is given in Table 10.1. Since these results are based on measurements in a wide variety of places and consequently include a wide range

[5] All the sound-level measurements were taken with a standard sound-level meter using the "40-db frequency-weighting network."

of noise levels, standard deviations [6] are given. A seasonal variation is also indicated. The noise level in residences is generally 3 db lower (in colder climates) in winter than in summer, since closed windows exclude a large part of the noise from outside. In factories the average level is generally lower in the

TABLE 10.1 *

SUMMARY OF ANNUAL AVERAGE NOISE LEVELS IN ROOMS IN DECIBELS

(D. F. Seacord)

Type of Location	*Average Noise Level*	*Standard Deviation*	*Excess of Summer over Winter*
Residence			
Without radio	43.0	5.5	3
With radio	50.0	8.0	4
Small store (1 to 5 clerks)	53.5	7.5	4
Large store (more than 5 clerks)	61.0	6.0	0
Small office (1 or 2 desks)	53.5	6.5	4
Medium office (3 to 10 desks)	58.0	6.5	1
Large office (more than 10 desks)	64.5	4.5	0
Factory office	61.5	9.5	−2
Miscellaneous business	56.0	7.5	1
Factory	77.0	12.0	−2

* The levels given in this table are "weighted"; that is, they are the levels measured with a standard sound-level meter incorporating a 40-db frequency-weighting network.

summer, probably because this season generally corresponds to one of lowered production and because open windows increase the effective absorption in the buildings. In order to reveal better the range of noises summarized in Table 10.1, distribution curves of the room noise for three broad classifications are given in Fig. 10.3. This graph shows the percentage of room locations that have noise levels less than the values indicated on the horizontal scale.[7] Thus in 90 per cent of the residences the level was less than 50 db with the radio off, and less than 60 db with the radio on.

[6] Standard deviation is defined as the root-mean-square of the deviations of a set of observations from the average value.

[7] Because an arithmetical probability scale is used in the vertical direction, "normal" distributions appear as straight lines.

This noise survey gives the average noise levels of diffuse sounds in rooms. Actually, the levels vary considerably, both with time (see Fig. 10.4) and from point to point, especially if there are

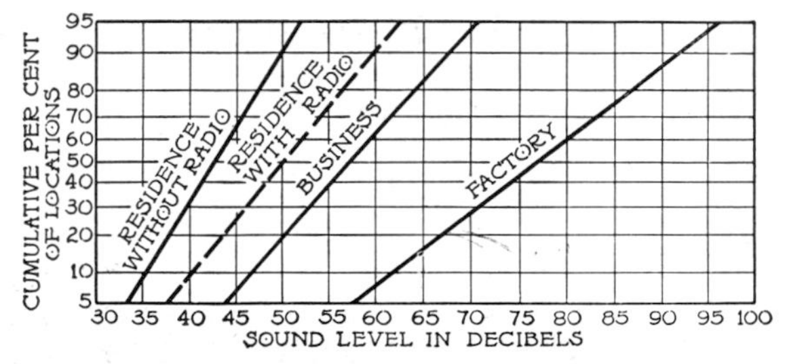

Fig. 10.3 Distribution of room noise levels measured in the Bell Telephone Laboratories survey. The curve shows the per cent of locations that have a noise level less than the values specified on the horizontal scale. (D. F. Seacord.)

fluctuating and prominent sources of noise within the room. In factories, these variations can be considerable, often more than 20 db for fluctuations with time. The noise levels of various

TABLE 10.2

Noise Levels of Various Machines at a Distance of 3 Feet

(H. J. Sabine and R. Wilson)

	Decibels
Lathes (average)	80
Cotton spinning machines	84– 87
Sewing machines	93–100
Looms	94–101
Riveter, riveting guns	94–105
Punch presses, various types	96–103
Ball mill	99
Drop hammers, bumping hammers	99–101
Wood planers	98–110
Wood saws	100
Wire-rope stranding machines	100–104
Headers	101–105

machines at a distance of 3 feet are given in Table 10.2. The average diminution in level with increasing distance from the machine depends on the room's size, shape, and acoustical treatment. Sabine and Wilson found the decrease to be of the order of 3 db for each doubling of the distance from the machine in *untreated* rooms; that is, at 40 feet the noise level was about 3 db lower than at 20 feet. In rooms having acoustical treatment on the ceiling, the decrease in level depends on the height of the ceiling and the absorption coefficient of the material. In such rooms, the drop in noise level is proportional to the distance from the machine and it is less in rooms with high ceilings than in rooms with low ceilings. For example, attenuations of

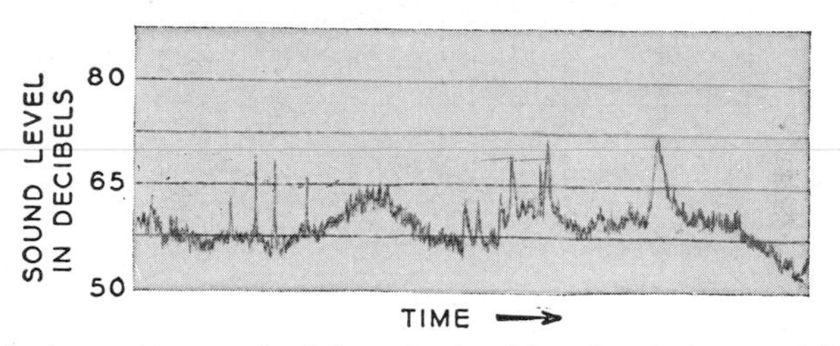

Fig. 10.4 A graphic record of the noise level in a hospital room, 6 feet from a window. Approximately 30 vehicles per minute pass on the street about 50 feet from the window.

the order of 0.4 db per foot were measured in rooms with ceiling heights averaging 10 feet, and 0.2 db per foot in rooms with ceiling heights about 20 feet (the ceiling materials in these cases had absorption coefficients of at least 0.70). These attenuations are in contrast to that in untreated rooms where the level dropped 3 db each time the distance was doubled.

A quantitative study of the frequency distribution of noise has been made in a large number of rooms, stores, offices, residences, restaurants, etc. Some of these data are presented in Fig. 10.5, which gives the average spectrum of room noise in 28 business locations. The curve represents the spectrum level defined by Eq. (2.1) for frequencies from 100 to 5000 cycles. The spectrum level diminishes almost uniformly with increased frequency at a rate of about 6 db for each octave. This is characteristic of many noises, especially those resulting from impacts; for example, noise from business machines, typewriters, rain striking a roof top, whitecaps on the surface of the ocean. The over-all noise level is obtained by summing the energy contributions from the entire spectrum.[8] For a noise having the spectrum shown in Fig. 10.5, a sound-level meter would yield a reading of 50 db (using the 40-db weighting network).[9] The results of the Bell Telephone Laboratories survey show that, although the over-all levels are different, the *shape* of the spectrum for room noise in most stores, offices, business locations, and residences is essentially

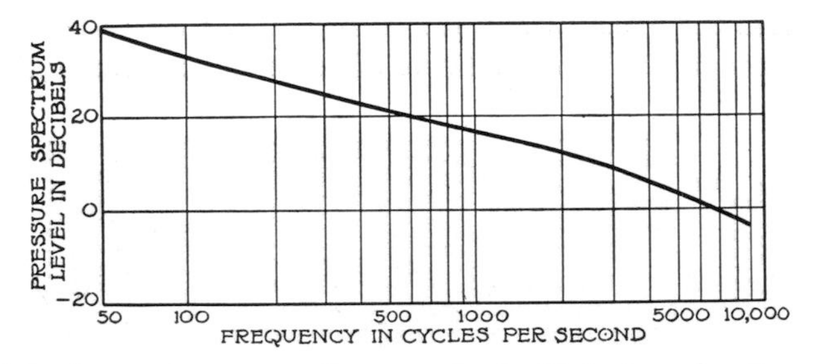

Fig. 10.5 Average spectrum of room noise in 28 business locations in and around Philadelphia. Spectrum adjusted to give a sound-level meter reading of 50 db, using the 40-db frequency-weighting network. (D. F. Hoth.)

the same. The curve may be raised or lowered to give the spectrum of room noise having a higher or lower over-all level.

Outdoor Noise

Sounds of outside origin are often the principal contributors to noise in offices, churches, and residences. The largest source of outdoor noise is generally automobile traffic. For this reason, it is desirable that all buildings in which quietness is an important factor, including churches, auditoriums, and hospitals, be not constructed near a busy, or potentially busy, street. In order to determine the sound-insulation needs of a building so that the planned insulation will meet future requirements, it is desirable to make a noise survey at the proposed site and to estimate the noise contributions from future traffic conditions.

[8] It should be recognized that the reading of a standard sound-level meter using a frequency-weighting network gives a "weighted" sound level in decibels, not the loudness level of the noise in phons. See "Sound-Level Meters," p. 15.

[9] The spectrum of average room noise for residences in which a radio is not playing is shown in Fig. 2.8. It has an over-all level of 43 db.

In this connection, Fig. 10.6 is of interest. It shows the relation between the average outdoor noise level (at the street curb), in decibels, *vs.* the flow of street traffic, in terms of vehicles per minute. These data indicate that the noise level varies directly with the logarithm of the number of passing vehicles. While automobiles, trucks, street cars, and subways account for the major sources of outdoor noise in most locations, airplane traffic is responsible for the peak noise levels in many areas and should be taken into account in estimating noise-insulation require-

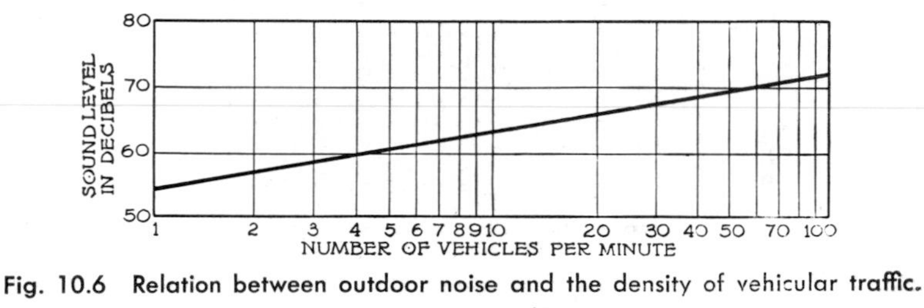

Fig. 10.6 Relation between outdoor noise and the density of vehicular traffic. (D. F. Seacord.)

ments in these localities. For example, in a noise survey at the proposed site of a theater in San Diego, California, where civilian and military planes can be seen and heard overhead during most of the daylight hours, the noise level from the passing planes exceeded 70 db twenty times during a typical one-hour interval. When no planes could be heard, the level was less than 61 db. The maximum level recorded during this typical one-hour period was 90 db, when 18 single-motor planes passed in formation at about 2000 feet elevation and at a minimum distance of about 1½ miles.

The average level of street noise varies with the time of day. In many business districts, the average noise level is lowest between 3 A.M. and 4 A.M. and rises rapidly after 5 A.M. to a maximum at about 10 A.M. In other areas, the maximum occurs later in the day, often during evening "rush" hours. Figure 10.7 gives the average result of measurements made at the street curb in a large number of business and residential locations. The dashed lines give data for New York City, but they are reasonably representative of the noise conditions in congested areas of other large cities. They show the per cent of the locations surveyed which have noise levels less than the decibel reading indicated along the horizontal axis. Thus, in the New York business district, 30 per cent of the locations had noise levels of less than 70 db. The solid lines give average data for business and residential districts in other cities, which include a wide range of areas, from congested citv districts to rural areas. As a result,

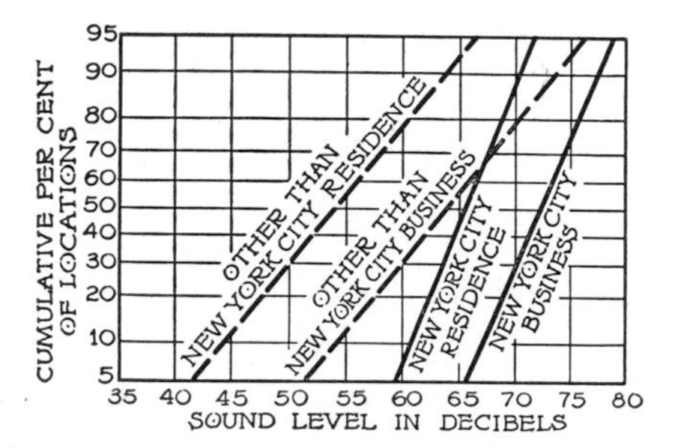

Fig. 10.7 Distribution of average street noise levels, measured in the Bell Telephone Laboratories survey, showing the per cent of locations having a noise level less than a value specified along the horizontal axis. (D. F. Seacord.) Peak levels of street noise are often 10 to 15 db above average levels. See Fig. 10.4.

the levels are lower than they are in New York City and the spread (range) of noise levels is greater.

Acceptable Noise Levels in Buildings

The highest level of noise within a building that neither disturbs its occupants nor impairs its acoustics is called the *acceptable noise level.* It depends, to a large extent, on the nature of the noise and on the type and customary use of the building. The time fluctuation of the noise is one of the most important factors in determining its tolerability. For example, a bedroom with an average noise level of 35 db, with no instantaneous peak levels substantially higher, would be much more conducive to sleep than would be a room with an average noise level of only 25 db but in which the stillness is pierced by an occasional shriek. Furthermore, levels that are annoying to one person are unnoticed by another. It is therefore impossible to specify precise values within which the noise levels should fall in order to be acceptable. It is useful, however, to know the range of average noise levels that are acceptable under average conditions. A compilation of such levels for various types of rooms in which noise conditions are likely to be a significant problem is given in Table 10.3. The recommended acceptable noise levels in this

TABLE 10.3 *

RECOMMENDED ACCEPTABLE AVERAGE NOISE LEVELS IN UNOCCUPIED ROOMS

	Decibels
Radio, recording, and television studios	25–30
Music rooms	30–35
Legitimate theaters	30–35
Hospitals	35–40
Motion picture theaters, auditoriums	35–40
Churches	35–40
Apartments, hotels, homes	35–45
Classrooms, lecture rooms	35–40
Conference rooms, small offices	40–45
Court rooms	40–45
Private offices	40–45
Libraries	40–45
Large public offices, banks, stores, etc.	45–55
Restaurants	50–55

* The levels given in this table are "weighted"; that is, they are the levels measured with a standard sound-level meter incorporating a 40-db frequency-weighting network.

table are empirical values based on the experience of the authors and others they have consulted. Local conditions or cost considerations may make it impractical to meet the high standards inherent in these relatively low noise levels. In more than 80 per cent of the rooms in some of the types listed, the prevalent average noise levels exceed the recommended acceptable levels. However, it should be understood that the acceptance of higher noise levels incurs a risk of impaired acoustics or of the comfort of the individuals in the room. The acceptable noise levels of Table 10.3 are useful in calculating the sound-insulation requirements of walls, partitions, and ventilation ducts under typical noise conditions.

Siting and Planning against Noise

The selection of the site for a building, the layout of the building itself, and the grading and landscaping of the site are indispensable parts of good planning against noise in buildings. The existence and persistence of quiet sites is dependent on zoning ordinances and their enforcement. Architects in every community should cooperate with the civic authorities in the segregation of noxious activities (including noisy industries, power stations, airports, traffic arteries) from buildings where quiet is an absolute necessity, such as schools, churches, hospitals, and residences. These buildings should be protected by civic planning. Interurban automobile and truck traffic should be routed *around, not through,* areas that have been zoned for schools, residences, and hospitals; express highways that must pass through zones requiring quiet surroundings should be isolated by means of embankments or parapets along the outer edges of the highway; trains should enter large metropolitan centers by underground routes; parks and landscaping should be planned to impede the propagation of noise into quiet zones; and approaches to airports, which are an increasing noise nuisance in all large cities, should be *from the outskirts* of the city, *not over it.* In these and similar zoning problems concerned with noise abatement and control, the architect should contribute his advice and make known the need for quiet.

CHOICE OF SITE. The architect is not always consulted about the selection of the sites for the buildings he designs. But the architect who appreciates the importance of noise control can render a most valuable service to his clients by advising them in the selection of a site. Building owners and committees are prone to overemphasize such features as convenience of access to the neglect of such matters as good acoustics, lighting, or air conditioning. Countless schools, churches, and hospitals are located on sites so noisy that those who are served by these buildings often refer to them as "the church where the sermon can't be heard because of the noise from passing trucks," or "the hospital where the noise keeps you awake all night."

Schools, churches, hospitals, and residences should not be located on noisy highways. Sites removed as little as a few hundred feet from such streets, especially if these streets are flanked with other buildings or dense planting, will be reasonably free from traffic noise. Quiet sites simplify the problems of elaborate sound insulation, reduce costs, and greatly enhance the utility and value of the completed structures. The architect and others who participate in the selection of the site for a building should always make use of the available data from noise surveys or have surveys made of the proposed sites. They also should anticipate as far as possible future trends that will affect the noise levels at these sites.

Grading and Landscaping

Anyone who has stood beside a railroad track and listened to the noise from a train as it enters a "cut" where there is an embankment between the listener and the train must have noticed the effectiveness of the embankment in reducing the noise. Even though *the low-frequency components* of the sound *are bent over and around the barrier, the higher frequency components are not;* for the latter, the barrier "casts a shadow" and the over-all noise at the position of the listener is reduced. Thus an earth embankment or a masonry garden wall often can be used to reduce the noise that impinges on a building and aid in the establishment of quiet conditions within the building without resorting to costly measures of sound insulation. It may reduce the level by as much as 5 db. If the surface of the barrier facing the source of noise is absorptive, such as a grassy turf, dense vines, other planting, or even leaf mold or peat moss, the over-all noise reduction may amount to as much as 8 or 10 db. Hedges or trees with dense foliage act as sound absorbers and reflectors, and their effectiveness increases with the extent (thickness, height, and density) of growth. A cypress hedge 2 feet thick has sound-obstructing value of about 4 db. Other types of dense planting can be used to attenuate noises to levels that will facilitate the architect's and builder's problem of providing adequate noise insulation for court rooms, hotels, residences, and other buildings.

Building Layout

The location of a building on its site, the arrangement of rooms, corridors, and vestibules, and the location of doors and windows, all have a bearing on the control of noise; they require careful consideration. For example, the noise level at the end of a room adjacent to a busy street may be at least 5 db higher than it is at the opposite end. In such a situation it is advantageous to place the speaker's platform at the end of the room adjacent to the street, which is the primary source of noise. This arrangement has two advantages: (1) the more distant parts of the audience are in the quieter section of the room, and (2) a speaker has a natural tendency to raise the level of his voice in the presence of noise. The sides of a building facing streets, playgrounds, or other sources of noise should house those activities that can tolerate the greatest amount of noise, and the sides of the building that face the quieter environment should be reserved for those rooms that require the quietest conditions. Windows should not open on noisy streets or yards. Doors which open on noisy streets should be supplemented by sound locks. While courts can be used to good advantage to shield certain rooms from street noise, they are usually serious offenders in reinforcing (by multiple reflections) the sounds that issue from windows opening on them. Many dwellers in city apartments attribute their sleepless nights to the disturbance from the neighbor's radio which blasts its strident noise into a court—a reverberant container that sustains the noise and aids in its efficient transmission from one open window to another.

A noisy room, such as a riveting shop, should be well removed from an office or room where quiet is valued. The doors and windows of adjacent rooms should be as far from each other as possible. It often is advisable to stagger the positions of the doors on the two sides of a hall or corridor so that no two doors face each other. Elevators, air-conditioning equipment, motors, and other noise-producing machinery should be removed and isolated from the sections of a building that can least tolerate such noises. Other suggestions and recommendations concerning the effective control of noise by building layout are presented in subsequent chapters dealing with special types of buildings.

11 · Reduction of Air-Borne Noise

Noise can ruin the acoustics of an otherwise well-planned auditorium. Failure to plan for adequate insulation against noise, in the design of buildings, is almost universal. Evidence of this failure is conspicuous in schools, churches, and civic buildings. The occupants of many residences and offices are victims of noise because basic principles of sound insulation were ignored or because fallacious principles were followed. One of the misconceptions inherent in many proposed methods for sound insulation is based upon the erroneous assumption that materials and methods which are effective for heat insulation are also effective for sound insulation. It is important that the architect and builder recognize that these are two separate problems, even though certain types of structures may be effective for both. In general, nearly all porous materials are good heat insulators and good *sound absorbers* as well. Nevertheless, they are usually poor *sound insulators.* Porous materials or other types of acoustical treatment can reduce, somewhat, the level of the noise transmitted into a room; but they do not provide the most effective means for insulation against outside noises. It is injudicious and uneconomical to construct a building without proper planning against noise and then to hope that absorptive material will solve the subsequent difficulties.

In this chapter we shall show how much reduction in the noise level in a room can be procured by insulation, and how much by absorption. We also shall show how the required noise-reduction can be predicted and attained.

How Sound Is Transmitted

Many of the different paths by which sound is transmitted from one room to another are illustrated in Fig. 11.1. Most

Fig. 11.1 Noise invades a room by many paths.

sounds that are communicated to a room, either from the outdoors or from elsewhere in the building, are included in one of the following classifications: [1]

[1] Sound can be transmitted *through* a partition, without causing the whole structure to vibrate, in a manner analogous to the transmission of light through a sheet of glass. However, the relative amount of sound transmitted by this means is usually negligible—except in the case of porous partitions, which are poor insulators unless they are very thick.

Sounds originating in the air which are transmitted:

(*a*) *Along a continuous air path through openings.* For example: through open windows, elevator shafts, doors, and transoms; through cracks around pipes, electric conduits, telephone outlets; through ventilating ducts. (Noise reduction in ventilation systems is the subject of Chapter 13.)

(*b*) *By means of diaphragmatic action of partitions.* Sound waves can force a partition to move back and forth as a diaphragm. By this means, sound from a source on one side of a wall can be communicated to the opposite side.

Sounds originating from direct impacts (for example, impulses produced by the dropping of objects on a floor, the scuffling of feet, footfalls, or the slamming of doors), *or from machinery vibrations* are transmitted through rigid structures with almost no attenuation. As a result of a direct impact in one room, large surfaces (such as walls) elsewhere in the building can be set into vibration, thereby radiating acoustical energy in a manner similar to the action of a sounding board on a piano.

These two types of sound transmission differ in many respects. Impacts generate impulses of short duration but large power; these are often propagated long distances. Sounds originating in the air generally are of much smaller power, persisting for a longer duration; they usually are greatly attenuated by divergence, and by intervening partitions, so that their disturbing influence is confined to regions near their origin. The methods of insulating solid-borne impact sounds are somewhat different from those of insulating air-borne sounds; a structure that is a very good insulator for one type may be a very poor insulator for the other. The general topic of the insulation of impact-generated sounds is discussed in the following chapter. The present chapter is restricted to methods of insulating air-borne sounds.

Transmission through Openings

Noise can be readily communicated from one portion of a building to another through openings such as windows or open doors. Frequently these openings limit the total amount of insulation which can be attained. Thus, if it is necessary to open windows for ventilation, the sound insulation between two adjacent rooms may be limited by the open windows to 20 db or even less. Under such circumstances, it would be profitless to provide separating partitions of relatively high insulation. Even very small openings, such as cracks around doors or windows, are effective in transmitting sound.[2] This is emphasized by the following example. Suppose that outdoor noise is transmitted into a room through an opening 10 inches wide and that the transmitted sound in the room attains a level of 60 db. By reducing the opening to a width of 1 inch, the acoustical power transmitted through it will be decreased tenfold (diffraction being neglected). In other words, the level of the transmitted sound will be lowered to 50 db. A further reduction in the width of the opening to 0.1 inch (diffraction being neglected again) would lower the noise level another 10 db. Thus, according to the above consideration, a reduction in the width to one one-hundredth of its initial opening decreases the level of the transmitted sound only 20 db. Actually, owing to the effect of diffraction, the reduction is not even this much. It is apparent, therefore, that a high degree of sound insulation requires the complete closing of cracks around windows. doors, pipes, and conduits.

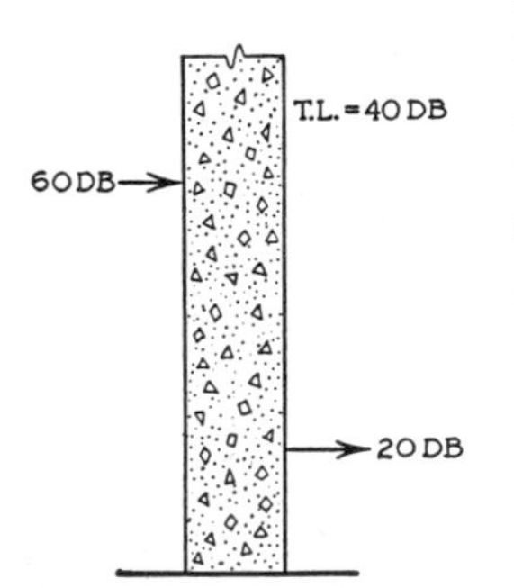

Fig. 11.2 A wall with a transmission loss of 40 db will reduce the level of a sound which is transmitted through it from 60 to 20 db, from 40 db to 0 db, etc.

Sound-Transmission Loss

The sound-insulative merit of a partition is generally expressed in terms of its *transmission loss* in decibels. The transmission loss is equal to the number of decibels by which sound energy which is incident on a partition is reduced in transmission through it. For example, the transmission loss (abbreviated T.L.) of the wall shown in Fig. 11.2 is 40 db. Thus incident sound energy which strikes this partition is reduced by 40 db in transmission through it; that is, only 1/10,000 of the incident energy

[2] See Chapter 4 for a discussion of "Diffraction."

is transmitted through the wall. The larger the value of T.L. for a partition, the greater the sound insulation it provides.

Rigid Partitions

The transmission of sound through a "rigid" partition, such as a brick, concrete, or solid-plaster wall, is accomplished principally by the forced vibrations of the wall; that is, the entire partition is forced into vibration by the pressure pulsations of the sound waves against it. The vibrating structure thus becomes a secondary source of sound and radiates acoustical energy in to the space on the opposite side of the partition. As one would suppose, the more massive the wall, the more difficult it is for sound waves to move it to and fro. This is a less technical, and less exact, way of saying that most partitions in buildings (walls, ceilings, and floors) are "mass-controlled" over the greater portion of the audible-frequency range. The insulative properties of a partition at low frequencies are partially dependent on its resonant frequencies, which are principally determined by its mass, stiffness, and internal damping. The lowest resonant frequencies of the structural partitions used in buildings are usually below 100 cycles, and the damping is usually high. For frequencies above 100 or 200 cycles, the mass is the important factor in determining the sound-insulation merit of a partition. The transmission loss of a wall increases with an increase in mass per unit area of wall section.

The manner in which the insulation value of a rigid partition depends on the mass per unit area of the structure is shown by the curve of Fig. 11.3. The transmission loss (averaged over the frequency range from 128 to 2048 cycles) is plotted against the weight in pounds per square foot of wall section. It will be noted that the curve approximates a straight line. The slope indicates that *the average T.L. of a rigid mass-controlled partition increases 4.3 db each time its weight is doubled.* Thus, the average T.L. for a rigid wall which weighs 10 pounds per square foot of wall section is about 37.4 db; the T.L. for a rigid wall which weighs 20 pounds per square foot is about 41.7 db; and the T.L. for a rigid wall which weighs 40 pounds per square foot is about 46 db. The data in Fig. 11.3 have been compiled from measurements taken by different laboratories on a large number of rigid partitions. Individual wall structures may depart considerably from the average values indicated. However, the average increase in T.L., for each doubling of weight of the partition, will generally be from 4 to 5 db. Another important aspect of insulation by rigid partitions is the variation of T.L. with frequency. The T.L. is low at low frequencies, and it increases with

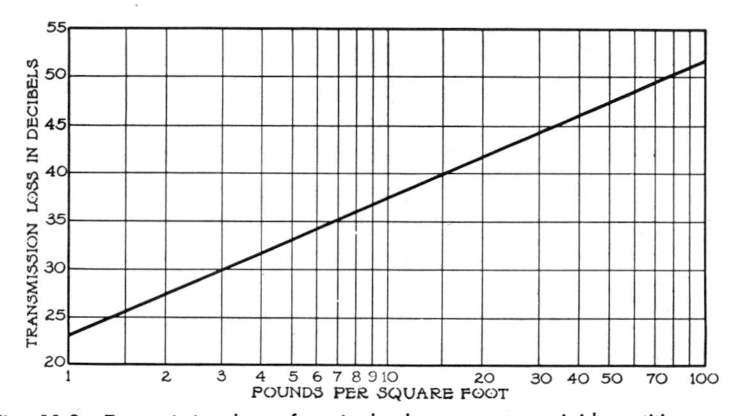

Fig. 11.3 Transmission loss of a single, homogeneous, rigid partition, averaged over the frequency range from 128 to 2048 cycles, *vs.* mass per square foot of the partition.

increasing frequency. The rate of increase of T.L. is not the same for all types of partitions; often it is quite irregular, especially at low frequencies, but it usually increases 3 to 6 db for each doubling of the frequency. For example, Fig. 11.4 shows the T.L. *vs.* frequency for a floor and ceiling partition constructed as shown in the figure. Since the T.L. of most partitions is small at low frequencies, these sounds are the most difficult to insulate. In the case of a *thin* flexible panel, such as a window pane, not only the mass but also the stiffness, the internal damping, the size of the panel, and the manner in which its edges are fastened contribute to its transmission-loss characteristics.

Because of the relatively slow increase in sound insulation with the increase in mass of a rigid partition, it is not always practical to secure a high degree of insulation by merely increasing the thickness of the wall. It would be necessary for a concrete wall to be nearly 2 feet thick in order for the wall to have an average T.L. of 60 db. On the other hand, two lightweight walls *isolated from each other* can be designed to provide this same amount of

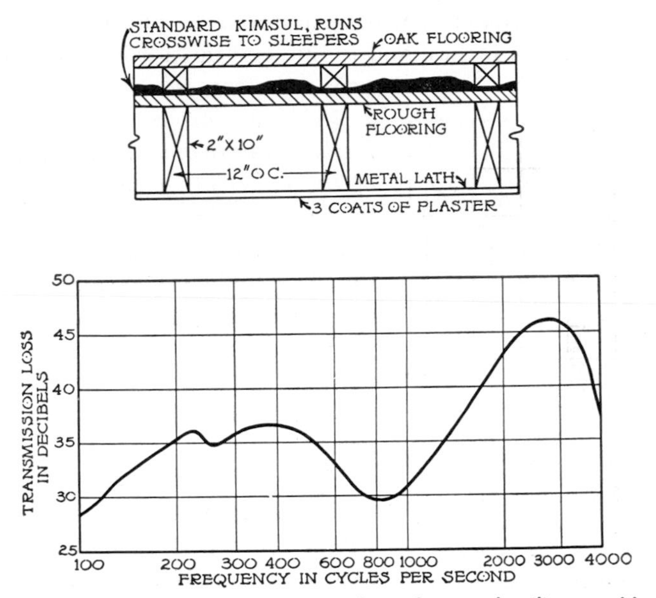

Fig. 11.4 Transmission loss vs. frequency for a floor and ceiling partition constructed as shown in the figure. (F. K. Harvey.)

transmission loss. *Where a large amount of sound insulation is required, it is most practical to employ discontinuous construction or compound partitions.*

Porous Materials

When sound is transmitted through a porous material, a portion of the sound energy is lost in the form of heat due to the motion of the air within the pores of the material; in non-rigid porous materials (such as rock wool), some energy is lost as a result of the motion of the component parts of the material. These losses are primarily responsible for the sound insulation provided by porous materials. If the material is uniform, the fractional loss of sound through the first inch of a porous blanket is about the same as the fractional loss through the next inch, etc. For example, suppose that sound having a frequency of 1500 cycles is transmitted through a layer of rock wool of 5 pounds per cubic foot density. The acoustical power will be reduced to approximately one half of its initial value in traversing a layer 1 inch thick; it will be reduced to one quarter of its initial value in 2 inches; and to one eighth in 3 inches. In general, therefore, *the T.L. provided by a uniform porous partition is directly proportional to the thickness of the material;* it can be expressed in decibel loss per unit thickness. This is in contrast to the transmission loss for a solid, rigid partition (usually much higher than for a porous partition of the same thickness) which increases approximately 4 to 5 db for each doubling of thickness.

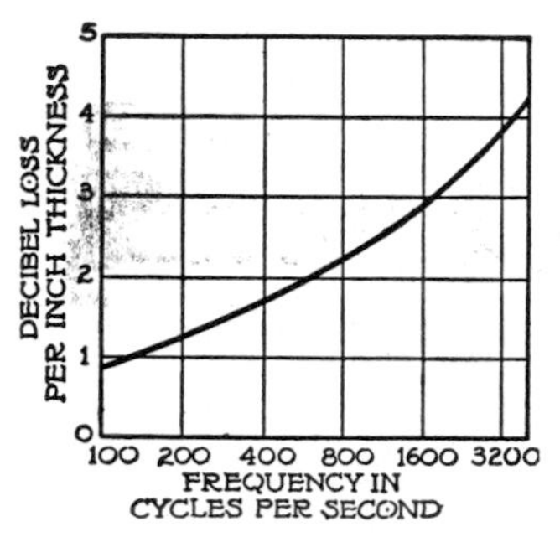

Fig. 11.5 Transmission loss through rock wool having a density of 5 pounds per cubic foot. The attenuation is expressed in terms of decibel loss per inch thickness of the porous material vs. frequency. (R. A. Scott.)

The results of attenuation measurements on non-rigid, porous materials show that such materials, if used by themselves as partitions, do not provide a large amount of sound insulation unless the partition is exceedingly thick. Figure 11.5 indicates that a blanket of rock wool 6 inches thick, having a density of 5 pounds per cubic foot, would provide a T.L. of but 14 db at 800 cycles and 19 db at 1500 cycles. Such a blanket is a good absorber of sound; it is a good heat insulator, but it is a poor sound insulator. Other measurements show that non-rigid porous materials having a higher density than this rock wool have correspondingly higher T.L.'s, but the increase in the T.L. is not sufficient to make such porous partitions, by themselves, practical structures for sound insulation. Although porous materials themselves are not good sound insulators, when used properly in conjunction with compound partitions, they can contribute to the total insulation supplied by the combined structure, as shown in the following section.

Compound-Wall Constructions

We have seen that the transmission loss in decibels of solid masonry or monolithic partitions increases directly with the weight per square foot of wall section. For walls having a mass of more than 15 pounds per square foot the increase in T.L. with an increase in weight is relatively small. It will be remembered that each successive doubling of the weight of the partition adds but 4 to 5 db. For this reason it is often economical to substitute two or more lightweight partitions that are isolated from each other for a heavy-masonry partition where relatively high value of transmission loss is required. Double-wall construction frequently offers the most practical means of obtaining high insulation at moderate cost and reasonable dead load. The separate partitions should be as completely isolated from each other as possible. *Structural ties between the separate partitions tend to convert the compound partition into a single rigid partition and thus reduce the sound insulation.* In this connection, it is important to make sure that pieces of wood, chunks of plaster, and other building materials are not dropped between the partitions of a double wall. The number of points at which the two partitions are tied together should be held to a minimum, and the ties should be of a flexible nature. Staggered-stud construction provides structural separation and therefore is often beneficial. The suspension of an absorptive blanket or fiberboard between double partitions, or between the wood studs or channel irons in staggered-stud partitions, may be a substantial aid to insulation if the two partitions are structurally separated. The absorptive material should not make a rigid or semi-rigid bridge between the two partitions, for then it may actually *lower* the sound insulation of the structure.[3] The effectiveness of the absorptive material depends somewhat on the absorption already present between the walls. In double walls which are not structurally separated, a slight additional insulation may be obtained by the introduction of an absorptive blanket, especially in lightweight partitions. However, this added insulation may not be economically worth while. Several compound partitions that

[3] In double partitions that are not structurally separated, a wall fill may increase the T.L. by increasing the mass per unit area of wall section.

have been used in motion picture studio construction are shown in Fig. 11.6. Examples of other double-wall constructions, and the values of their T.L.'s at different frequencies, are listed in the tables in Appendix 2.

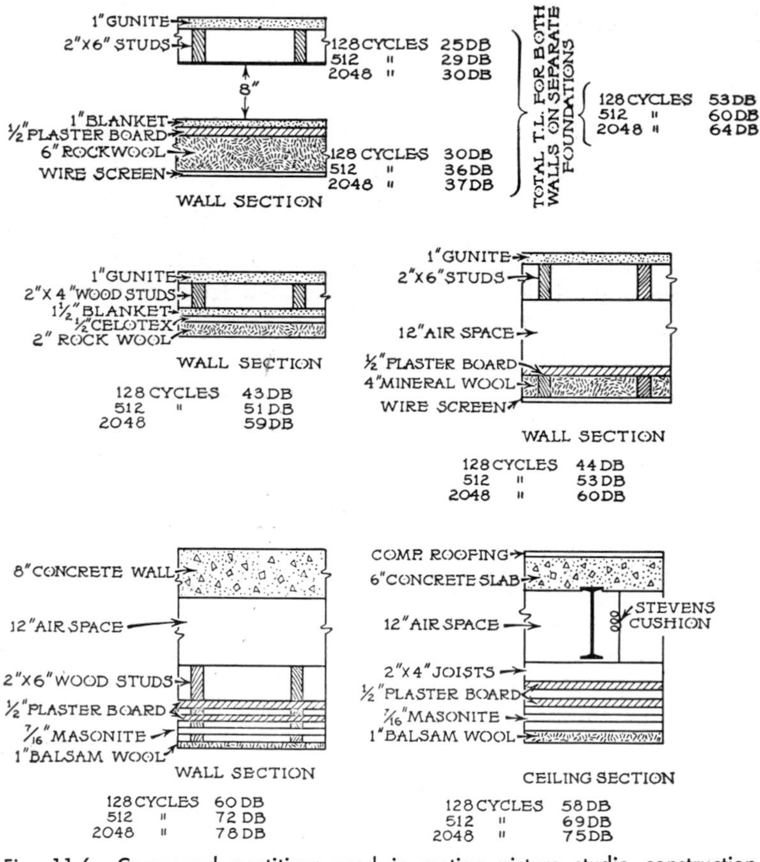

Fig. 11.6 Compound partitions used in motion picture studio construction. The transmission loss for several frequencies is given in decibels. In the upper figure the T.L. is given for each of two walls on separate foundations and for the compound structure.

The principles of insulation of air-borne sounds by floors and ceilings are essentially the same as those for walls. However, an additional important factor must be taken into account in planning: floors and ceilings should be designed to isolate sounds originating from impacts. The subject of noise isolation by floors and ceilings is treated in the following chapter.

Windows and Doors

Windows and doors are usually the "paths of least resistance" in the over-all sound insulation of a room. Therefore, they should be given careful consideration for buildings in which good insulation is required. The positions of windows ought to be carefully planned so that, without sacrificing utility or beauty, they are removed as far as possible from other windows in adjacent noisy rooms or from street noise. It is even helpful to hinge the frames of a window which swings outward, so that the panes reflect, in a direction away from the window of the adjacent room, sound passing through the open window. Good sound insulation is virtually impossible unless doors and windows fit tightly in their frames; and threshold cracks must be eliminated.

The average transmission loss of a window depends primarily on the thickness of the pane; the heavier the glass, the more insulation it provides. At frequencies well above the lowest resonant frequency of the pane, the average sound insulation increases with increasing frequency. It is not possible to give precise values for the transmission loss of various windows because the T.L. depends, to a considerable extent, on the area of the panes and on the manner in which they are set. For these reasons, *average* values are listed in Table A.11, Appendix 2. The values listed are the T.L.'s averaged over the frequency range from 128 to 2048 cycles. Although not precise, these values are indicative of the *relative* sound insulation provided by the windows listed.

Double panes are frequently used to increase the sound insulation of a window. To be most effective, the panes should be structurally isolated. However, this is usually impractical except in double-wall constructions. The transmission losses for several types of double windows are listed in Table A.11, Appendix 2. If the spacing between rigidly mounted panes is as little as $\frac{1}{4}$ inch, the transmission loss through the multiple structure will be about the same as it would be through a single pane having a weight per unit area equivalent to that of the combination. Increasing the air space increases the transmission loss considerably. Thus, enlarging the spacing between double-strength (1/4-inch) plate-glass panes from 1/4 inch to 1/2 inch will add about 3 db of sound insulation. If the two panes are not rigidly joined at the edges but are set in felt or rubber, and if the panes fit tightly against the felt or rubber so that there are no "leaky" threshold cracks, the T.L. will be increased at least 4 or 5 db. Increasing the separation between the two panes from 1/2 inch to 6 inches may add as much as 10 db to the transmission loss.

Double-pane windows, and sometimes triple-pane windows, with 6 inches or more separation between the sheets of glass, are required in special rooms where excellent sound insulation is needed—for example, in radio studios. In such rooms the periphery of the space between panes should be lined with sound-absorptive material. This may add as much as 5 db to the transmission loss of the window. Each pane should be set in felt or rubber, and it is advisable to tilt one pane with respect to the other. A tilt of as little as 1 inch in 12 inches will suffice to suppress high transmission of certain resonant frequencies. Also, by selecting window panes of different thicknesses (so their resonant frequencies are different), a more uniform insulation *vs.* frequency characteristic will be obtained.

The transmission loss of a door increases with increased weight; the T.L. also increases with frequency. Most doors of ordinary construction have an average transmission loss of 20 to 25 db; some specially manufactured doors have T.L.'s as high as 40 db. The general trend of the T.L.-frequency characteristics follows approximately that of the corresponding curve for rigid partitions (see Fig. 11.3). The effectiveness of any door in providing sound insulation depends largely on the seal around the edges. For example, tests on one steel door showed that the placement of a rubber strip on the outer step of the jamb increased the T.L. 4 db. A force of 400 pounds on the panel made still better contact at the edges, and the T.L. was increased another 4 db. The average T.L.'s for a number of different types of doors are given in Table A.11, Appendix 2.

Noise-Insulation Factor

In most rooms, the walls, ceiling, and floor have greatly different sound-insulative properties. The walls often consist of a variety of constructions, and they usually include doors and windows. In order to obtain an over-all rating of the insulation that a room provides against noise of outside origin, the area of each boundary surface together with its transmission loss must be taken into account. The rating also depends on the total amount of absorption in the room. In this section we shall show how these physical properties are related in establishing a *noise-insulation factor* for the room.

The fraction of incident sound energy transmitted through a partition is called its *transmission coefficient* τ. It is related to the T.L. of the partition by the equation

$$\text{T.L.} = 10 \log_{10} \frac{1}{\tau} \quad \text{db} \tag{11.1}$$

Suppose that part of the total area of a partition has one T.L. and another part, constructed of different material, has a different T.L. Then the average T.L. for the partition *cannot* be obtained by averaging the T.L.'s. Instead, we must average the τ's for the two parts. Thus, suppose that a wall is constructed of two panels of equal area, one having a T.L. of 40 db ($\tau = 0.0001$), and the other having a T.L. of 20 db ($\tau = 0.01$). Then, the average T.L. for this partition is only slightly greater than 20 db—practically that of the less insulative panel. Suppose that the boundaries are made of a number of sections having different areas. Then, if it is assumed that each element of construction has sound of the same level incident upon it, the average transmission coefficient τ_{av} is given by

$$\tau_{av} = \frac{\tau_1 S_1 + \tau_2 S_2 + \tau_3 S_3 + \cdots}{S} = \frac{T}{S} \tag{11.2}$$

where τ_1, τ_2, and τ_3 are the transmission coefficients of the different parts of the boundary; S_1, S_2, and S_3 are their corresponding areas; and S is the sum of all these areas. If a wall is constructed of two sections having different T.L.'s, and if the area of the section having the larger T.L. is less than twice that of the other section, the assumption that the whole wall has a transmission loss equal to that of the less insulative section is a good approximation.

The numerator of Eq. (11.2) is called the *transmittance* and is designated by T. The transmittance and the total number of units of absorption a in a room are the principal factors in establishing a figure of merit for the noise-insulative properties of its boundaries. Such a rating is given by the *noise-insulation factor* of a room. This factor is expressed by:

$$\text{Noise-insulation factor} = 10 \log_{10} \frac{a}{T} \quad \text{db} \tag{11.3}$$

This relationship is often of considerable importance in problems of design since it can be used to determine whether it is more economical to increase the noise-insulation factor of a room by altering the insulative properties of the partitions or by adding more absorption.

When the average values of τ are used, the noise-insulation factor gives, very roughly, the difference in decibels between the level of noise outside a room and the level of the transmitted noise that will be established in the room. Thus, if the noise level outside is 60 db and the noise-insulation factor is 25 db, the level inside is of the order of 35 db. A single figure resulting from this calculation is obviously only an approximation, because the sound insulation of the panel, the spectrum level of the noise, and the sensitivity of the ear vary with frequency. Consequently, the noise-insulation factor varies with frequency. However, for many problems of routine design, an application of Eq. (11.3) at a single frequency, 512 cycles, is extremely useful, especially in determining the relative merit of increasing the transmission loss of different boundary surfaces.

An example of the use of Eq. (11.3) is given here to illustrate the importance of eliminating small areas which have relatively large transmission coefficients if good noise insulation is required. Consider a small lecture room with a total absorption, including the audience, of 2400 square-foot-units (sabins). The table below describes the boundaries and the transmittance through them.

Material	*Area in Square Feet*	τ	τS *in Square Feet*
4-inch concrete slab ceiling plus 1-inch acoustical tile	2500	0.000025	0.063
4-inch concrete slab floor plus floor covering	2500	0.000025	0.063
8-inch brick walls plus ½-inch plaster	4500	0.000008	0.036
3/16-inch glass windows, closed	400	0.0011	0.440
1½-inch hardwood doors, good closure	100	0.001	0.100
Total transmittance T, in square feet			0.702

Hence,

$$\frac{a}{T} = \frac{2400}{0.702}$$

Therefore,

$$10 \log_{10} \frac{a}{T} = 35$$

Thus the noise-insulation factor is 35 db. If the windows were replaced by double windows with a transmission coefficient of 0.00027, the noise-insulation factor would be increased to 39 db. On the other hand, if the windows were opened ($\tau = 1.0$), the value of a/T would be 2400/200. In this case the noise-insulation factor would be only 11 db. It is obvious from these examples that little would be gained by an increase in the insulation value of the walls since most of the sound is communicated through the windows and doors.

Noise-Insulation Requirements

Most cities or counties have building codes designed to protect the general public against fire, electrical hazards, building collapse, etc., but there are few codes that specify protection against noise. Architects and builders who have an understanding of noise-insulation requirements, and who know how these requirements can be fulfilled, will be in a position to give advice in setting up noise-insulation regulations. Some progressive municipalities throughout the world are already considering the adoption of such ordinances. In Sweden requirements regarding sound insulation have been in force since 1946. The Swedish regulations specify the minimum transmission losses against air-borne sound for partitions and floor-ceiling constructions given below. These noise-insulation requirements for the partitions of

AVERAGE MINIMUM INSULATION REQUIREMENTS IN DECIBELS *

	Frequency Range		
Type of Room	*100–500 cycles*	*550–3000 cycles*	*100–3000 cycles*
Hospitals	44	56	**50**
Dwelling rooms	42	54	**48**
School rooms	36	48	**42**
Work rooms	34	46	**40**

* Transmission loss measured in decibels with sound-level meter incorporating an appropriate frequency-weighting network.

a room are based primarily upon average noise conditions which were observed at typical outdoor locations and on the noise levels which are acceptable within the different types of rooms. Even the results of the preceding simple considerations can be exceedingly useful in acoustical planning if they are used with a comprehension of their limitations.

Noise-insulation requirements for rooms and buildings should be calculated just as routinely as are reverberation requirements; *often they are much more important.* The nomogram in Fig. 11.7 is an aid in the determination of the approximate *minimum* insulation requirements. Average noise conditions which may exist outside the room are listed in the column at the left of the chart. The acceptable noise conditions are listed in the column to the right. Insulation needs are estimated in the following way. After an estimate or survey of the exterior noise conditions, the appropriate level on the scale to the left is selected. This point is connected by a straight line through a point on the scale to the right which corresponds to the *desired* noise conditions. Then the point of intersection of this straight line with that of the center scale determines the *approximate* minimum insulation requirement. The weight (in pounds per square foot of wall section) of a single rigid partition that will provide this in-

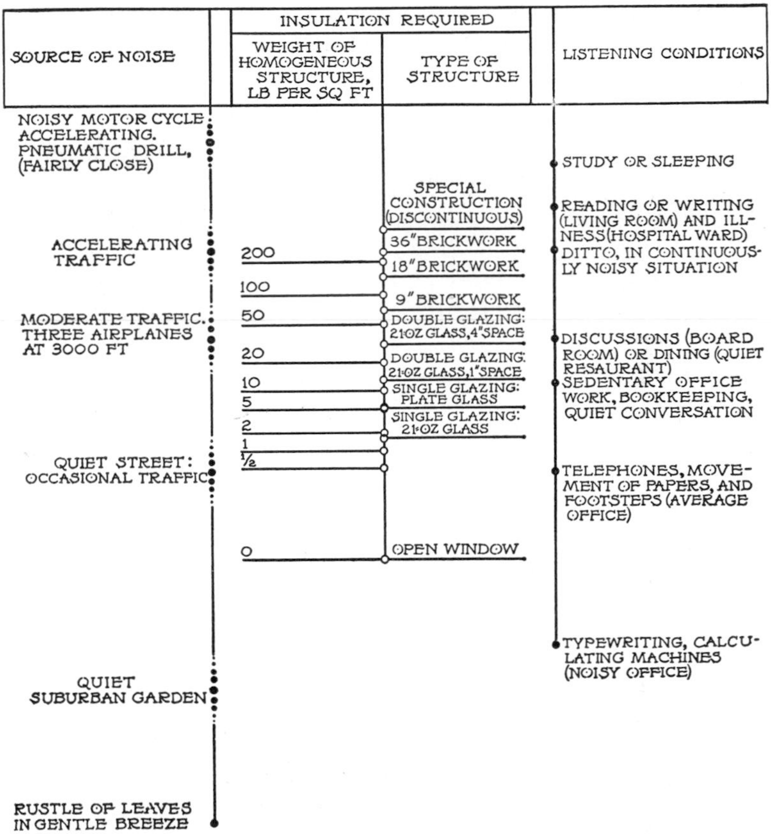

SOURCE OF NOISE
INSULATION REQUIRED
WEIGHT OF HOMOGENEOUS STRUCTURE, LB PER SQ FT
TYPE OF STRUCTURE
LISTENING CONDITIONS

NOISY MOTOR CYCLE ACCELERATING. PNEUMATIC DRILL, (FAIRLY CLOSE)
ACCELERATING TRAFFIC
MODERATE TRAFFIC. THREE AIRPLANES AT 3000 FT
QUIET STREET: OCCASIONAL TRAFFIC
QUIET SUBURBAN GARDEN
RUSTLE OF LEAVES IN GENTLE BREEZE

200
100
50
20
10
5
2
1
½
0

SPECIAL CONSTRUCTION (DISCONTINUOUS)
36" BRICKWORK
18" BRICKWORK
9" BRICKWORK
DOUBLE GLAZING: 21-OZ GLASS, 4" SPACE
DOUBLE GLAZING: 21-OZ GLASS, 1" SPACE
SINGLE GLAZING: PLATE GLASS
SINGLE GLAZING: 21-OZ GLASS
OPEN WINDOW

STUDY OR SLEEPING
READING OR WRITING (LIVING ROOM) AND ILLNESS (HOSPITAL WARD)
DITTO, IN CONTINUOUSLY NOISY SITUATION
DISCUSSIONS (BOARD ROOM) OR DINING (QUIET RESAURANT)
SEDENTARY OFFICE WORK, BOOKKEEPING, QUIET CONVERSATION
TELEPHONES, MOVEMENT OF PAPERS, AND FOOTSTEPS (AVERAGE OFFICE)
TYPEWRITING, CALCULATING MACHINES (NOISY OFFICE)

Fig. 11.7 Nomogram for determining the approximate minimum sound-insulation requirements. (See text.) (Reproduced by permission of the Comptroller of His Britannic Majesty's Stationery Office and the Director of Building Research.)

sulation is indicated. In many cases, it is advisable to utilize a compound partition having an equivalent transmission loss. The use of Fig. 11.7 will yield the minimum average requirements for insulation against specified conditions of noise. This nomogram should be regarded only as an *aid* in establishing an *approximate lower limit* of the amount of sound insulation required to meet specified conditions—not as the final means for determining the types of wall construction, doors, windows, etc., that will give satisfactory sound insulation.

It is obvious, from a consideration of the numerical example given in the preceding section, that a determination of the overall insulation requirements must take into account the most probable "sound leaks," such as windows and doors. For this reason, it is helpful to use the noise-insulation factor, which includes a consideration of all bounding surfaces, as a guide in estimating insulation needs. The following working rule has proved satisfactory: *Subtract the acceptable noise level from the average level (averaged over time) of the outside noise; to this difference add 10 db. The result is the noise-insulation factor required to furnish adequate sound insulation.* The additive term of 10 db is included (1) to provide some protection against disturbances from the usual surges of outside noise that are above the average level, and (2) to allow for unavoidable differences between the sound insulation provided by the actual structures and those determined by laboratory tests on model partitions. For example, if the average level of outside noise is 70 db, and a noise level of 35 db is acceptable in the room, the noise-insulation factor should be about 45 db. In using this rule for estimating insulation requirements, one usually assumes values for the noise levels and transmission loss that are averaged over frequency (as well as time), although each has a characteristic which varies with frequency. This procedure gives an approximate solution. When a more exact solution is required, it is preferable to make calculations at a number of frequencies, particularly below 1000 cycles, and to take into account the frequency-sensitivity characteristics of the ear. When the insulation needs are especially important, as for radio studios and soundproof test rooms, the method of the next section can be used as a means for determining the most effective and practical types of partitions and constructions for required noise-insulation factors.

Calculation of Loudness Reduction Resulting from Sound Insulation

In order to rate the effectiveness of sound insulation in terms that correspond more closely to the judgments of listeners than does the "average transmission loss," one must consider the aural characteristics of the ear, the frequency characteristics of the outside noise, the T.L.-frequency characteristic of the insulation, and the frequency characteristic of the noise generated within the room. All these factors are taken into account in the following method of calculating noise reduction, which is based on the procedure for loudness calculation [4] (see p. 32). In computing the per cent loudness reduction due to sound insulation, the following steps are taken:

1. Determine the pressure spectrum (L_{ps} *vs.* frequency) of the outside noise; see Eq. (2.1).
2. In the following way determine the predicted pressure spectrum of the noise transmitted into the room: (*a*) Calculate the noise-insulation factor for the room at each of a number of frequencies spaced at intervals no greater than an octave; see Eq. (11.3). (*b*) At each of these frequencies subtract the noise-insulation factor from the pressure spectrum of the outside noise.
3. Determine the pressure spectrum of the noise generated within the room.
4. Determine the pressure spectrum of the combination of transmitted noise from outside and noise generated within the room by means of Fig. 10.1; this is a combination of the spectra of steps 2(*b*) and 3.
5. By the use of Fig. 2.9, calculate the masking spectrum (M *vs.* frequency) for pressure spectra of steps 1 and 4.
6. Plot the two masking spectra, determined in step 5, on the Loudness Computing Chart of Fig. 2.10.
7. The per cent difference between the areas under these two curves on the Loudness Computing Chart represents the per cent loudness reduction.

The following example illustrates the use of this kind of calculation. Suppose that a room has a noise-insulation factor that varies with frequency in accordance with the dashed curve (*a*) of Fig. 11.8, and assume that the pressure spectrum of the outside noise is the one indicated by

[4] See footnote, p. 34, regarding the calculation of the loudness reduction for pure tones and the use of equal-loudness contours for this purpose.

curve (*b*) in Fig. 11.8.[5] This spectrum corresponds to an over-all noise level of approximately 80 db. Then the predicted pressure spectrum of the room noise, shown on the graph, is obtained by subtraction. The masking spectra corresponding to these pressure spectra are plotted on

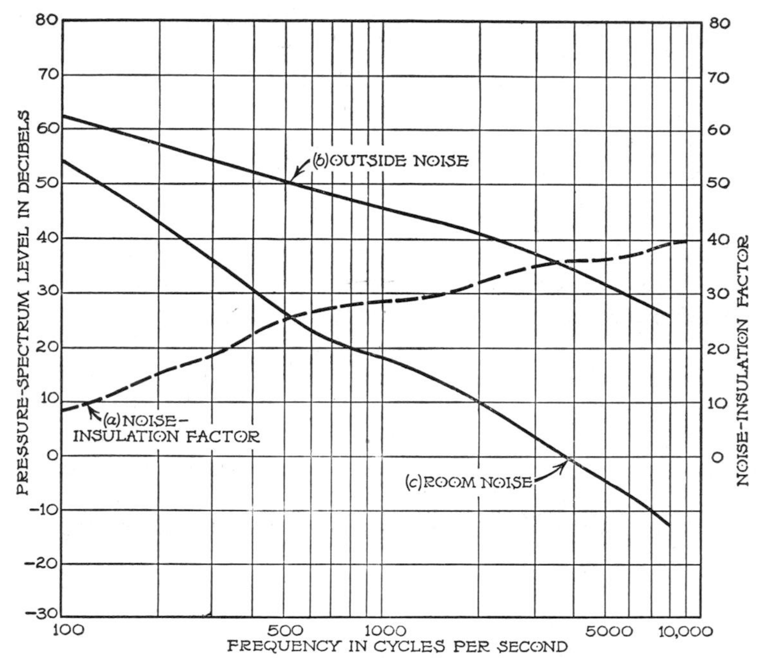

Fig. 11.8 Example showing (*a*) the spectrum of noise outside a room corresponding to a total noise level of 80 db, (*b*) the noise-insulation factor *vs.* frequency characteristic for the room, and (*c*) the predicted pressure spectrum of the room noise.

a Loudness Computing Chart; see Fig. 11.9. The area under these curves corresponds to 113 sones and 14.8 sones, respectively. Hence, the difference in loudness between outside noise and room noise is 87 per cent; or the loudness of the room noise is 13 per cent of that of the outside noise.

[5] If the noise level is not the same at all the boundaries of the room, appropriate corrections can be made. However, it usually is better to make the assumption that the noise level at all boundaries is equal to that on the noisiest side of the building; for the highest level is usually street noise, which often is adjacent to the poorest insulating surfaces, the windows. Furthermore. this assumption allows for a "factor of safety."

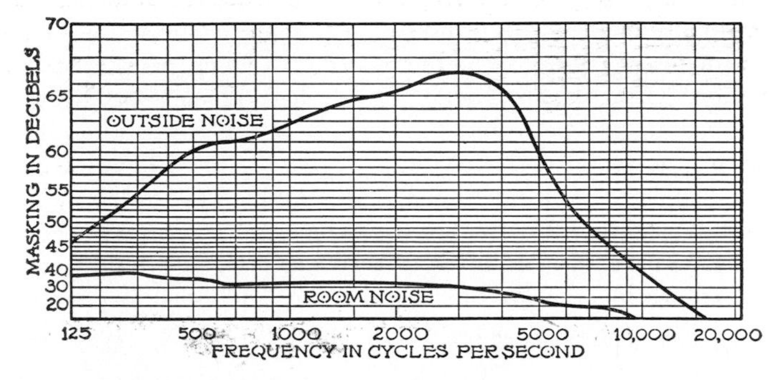

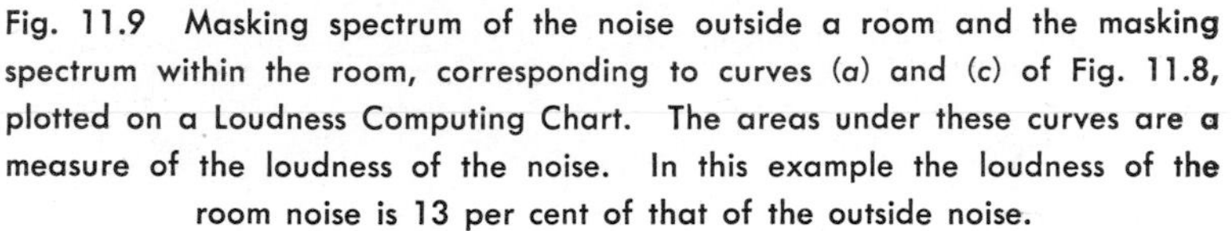
Fig. 11.9 Masking spectrum of the noise outside a room and the masking spectrum within the room, corresponding to curves (*a*) and (*c*) of Fig. 11.8, plotted on a Loudness Computing Chart. The areas under these curves are a measure of the loudness of the noise. In this example the loudness of the room noise is 13 per cent of that of the outside noise.

Tables of Sound-Insulation Data

Tabulated sound-insulation data on a large number of wall, ceiling, and floor constructions (and other partitions or panels) are given in Appendix 2. These results have been arranged according to groups that correspond essentially to the following classification, which follows that of the U. S. Bureau of Standards: [6]

Wood-Stud and Steel-Stud Partitions (Table A.7). This type includes wood-stud and steel-stud partitions faced with plaster on wood, metal, or gypsum lath, or faced with fiberboard, plywood, or similar materials. These partitions generally provide greater insulation than do rigid masonry walls of the same weight; but cracks or other openings will greatly reduce their insulation value. Sound is transmitted through such walls principally by setting one facing into vibration (except for the sound transmitted through holes, cracks, etc.); this vibration is then communicated to the second facing, mainly through the studs, and to a lesser extent by the air between these facings. Staggered-stud construction, therefore, is better than single-stud construction.

[6] These sound-insulation data of wall and floor constructions, summarizing the results of measurements made by the U. S. Bureau of Standards, can be obtained from the Superintendent of Documents, Washington 25, D. C.: Bureau of Standards Report BMS 17 (1939), 20 cents; Bureau of Standards Supplement No. 1 to Report BMS 17 (1940), 5 cents; Supplement No. 2 to BMS 17 (1947), 10 cents.

Suspended absorptive blankets between the staggered studs, as in Fig. 11.10, usually increase the transmission loss—especially if they are covered with heavy paper, or similar material, on one or, preferably, on both sides. The blanket should provide a complete septum over the extent of the entire partition. The use of such blankets as a means of increasing the sound insulation in walls is usually economical only in staggered-stud or in structurally isolated double-wall constructions.

BRICK, TILE, MASONRY, AND POURED-CONCRETE PARTITIONS (TABLE A.8). In this group are included brick, poured concrete, glass blocks, concrete blocks, cinder blocks, clay tile, gypsum tile, and other masonry partitions. Their insulation characteristics are approximately those of rigid

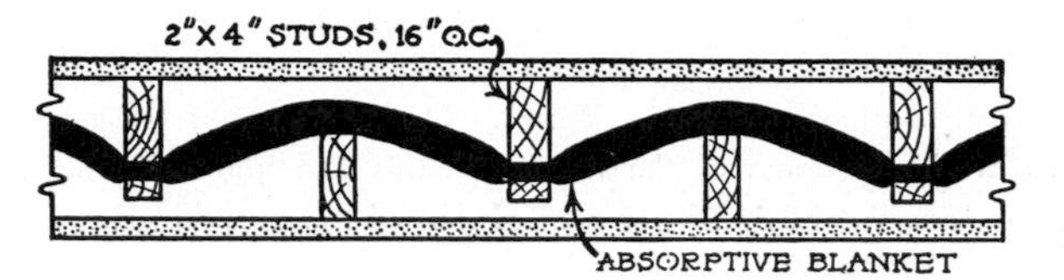

Fig. 11.10 Staggered-stud wall construction with absorptive blanket.

partitions, as given in the curve of Fig. 11.3. Porous blocks, when plastered, usually have a transmission loss which is 3 or 4 db greater than that predicted by this curve. If the porous blocks are not plastered, they may be very poor insulators; sound "leaks" through the interstices of the blocks. If porous materials are used in wall constructions, the interstices should be of the non-communicating type; that is, each cavity should be completely enclosed. The Bureau of Standards has conducted tests showing that 4-inch hollow tile, although not adequate for most purposes of sound insulation, can give relatively high insulation if plastered surfaces are furred out from the tile. The more *isolation* between the plastered surfaces and the tile, the better is the insulation. For example, measurements show that one construction of plaster on lath, which was tied [7] to 4-inch hollow clay tile, rated slightly better than a solid 8-inch brick wall.

PARTITIONS WITH RESILIENT ATTACHMENT OF LATH (TABLE A.9). These are constructed of lath fastened to studs by special nails or resilient clips; see Fig. 11.11. Such walls provide somewhat more insulation than do those in which the lath is nailed to the studs in the usual manner. The special nail or clip reduces the "coupling" between the plastered lath and the studs, thus reducing the vibrational (acoustical) energy communicated to the opposite side of the partition.

[7] Tied with wires which had been embedded in the mortar joints.

SOLID-PLASTER PARTITIONS (TABLE A.10). Weight is the primary factor controlling the insulation of solid-plaster-between-stud partitions and of studless plaster partitions. The proportions and kind of plaster generally are not significant, with one apparent exception: lime plaster gives slightly higher insulation than do other types.

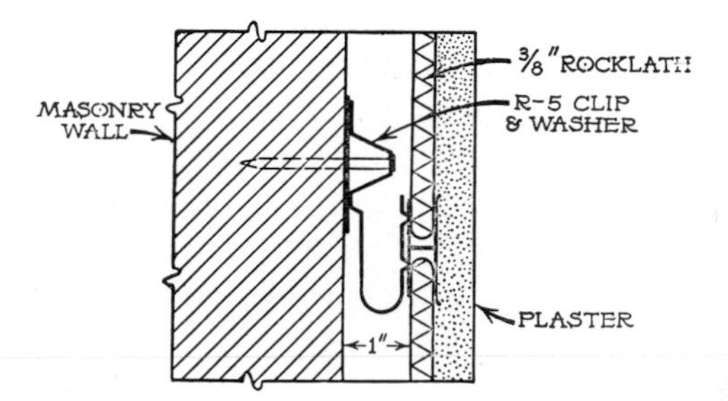

Fig. 11.11 Partition with resilient clip attaching plaster on Rocklath to wall. (U. S. Gypsum Co.)

MISCELLANEOUS PARTITIONS (TABLE A.11). This group includes a wide variety of partitions, doors, window structures, and single panels of porous materials, fiberboard, plywood, etc. It should be remembered that the transmission loss of a thin, rigid partition depends, to a large extent, on its stiffness and therefore on its dimensions. Hence, the values given in the tables in Appendix 2 should be regarded as rough approximations.

Appendix 2 also includes sound-insulation data, in Table A.12, for the following types of floor construction: wood and steel joist floors, masonry and concrete floors, miscellaneous floors. The insulation of floor constructions against impact is discussed in the next chapter.

In the tables of sound-insulation data given in Appendix 2, the following information is listed: the composition of the partition, its mass per square foot of wall section, its thickness, and its transmission loss (T.L.) at representative frequencies. In addition, cross-sectional diagrams are shown for many of the partitions. The authority for each measurement is listed. Because of differences in methods of measurement (and other uncertainties), differences of less than 3 to 4 db in the values of the T.L.'s from different laboratories should not be regarded as significant.

It is important to keep in mind the previous discussion of noise-insulation requirements when using these tables, for the panels which show the highest *average* values of T.L. may not necessarily supply the greatest amount of insulation for all types of noise. It is usually important, in determining the best type of partition for each problem that arises, to consider the insulation values at different frequencies as well as the average value of T.L. for all frequencies. Thus, if a particular noise has much of its energy concentrated in a certain frequency range, it usually will be advantageous to select or design structures and materials that will provide high sound insulation for this frequency range, especially if this range is one for which the ear is most sensitive.

Noise Reduction by Sound-Absorptive Treatment

The level of noise which is transmitted into a room is reduced relatively little by sound-absorptive treatment, as Eq. (11.3) indicates. Acoustical materials are not a cure for poor sound insulation. However, they are extremely useful, and frequently indispensable, in controlling noise generated within a room and in reducing the transmission of noise through corridors, from one room (or one part of a room or building) to another. The value of such treatment is not adequately appraised by Eq. (11.3). If the noise level is reduced 3 db, the sound level of speech can be reduced about the same amount. Thus the acoustical power expended in speaking can be reduced by a factor of 2.

The installation of acoustical materials in a room has the following beneficial effects: (1) it reduces the reverberation time, usually several fold; (2) it reduces the over-all noise level; and (3) it tends to localize noise to the region of its origin (a distant source is attenuated relatively more than one near-by). Since unexpected noises are particularly annoying, this reduction of remote sources of sound is especially helpful. All three factors have important effects on the general fatigue of individuals in the room.

Calculation of Noise-Level Reduction. It follows from a consideration of Eq. (8.6) that, if the output of the sources of noise within a room remains constant, the reduction of the noise level (of diffuse sound) due to the installation of absorptive material is given by

$$\text{Noise reduction} = 10 \log_{10} \frac{a_{\text{after}}}{a_{\text{before}}} \text{ db} \tag{11.4}$$

where a_{after} = total number of square-foot-units (sabins) of absorption in a room after treatment, and a_{before} = total number of sabins of absorption in room before treatment. Thus, if the absorption in a room is doubled, the average noise level will be decreased by 3 db. Equation (11.4) is expressed graphically in Fig. 11.12. Since the absorption coefficient of acoustical mate-

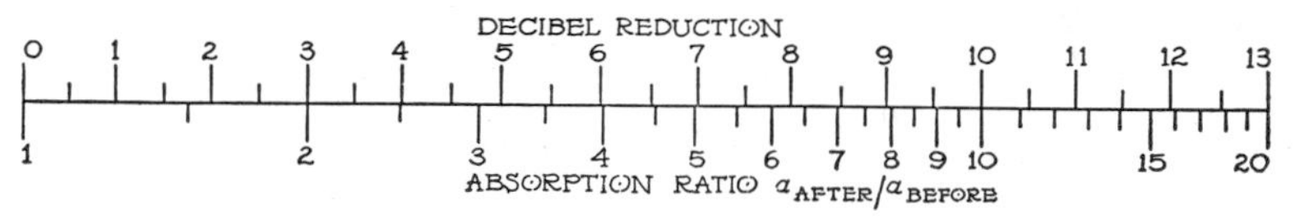

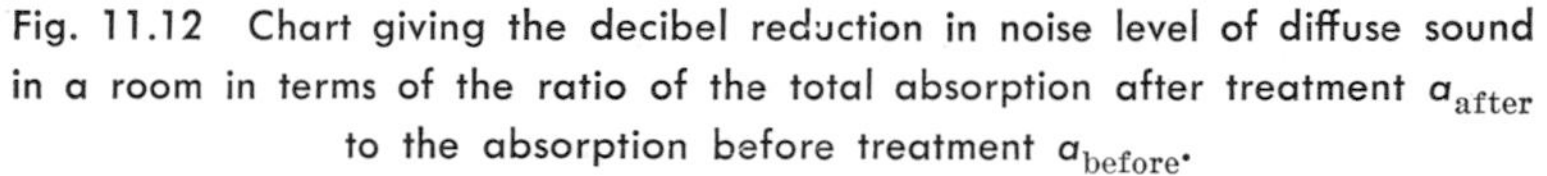

Fig. 11.12 Chart giving the decibel reduction in noise level of diffuse sound in a room in terms of the ratio of the total absorption after treatment a_{after} to the absorption before treatment a_{before}.

rials and the pressure spectrum level of room noise both vary with frequency, the reduction will be different at different frequencies. However, to arrive at a single figure which is somewhat representative of the noise-level reduction, it is customary to employ *noise-reduction coefficients* in such computations. The noise-reduction coefficient (N.R.C.) of a material is the average, to the nearest multiple of 0.05, of the absorption coefficients at 256, 512, 1024, and 2048 cycles.

As a numerical example, we shall compute the noise reduction in an office 35 feet by 49 feet by 10 feet which results from the treatment of the entire ceiling with a material having an N.R.C. of 0.55. The untreated ceiling and walls have a plaster surface whose N.R.C. is 0.03; the floor has an N.R.C. of 0.04; in addition, there are 39 sabins of absorption due to desks, chairs, and miscellaneous items. Hence, the quantity a_{before} is computed as follows:

	Area in Square Feet		*N.R.C.*		*Units of Absorption in Square Feet*
Floor	1715	×	0.04	=	69
Walls	1700	×	0.03	=	51
Miscellaneous				=	39
Ceiling	1715	×	0.03	=	51
Total			a_{before}	=	210

If the ceiling is covered with a material having an N.R.C. of 0.55, the *increase* in absorption is $1715 \times 0.52 = 892$ square-foot-units. Then, the value of a_{after} is $210 + 892 = 1102$ square-foot-units. From Eq. (11.4), the noise reduction attributable to the installation of the acoustical material would be approximately 7 db. The reverberation time would be reduced from 3.9 seconds to 0.7 second.

Calculation of Per Cent Loudness Reduction. The value of sound-absorptive materials in decreasing room noise is underrated when it is expressed in terms of the decibel reduction in noise level. The reasons for this are easy to understand when it is recalled that, in such an evaluation, the characteristics of the ear are neglected and little or no account is taken of the frequency characteristics of the source of noise and of the absorptivity of the acoustical material. A more significant appraisal can be made if all these factors are taken into account and the decrease in noise level is expressed in terms of per cent loudness reduction, which can be computed as follows:

1. Determine the pressure spectrum (L_{ps} *vs.* frequency) of the room noise by the methods of Chapter 2; see Eq. (2.1).
2. Determine, at each of a number of frequencies spaced at intervals no greater than an octave apart, the ratio of the total absorption in the room after treatment to the total absorption before treatment. (In large rooms, the effect of air absorption should be included at high frequencies.)
3. At each of these frequencies compute the decibel reduction in the pressure spectrum level by means of Eq. (11.4) and the ratios of step 2.
4. Subtract these decibel reductions from the original pressure spectrum of step 1 to obtain the predicted pressure spectrum of the room noise.
5. By the use of Fig. 2.9, calculate the masking spectrum (M *vs.* frequency) for the pressure spectra of steps 1 and 4.

6. Plot the masking spectra, determined in step 5, on the Loudness Computing Chart of Fig. 2.10.

7. The per cent difference between the areas under these two curves on the Computing Chart represents the per cent loudness reduction.

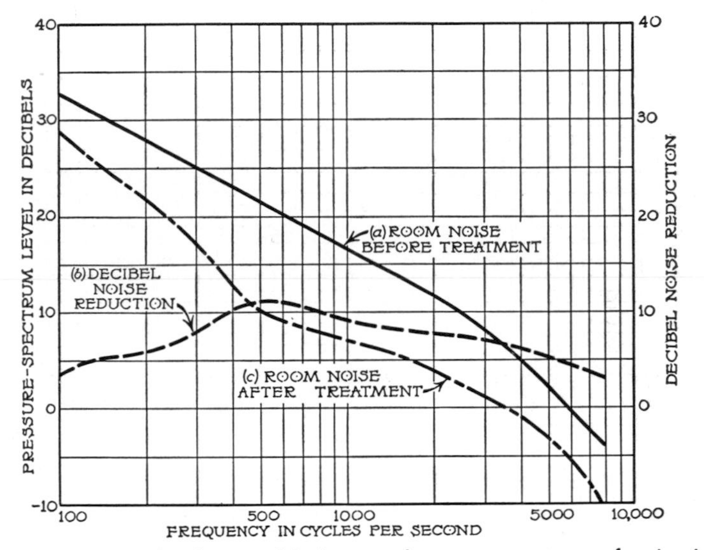

Fig. 11.13 Example showing (*a*) the sound-pressure spectrum of noise in a room corresponding to a total noise level of 50 db, (*b*) the decibel reduction in spectrum level due to the installation of absorptive material, and (*c*) the predicted spectrum of the room noise after treatment.

As an illustration of the use of this procedure, we shall calculate the per cent loudness reduction due to an installation of acoustical material in an office where the noise has the pressure spectrum shown in Fig. 11.13. This pressure spectrum has an over-all level of 50 db. The masking spectrum corresponding to this noise is plotted on a Loudness Computing Chart, Fig. 11.14. Suppose that the ratios of the total absorption in the room after acoustical treatment to the total absorption before treatment, and the corresponding decibel noise reductions in the pressure spectrum level, are as follows:

Frequency in cycles	128	256	512	1024	2048	4096
$a_{\text{after}}/a_{\text{before}}$	3.2	5.0	12.9	7.9	6.3	4.0
Decibel noise reduction	5	7	11	9	8	6

From these data and the original pressure spectrum of the noise, the *predicted pressure spectrum* (Fig. 11.13) is calculated. The masking spectrum is then determined and plotted in Fig. 11.14. The areas under the curves on this Loudness Computing Chart correspond to

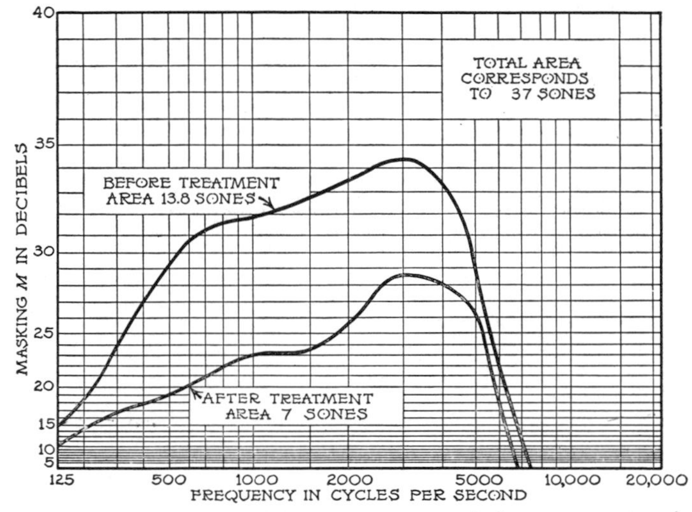

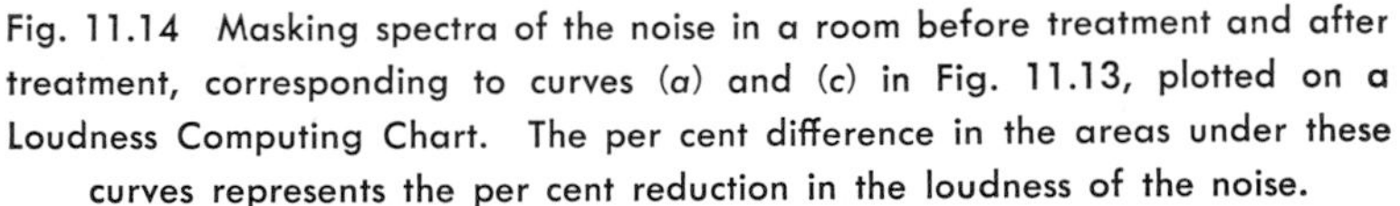

Fig. 11.14 Masking spectra of the noise in a room before treatment and after treatment, corresponding to curves (*a*) and (*c*) in Fig. 11.13, plotted on a Loudness Computing Chart. The per cent difference in the areas under these curves represents the per cent reduction in the loudness of the noise.

13.8 sones (before treatment) and 7 sones (after treatment). Therefore, the installation of the acoustical material has reduced the loudness of the room noise by about 50 per cent.

Application of Acoustical Materials

The use of acoustical materials for noise-reduction purposes exceeds by far all other uses of such materials. We already have considered (in Chapter 6) factors which should govern the selection of these materials, and the best methods of mounting and decorating them. In the present section we are concerned with the means of determining how much absorptive treatment to apply in a room, and where best to place it in order to provide the most effective and practical job of noise reduction. The principal considerations which decide these questions are the shape of the room, how much absorption is already in the room, and where this absorption is located. In rooms with low ceilings (less than about 12 feet) and with the other dimensions comparably large, as in large offices, work rooms, etc., the answer usually is simple: Apply a material having a noise-reduction coefficient (N.R.C.) of at least 0.6 over the entire ceiling. This will generally give satisfactory results. The higher the ceiling, the higher must be the value of the N.R.C. required for a good job.[8,9] *In rooms where the ceiling height is not small compared with the other dimensions, and especially in rectangular rooms with smooth, reflective walls, treatment of the ceiling only will not yield satisfactory results no matter how absorptive the ceiling treatment is.* Sustained reflections between the hard side walls will take place; these reflections often result in a disappointingly low reduction of noise, in room resonance (certain modes of vibration are too little damped), in excessive reverberation, and sometimes in flutter echoes. Here, the usual solution consists of the application of some acoustical treatment to the *side walls* as well as to the ceiling. It need not cover the entire side walls; application of the absorptive material in patches, panels, or bands (strips) covering about one half of the wall surface usually suffice. Where patches of material are used it is advantageous to place the patches on opposite pairs of walls so that reflective surfaces of the walls are not directly opposite each other. Figure 17.1 is a good example of an installation of acoustical material in a room which has a high ceiling. The absorptive treatment has been applied to the available areas of the side walls down to the wainscot.

Figure 11.12 shows that the decibel reduction of noise is not directly proportional to the absorption added to the room; instead, the first few units of absorption do the most good, and

[8] These guiding principles also apply to the acoustical treatment of corridors. The greater the height of the corridor, the larger should be the N.R.C. of the acoustical treatment. If the height is greater than the width, use an N.R.C. of at least 0.70 and apply some material to the side walls.

[9] In the section "Noise in Buildings," Chapter 10, measurements were included which indicated how the sound level decreased with increasing distance from a noise source: In rooms with ceiling treatments having an absorption coefficient of at least 0.70, the attenuation was of the order of 0.4 db per foot for ceiling heights of about 10 feet, and 0.2 db per foot for ceiling heights of about 20 feet.

successive increments follow "the law of diminishing return." Hence, if a room has a great deal of absorption in it, provided by its boundaries and furnishings, then the addition of absorptive treatment on the ceiling may not produce a marked effect in the reduction of the noise level in the room. An estimate of the number of units of absorption required in a room for effective control of noise by absorption can be obtained by the use of the optimum reverberation-time curve for speech together with the reverberation-time nomogram, Fig. 8.11. First, by means of the curve in Fig. 9.11, find the optimum reverberation time for speech in the room. *For noise-reduction purposes, the reverberation time should not exceed the optimum time for speech, and, for large rooms, noisy rooms, or rooms in which quiet is a prime objective, the reverberation time should be not more than two thirds to three fourths of the optimum time for speech;* in general, in all rooms except very large ones, it should not exceed 1 second. The nomogram can then be used to give the units of absorption corresponding to the appropriate reverberation time and volume. The value of total required absorption determined by this rule is not critical; however, the rule is a practical guide for estimating the required absorption for satisfactory acoustical treatment. The absorptive material should then be distributed in accordance with previous recommendations.

12 · Reduction of Solid-Borne Noise

The means of transmission and suppression of solid-borne sound are considerably different from those of air-borne sound, described in the preceding chapter. Solid-borne noise usually originates from impacts or machinery vibration; for example, the "hammering" of a water pipe, the dragging of a chair across an uncarpeted floor, or the shaking of an electric refrigerator caused by the unbalance of its motor. Here, the amounts of instantaneous vibratory power involved are tremendous compared with those of the usual sources of air-borne noise. For this reason, and because vibration is communicated through continuous structures with little attenuation, it often is transmitted great distances in a building. Hence, solid-borne noise should be suppressed at its source wherever it is practical to do so. Measures for accomplishing this include the use of: heavy carpeting, cork tile or linoleum-on-felt to reduce impact transmission to the floor; a segment of flexible metallic or rubber hose in a pipe to lessen the propagation of impulses along it; flexible mountings for motors and other types of machinery to suppress the communication of vibration to the floor; and a short section of canvas in a ventilation duct to prevent the passage of vibration along the duct.

Once vibration is transferred to a solid building structure, such as a concrete slab, it travels through the structure with a speed of about ten times that of sound in air. If it reaches a flexible partition, a floor, or a wall, the vibration may force the partition into oscillation, and annoying sounds may then be radiated. The efficiency of radiation depends on the ratio of the dimensions of the partition to the wavelength of the sound (the greater this ratio, the greater the efficiency) and on the internal damping of the partition. For these reasons, a panel that is subject to vibration is sometimes cross-braced to divide it into smaller areas, and damping material (pugging or "dum-dum") may be applied to one side of the panel.

Although massive, rigid partitions (for example, concrete walls) provide effective insulation against air-borne noise, they offer poor protection against solid-borne vibration. On the other hand, porous materials, such as blankets, which are relatively poor insulators of air-borne sound, can be used in such a way as to present an effective barrier against the transmission of solid-borne vibration. The most effective type of structure for prevention of propagation of solid-borne noise is that of "discontinuous construction," whereby the transmission path is severed or contains marked discontinuities in density and elasticity. This type of construction and other means of protection against solid-borne noise are discussed in the following sections.

Floors and Ceilings

Well-designed floor systems must provide adequate insulation against both air-borne and solid-borne noise. A construction that is excellent for one may be poor for the other. For example, a bare concrete slab 1 foot thick has a high transmission loss for air-borne sound, yet it propagates impacts readily. Impacts must be prevented from imparting much energy to a floor if their transmission through the structure is to be suppressed. This can be accomplished by means of a resilient covering, such as carpet or cork tile which will absorb some of the impact. Since the portion of the vibratory energy communicated to the floor will be propagated with little attenuation unless there are structural breaks or discontinuities in the construction, staggered joist construction is superior to ordinary joist construction.

Ratings of partitions for the insulation of air-borne sounds, in terms of transmission loss, are given in Table A.12, Appendix 2. A similar type of rating for the isolation [1] of impacts against a

[1] The term *isolation* is usually applied to solid-borne sound; *insulation* is applied to air-borne sound.

floor would be much more complicated; it would depend on many factors, such as the type of blow imparted to the surface (for example, whether it is from a hammer or shoe) and the frequency distribution of the impact noise. However, a useful single figure of merit can be obtained in a laboratory as follows: In a room beneath the test floor, compare (1) the (weighted) noise level created by impacts on a floor construction used as a standard of reference with (2) the noise level created by impacts on the floor construction under test. Then, the *impact-noise reduction* of the floor under test is the number improvement in decibels the test floor construction provides over the floor used as a standard of reference in isolating impact noise transmitted to a room beneath. For example, suppose that the sound level in a room due to a continuous series of impacts on a concrete floor above is 50 db. Then, if carpet on that concrete slab reduces the level of the noise in the room to 45 db, the impact-noise reduction is said to be 5 db. Tables 12.1 and 12.2 give the ratings of a number

TABLE 12.1

IMPACT-NOISE REDUCTION FOR CONCRETE FLOORS

Impact Noise Transmitted to Room beneath Various Concrete Floor Constructions Compared with That for the Bare Concrete Slab

(R. Lindahl and H. J. Sabine)

Nature of Floor Construction Laid on Concrete Slab	*Impact-Noise Reduction in Decibels*	*Comment of Observer*
None (bare concrete)	0	Bad
Asphalt tile (5/32-inch)	0	Nearly as bad
Asphalt saturated felt (1/8-inch)	2	Nearly as bad
Rubber tile (3/16-inch)	7	Better than concrete
Heavy carpet (no pad)	10	Good
Linoleum (3/16-inch) on felt	12	As good as carpet
Asphalt-saturated fiberboard (1/2-inch)	12	As good as carpet
Hardboard (3/16-inch) over fiberboard (1/4-inch)	17	Better than carpet
Wood floor (3/4-inch) on sleepers (2-inch by 3-inch)	19	Very quiet
Cork tile (1/2-inch)	20	Very quiet
Wood floor (3/4-inch) on sleepers (2-inch by 3-inch), rock-wool fill	20	About same as construction without rock wool

TABLE 12.2

IMPACT-NOISE REDUCTION FOR CONCRETE FLOORS

Impact Noise Transmitted to Room beneath Various Concrete Floor Constructions Compared with That for the Bare Concrete Slab

(Building Research Station at Garston)

Nature of Floor Construction Laid on Concrete Slab	*Impact-Noise Reduction in Decibels* *
None (bare concrete)	0
Carpets, etc.:	
Linoleum (1/8-inch) and linoleum (1/8-inch) on roofing felt	5
Wood blocks, thin carpet, rubber	5–10
Carpet (1/8-inch) on underfelt (1/8-inch), hard rubber-cork composition (1/4-inch)	10
Sheet rubber (1/16-inch) on sponge rubber (1/4-inch)	20
Screeds (concrete slab), 2-inch, on following underlays:	
Clinker	5–10
Granulated cork (1-inch)	10–15
Slag-wool quilt or eelgrass quilt	15–20
Glass-silk quilt, single-layer, or eelgrass quilt, double-layer	20
Glass-silk quilt, double-layer	25
Boarding on battens on following underlays:	
Clips	5–10
Asbestos pads or felt pads (1/2-inch)	5–10
Fiberboard pads (1/2-inch)	10
Felt pads (1-inch) or rubber pads (1/2-inch)	10–15
Eelgrass quilt or slag-wool quilt (1/2-inch)	10–20
Glass-silk quilt or rubber pads (1/2-inch)	15–20
Suspended ceilings:	
Plaster (1/4-inch) on fiberboard (1/2-inch) on battens in clips (2-inch by 2-inch)	5–10
Plaster (3/8-inch) on plasterboard (3/8-inch) on battens in felt-lined clips	10–15

* Frequency-weighted sound-pressure level, averaged to the nearest 5 db.

of floor constructions laid on a concrete slab, compared with the rating of the concrete slab itself.

In Table 12.3 is given the impact-noise reduction for wood floors; a normal board and joist floor, together with a lath and plaster ceiling, is used as a basis of comparison. The addition of 20 pounds per square foot of sand or ashes improves the impact-noise reduction by 10 db. The reason is that the whole floor, being of light weight, tends to vibrate under impact. In-

TABLE 12.3

IMPACT-NOISE REDUCTION FOR WOOD JOIST CONSTRUCTION

Impact Noise Transmitted to Room beneath Various Wood Floor Constructions Compared with That for an Ordinary Board and Joist Floor, Lath and Plaster Ceiling

(Building Research Station at Garston)

Nature of Floor Construction	*Impact-Noise Reduction in Decibels* *
Normal board and joist floor, lath and plaster ceiling	0
Plaster on plaster board ceiling	0
Pugging (sand or ashes, 10 lb per sq ft, or slag wool, 2 lb per sq ft)	5
Separate joists to carry ceiling	5
Floor boards on cross battens on glass-silk quilt, not nailed	5
Floor boards on fiberboard, on sub-boarding	5
Floor boards on glass-silk quilt, on sub-boarding	10
Sand or ashes, 20 lb per sq ft	10
Carpet on underfelt	10

* Frequency-weighted sound-pressure level, averaged to the nearest 5 db.

creasing the mass of the structure reduces this tendency. The added material also increases the damping of the structure.

If good isolation against impacts is to be obtained, the ceiling should not be rigidly connected to the floor joists. If the ceiling is carried on independent joists, as in Fig. 12.1, the impact-noise

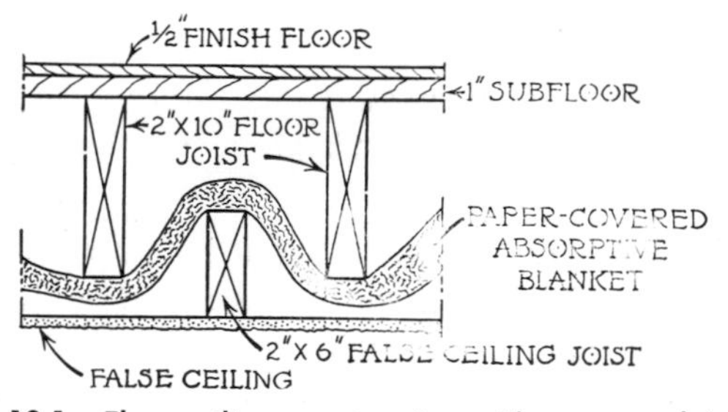

Fig. 12.1 Floor-ceiling construction with staggered joists.

reduction will be improved by at least 5 db. (The principal benefit of the addition of a paper-covered absorptive blanket, as shown, is in the reduction of air-borne sounds.) Resilient clips provide a convenient means for attaining good isolation between a ceiling and the structural floor from which it is hung. A method for suspending acoustical tile ceilings is illustrated in Fig. 6.11. Similar methods are effective also for suspending ordinary lath and plaster ceilings.

Consideration must be given to the floor surface in the acoustical design of a floor construction. It is frequently important to choose a wearing surface such that impacts against it will be muffled and thus will generate little noise in the room where they are produced. The benefits resulting from the use of carpeting in the aisles and lobby of an auditorium have already been stressed in Chapter 9. The "relative noisiness" values of a number of floor surfaces are listed in Table 12.4. These data give the

TABLE 12.4

IMPACT NOISE CREATED IN ROOM ABOVE COMPOUND FLOORS

Various Floor Constructions Laid on Concrete Compared with Bare Concrete Slab

(R. Lindahl and H. J. Sabine)

Nature of Floor Construction Laid on Concrete Slab	*Noise Level Reductions in Decibels*
Wood floor (¾-inch) on sleepers (2-inch by 3-inch)	−10
Same, with rock-wool fill	−7
None, bare concrete	0
Hardboard (3/16-inch) on fiberboard (½-inch)	2
Asphalt tile (5/32-inch)	2
Rubber tile (3/16-inch)	3
Heavy carpet (no pad)	4
Cork tile (½-inch)	5
Asphalt-saturated felt (⅛-inch)	5
Asphalt-saturated fiberboard (½-inch)	5
Linoleum (3/16-inch) on felt	7

relative noise levels created in a room by impacts on different floor constructions laid on a concrete slab. Compare the order in which these structures rank with the order of the corresponding impact-noise reductions for these same structures (see Table 12.1). Note that floors which provide good isolation against the transmission of impact noise to the room beneath may not necessarily provide a "quiet" wearing surface in the room where the impacts are produced; for example, wood flooring.

FLOATING FLOORS. The isolation provided by a floor system against mechanical impact can be greatly improved by the use of a "floating" floor which rests on the structural floor but is separated from it by a resilient support or quilt, as in Fig. 12.2.

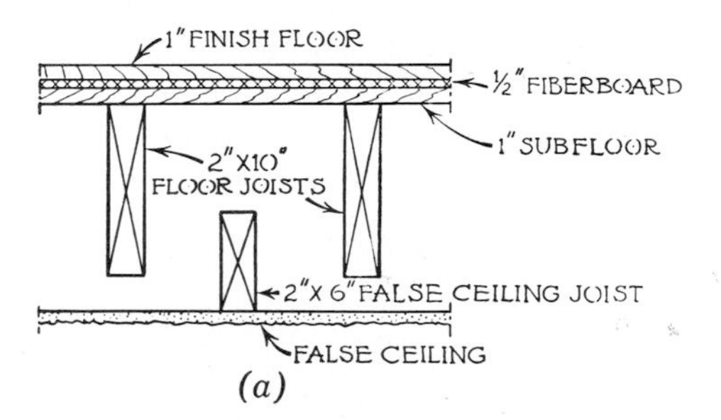

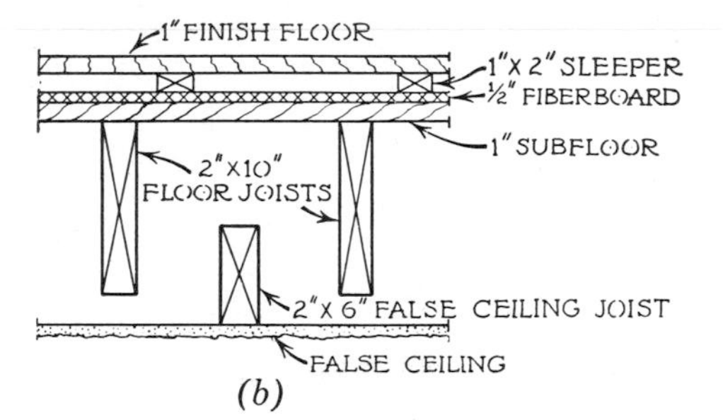

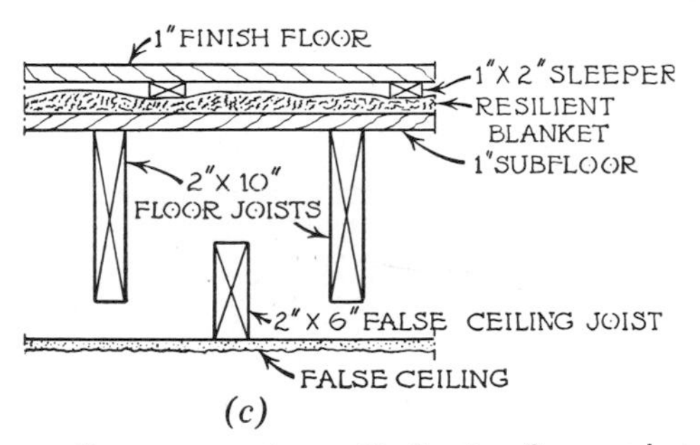

Fig. 12.2 Floor-ceiling constructions with floating floor and staggered joists, finish floor on (*a*) fiberboard; (*b*) sleepers on fiberboard; (*c*) sleepers on resilient blanket.

Not only is the impact-noise reduction improved by such floating floors, but also the transmission loss for air-borne sound is increased slightly.

In choosing a resilient support for a floating floor, one must consider the safe amount of loading the support can withstand without being compressed to the extent that its resilience is practically lost. It is important to select a material that will have a long life and will not continue to pack or settle significantly. In general. the most satisfactory materials are very resilient ("springy"), and they return to their initial condition when the load is removed. Slab cork, granulated cork, rubber, fiber-

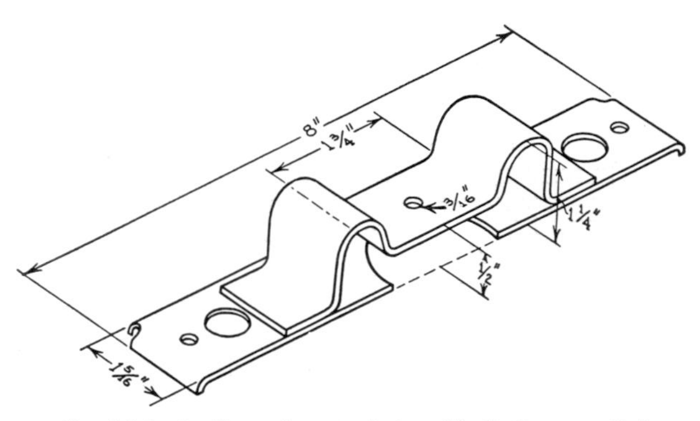

Fig. 12.3 Resilient sleeper chair. (U. S. Gypsum Co.)

board, felt, wood-wool and certain types of mineral-wool blankets are among the available materials that meet these requirements. They must be properly loaded, but not overloaded. All these materials continue to settle and to become less resilient with age. Flexible steel supports and clips are less subject to such defects. The effectiveness of a floating floor system is dependent on the extent of isolation provided by the resilient supports between the floated finish floor and the structural floor. Care should be taken that this isolation is not "shorted" by nails or by solid connections anywhere–including the junctions between the floating floor and the walls. Thus, in Fig. 12.2(*c*), a resilient blanket is laid on a subfloor; the finish floor is nailed to sleepers which rest on the blanket. The sleepers must not be rigidly connected to the subfloor; they should float on the blanket or be fastened by special resilient chairs (one such chair is shown in Fig. 12.3). The blanket should have a paper, or similar, covering on at least one side to improve the insulation against air-borne sounds.

The impact-noise reduction of a floating wood floor, separated, by means of different types of resilient isolators, from a maple subfloor on a concrete slab, is given in Table 12.5.

TABLE 12.5

IMPACT-NOISE REDUCTION FOR FLOATING FLOORS

Isolation Provided by Floating a Wood Floor, Separated from a Wood-Covered Concrete Slab, by Means of Various Types of Resilient Isolators

Type of Isolator	*Impact-Noise Reduction in Decibels*
None	0
Wool (1-inch) and fiberboard (½-inch)	8
Fiberboard (1-inch)	11
Cork (1½-inch)	12
Commercial flexible steel "chair"	24

For isolation of heavy impacts, wood floor constructions are very poor. Since they are relatively light, the whole floor easily can be set into vibration. The floor construction shown in Fig. 12.4 is a superior one. For example, a resilient quilt is laid on

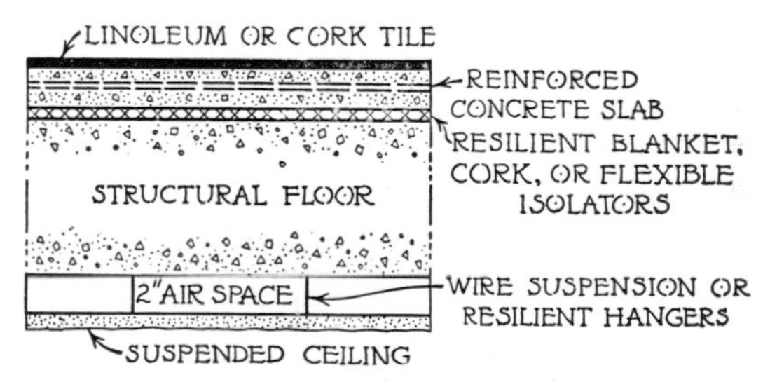

Fig. 12.4 Section of floor and ceiling showing floating floor and flexibly suspended ceiling.

a concrete supporting floor; a concrete slab 2 inches thick is then poured directly on the blanket. The wet concrete must not be allowed to leak through the quilt to the supporting floor and thus "short" the isolation. In a technique employed in Sweden the edges of the paper liner are opened on the exposed side of the quilt, lapped over the edges of the adjacent section of the quilt, and then sealed with an adhesive. The blanket should be turned up along the walls to prevent contact of the partitions

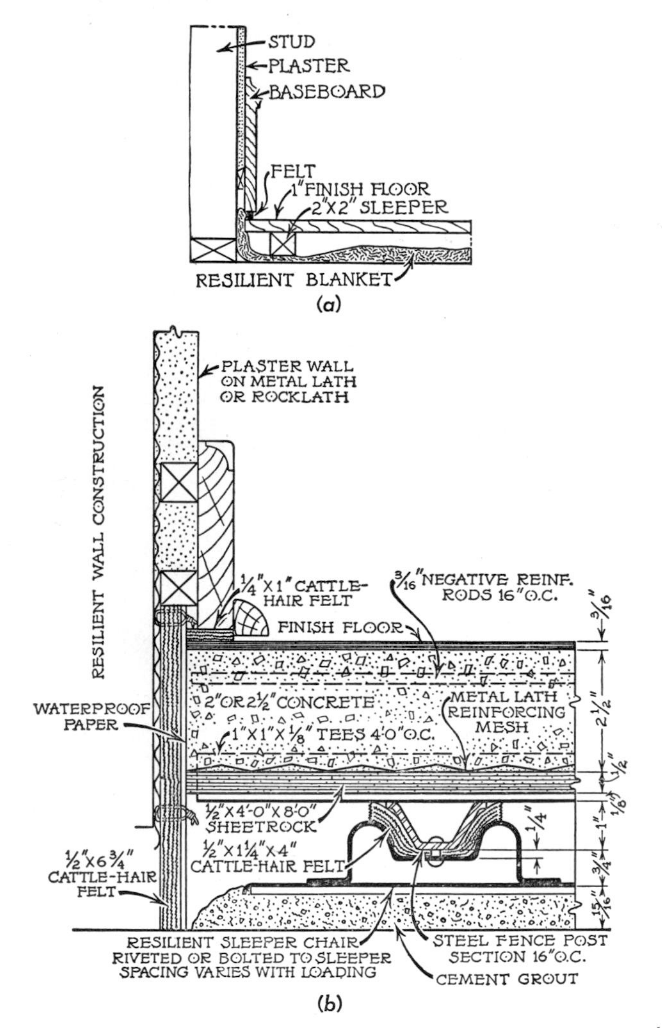

Fig. 12.5 Detail of junction between floating floor and wall for (*a*) wood floor and (*b*) concrete floor. (U. S. Gypsum Co.)

with the floating slab. Figure 12.5(*a*) shows a detail of the junction between a floating wood floor and a wall. Felt is used to obstruct the flow of sound through the crack between the finish floor and the baseboard. Figure 12.5(*b*) shows a detail of the junction between a floating concrete slab and a resilient wall construction. This type of floating floor is sometimes used in broadcast studios or in buildings having similar sound-isolation requirements.

Discontinuous Construction

We have already indicated that air-borne sound can be effectively impeded by certain types of multiple wall structures and that solid-borne sound can be suppressed by discontinuities in the transmission path. The advantages of good insulation against both types of noise have been incorporated in *discontinuous construction,* whereby the rooms in a building are treated essentially as a suspended "box within a shell." The basic elements are shown in Fig. 12.6.[2] The walls of a room are built on a floating floor. Ties between the walls and the continuous construction are avoided, but, where they are absolutely necessary, special resilient isolators are employed. The ceiling is suspended from the structural floor by resilient hangers, enough space being left above the false ceiling for pipes and other services. Because of its cost, discontinuous construction has not had extensive use. In many instances where economic considerations do not permit such construction throughout an entire building, it may be practical to isolate a few rooms within which quiet is especially desirable. On the other hand, if a building is in a quiet location and if, with one or two exceptions, the rooms within it are quiet, it may be most practical to isolate the noisy rooms from the rest of the building by means of discontinuous construction.

The benefits of discontinuous construction can be almost entirely lost if proper treatment of the details is neglected. Windows and doors should not form a rigid link between a detached room and the surrounding continuous construction, nor should pipes, ducts, etc., be allowed to present a solid bridge between these elements.[3] Pipes should be suspended from the structural

[2] For a comprehensive treatment of the subject of discontinuous construction, see R. Fitzmaurice and W. Allen, "Sound Transmission in Buildings," His Majesty's Stationery Office, London (1939).

[3] For the treatment of details of walls, windows, doors, chimneys, balconies, plumbing, etc., see "Sound Transmission in Buildings," *loc. cit.*

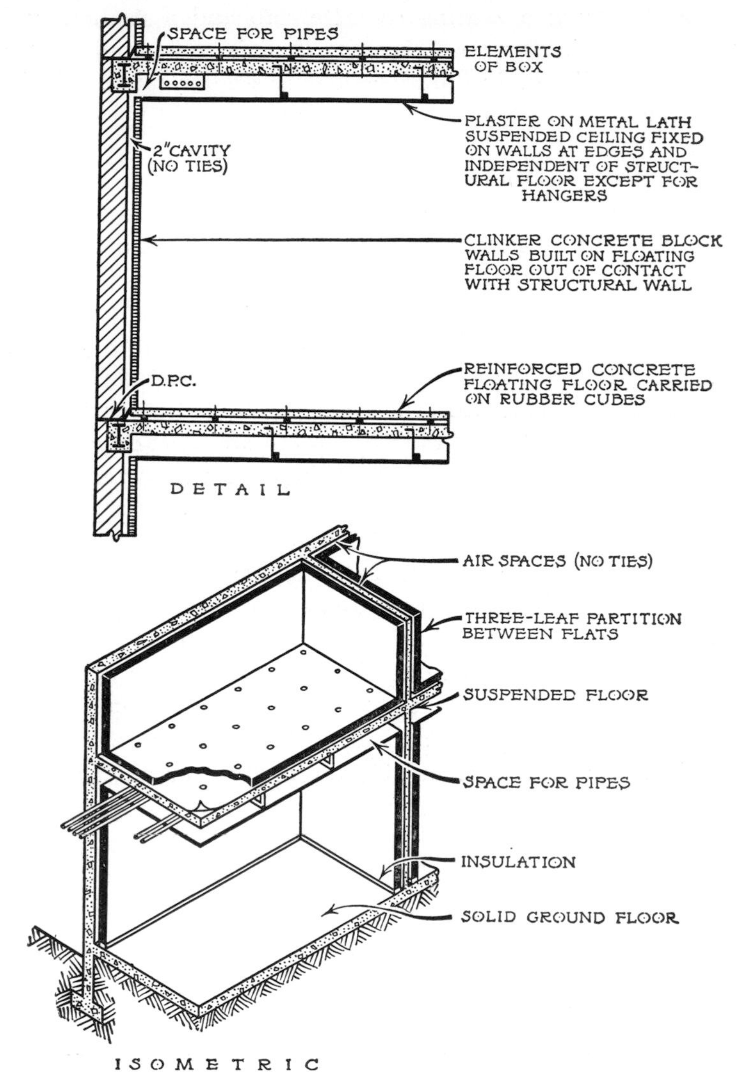

Fig. 12.6 Example of discontinuous construction. (Reproduced by permission of the Controller of His Brittanic Majesty's Stationery Office and the Director of Building Research.)

floor by resilient supports. Where they penetrate walls, pipes should be isolated from the partitions by rubber, felt, or other compliant material. Care should be exercised to prevent cracks at these junctions, otherwise the insulation against air-borne sounds will be reduced. The insertion of a short length of flexible hose, 6 to 12 inches long, will be found to be helpful in

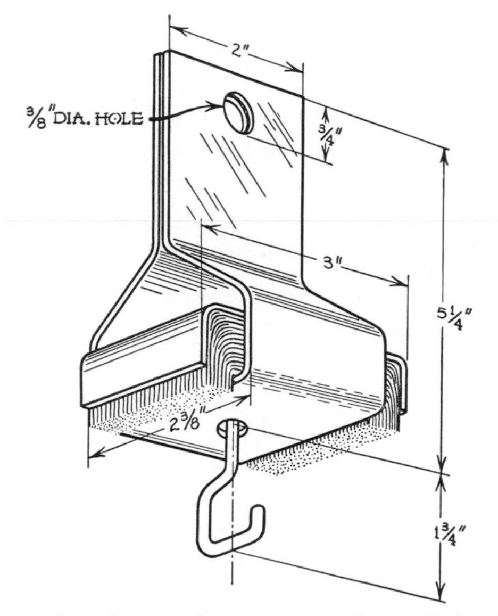

Fig. 12.7 Resilient hanger for isolating ceiling. (Johns-Manville.)

providing a substantial reduction in the vibration transmitted along a pipe.

A practical method of securing better isolation than that obtained from ordinary continuous construction is that of "semi-discontinuous construction." Although this method is not so effective as that of discontinuous construction, its cost is somewhat lower. Here a floating floor is employed, but, unlike discontinuous construction, the walls of the room are not completely isolated from the structural walls. Instead, resilient "chairs" or clips such as those shown in Fig. 11.11 are used to support the inner walls. The ceiling is suspended from the structural floor by resilient hangers. One type of resilient hanger is shown in Fig. 12.7.

Isolation of Machinery Vibration

Proper control of machinery vibration can greatly reduce the noise in many factories and buildings. Frequently, the cause of such vibration can be removed at its source, for example, by tightening loose parts, by improving the balance of rotating parts, or by increasing the accuracy with which the moving parts fit. Wherever practical, such measures should be taken; but sometimes they are either impossible or costly. Under these conditions, a resilient mount for the source, made of rubber, cork, felt, springs, or similar products, can be useful in preventing vibration from being transmitted to large surfaces, such as partitions, that radiate sound efficiently.

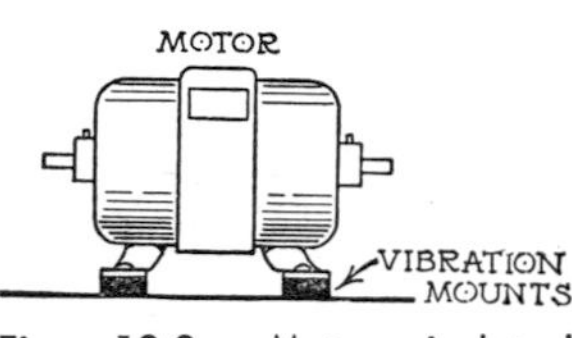

Fig. 12.8 Motor isolated from floor by means of vibration mounts.

Consider the motor shown in Fig. 12.8. While this machine is in operation, there are forces that tend to make it oscillate in both the vertical and the horizontal directions. These oscillatory forces would be transmitted to the floor if the motor were bolted directly to it. However, if resilient supports, usually called *vibration mounts,* permit the machine to move up and down, the inertia of the moving system can be used to oppose these forces so that only a slight force is transmitted to the floor. If the mounting system is properly designed, little vibration will be communicated to the floor even though the machine may oscillate up and down with a relatively large amplitude. In many types of vibration problems substantial benefit can be derived from mounts which yield principally in the vertical direction. Machinery vibration often is very complex, and, in general, a complete mathematical account of this vibration (and means of isolating it) requires formulation in terms of six possible degrees of freedom—three degrees of translation, and three degrees of rotation. A better job of isolation may be obtained when regard is given to all possible modes of vibration and rotation than when only selected modes are considered. This is taken into account in the design of certain commercial flexible mounts, clips, and chairs, where such materials as felt and rubber are used to increase the isolation of the two horizontal modes of vibration and all three modes of rotation.

In this section we shall discuss vibration systems in which the displacement is principally in the vertical direction. The mounting system on which the machine shown in Fig. 12.8 rests has a static displacement (that is, it sags) due to the weight on it. If the machine is given an additional downward displacement and then released, it will vibrate up and down about its equilibrium position with an amplitude that decreases with time. A simplified portrayal of this motion is shown in Fig. 12.9. The number of these complete oscillations per second, f, is its resonant frequency. A resilient mounting system is effective in isolating vibration *only* when f is at least two times lower than the disturbing frequency of the vibration, F. The smaller f is compared to F, the better the isolation. This is expressed analytically in the following way. The ratio between the force transmitted to a floor when a mounting is employed and the force which would be transmitted if the vibrating mass were rigidly attached to the floor is called the *transmissibility*, T. It is given, approximately, by the expression

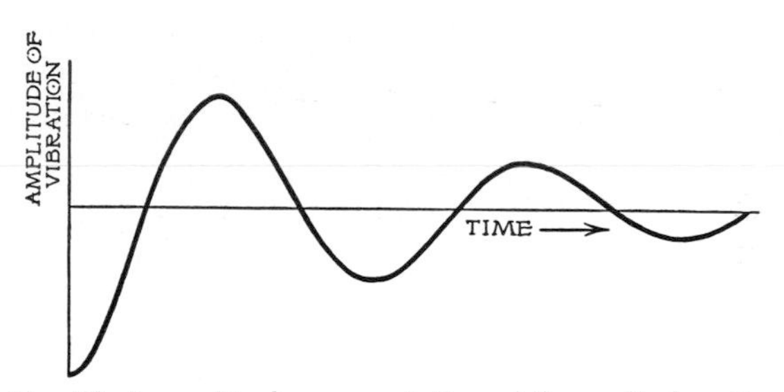

Fig. 12.9 Simplified graphical representation of the vertical motion of a mass which has been given a downward displacement on a resilient support.

$$T = \frac{1}{\left(\frac{F}{f}\right)^2 - 1} \tag{12.1}$$

Mechanical damping of the mounting system has been neglected in Eq. (12.1); the presence of such damping is particularly important in the frequency region of resonance, $F = f$, where it prevents the transmissibility from becoming very large. When the magnitude of T is less than 1, the mounting system will be beneficial. Under these conditions, the percentage reduction of the transmitted force, frequently referred to as "the per cent reduction of vibration," is given by the equation

$$\text{Vibration reduction} = 100(1 - \mathrm{T}) \quad \text{per cent} \qquad (12.2)$$

and is expressed graphically in Fig. 12.10. If the vibration mounting system supporting the machine in Fig. 12.8 has a static displacement of 0.15 inch, and if the motor rotates at 1200 revolutions (cycles) per minute, then, according to this chart, the vertical component of transmitted force is reduced about 80 per cent. If the motor in this example runs at speeds *less than* 685 revolutions per minute, this mounting system will *increase* the vibration energy communicated to the floor instead of reducing it!

The resonant frequency f can be computed in the following way. Let K equal the total stiffness constant (the force required to produce unit displacement of the mounting system), and let M be the mass which the mounting system supports. Then, the resonant frequency is given by

$$f = \frac{1}{2\pi\sqrt{M/K}} \quad \text{cycles/sec} \qquad (12.3)$$

An alternate expression for Eq. (12.3) is

$$f' = \frac{188}{\sqrt{d}} \quad \text{cycles/min} \qquad (12.4)$$

where d is the static displacement of the mounting system in inches, due to the load on it, and f' is the frequency in cycles per minute.

Frequently, a number of identical vibration mounts are used to support a single load. If n mounts are connected together so that the same load is transmitted from one mount to the next (this is called a *series mounting*), the total stiffness is equal to

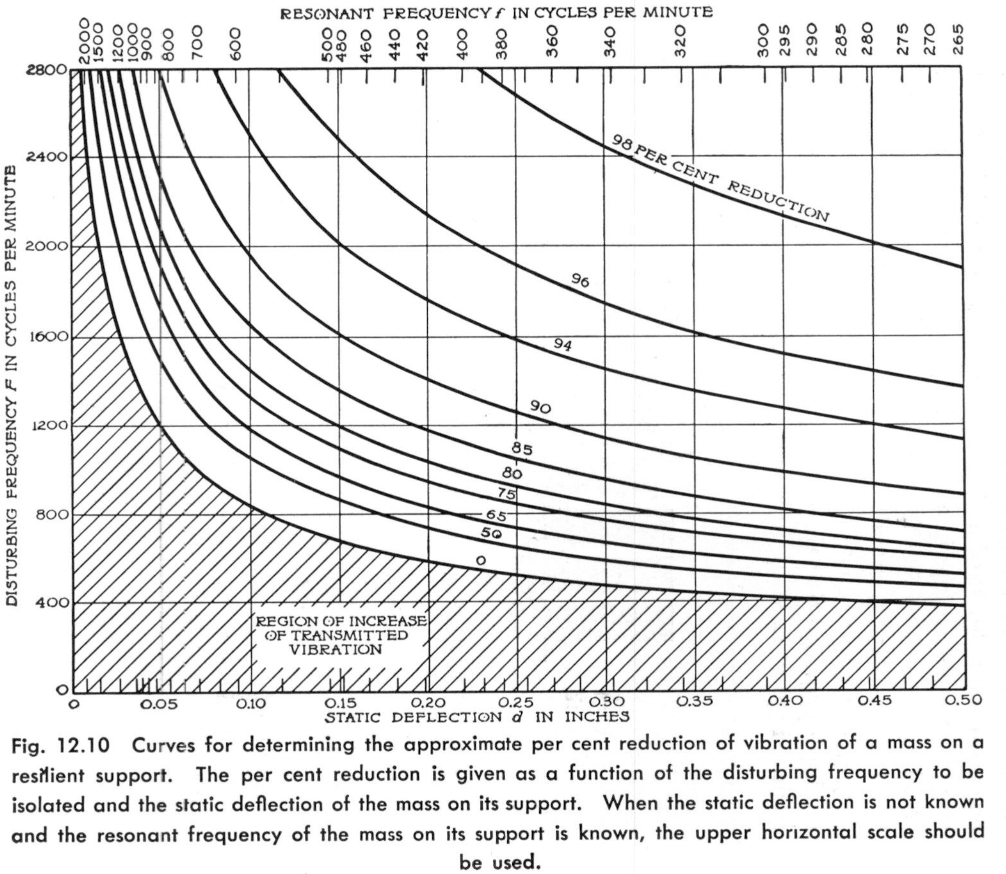

Fig. 12.10 Curves for determining the approximate per cent reduction of vibration of a mass on a resilient support. The per cent reduction is given as a function of the disturbing frequency to be isolated and the static deflection of the mass on its support. When the static deflection is not known and the resonant frequency of the mass on its support is known, the upper horizontal scale should be used.

$1/n$ times the stiffness of an individual mount. For example, a combination of three identical springs in series, Fig. 12.11(*c*), has one-third the stiffness of a single spring. Thus a weight of 200 pounds on three springs in a series will have a static displacement equal to three times that for the same weight on a single spring. Similarly, if a load is equally distributed over a cork slab, the stiffness will be approximately halved if the thickness of the slab is doubled, provided that the bearing area of the slab remains constant. Figure 12.11(*b*) illustrates the case where the load is

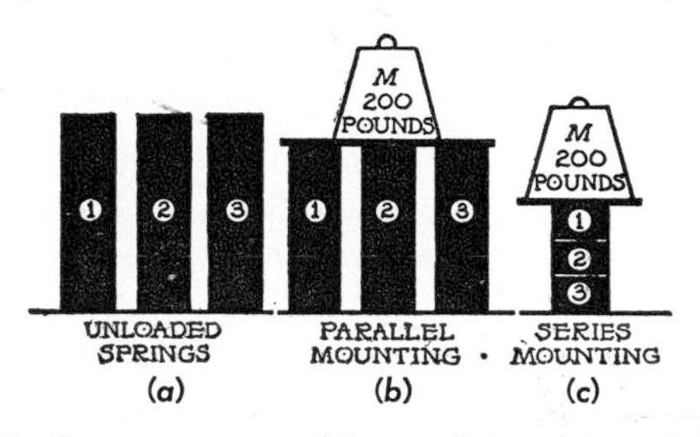

Fig. 12.11 Resilient supports used in parallel and in series. (See text.)

divided among n similar mounts (this is called a *parallel mounting*). Here, the total stiffness is equal to n times the stiffness of a single mount. Therefore, if three springs in parallel are identical, the total stiffness of the combination is three times that of a single spring. Similarly, for a given thickness of a cork mounting, an increase in the bearing area of the cork slab under a mass increases the total stiffness (in the same proportion) and thus increases the resonant frequency of the combination.

Resilient supports must be selected with great care, especially when the frequency of vibration is quite low; otherwise, the results may be worse than if the machine were rigidly fastened to the floor! Many commercial vibration mounts are available. Others can be designed from cork, rubber, felt, etc. *In all cases, one should make sure that the mounting system is neither overloaded nor underloaded, and that it provides a resonant frequency several times lower than the lowest frequency of vibration to be isolated. When cork, rubber, or resilient blankets are employed, they should be loaded to about 50 to 75 per cent of the upper safe limit of loading.*

13 Control of Noise in Ventilating Systems

A well-designed ventilating or air-conditioning system should operate quietly enough to be free from annoying noise generated by or transmitted through the system, and it should not impair privacy by communicating conversation or noise from one room to another connected to it by means of air ducts. The noise contributed to a room by a ventilating system usually can be reduced to an amount that is negligible compared to other noise in the room by: (1) the suitable choice and installation of motors, fans, and grilles; (2) streamlining of the air-transmitting system so that turbulence is avoided; and (3) application of absorptive treatment within the duct system and in the equipment rooms. In systems where the noise level in the room must be very low, more elaborate measures may be needed, such as the use of special acoustical filters. In this chapter we shall discuss the means by which noises originate in air-conditioning or ventilating systems, and how they can be controlled.[1] Although the responsibility for controlling these noises usually is assumed by the heating and ventilating contractor, the architect should be familiar with the principles and procedures discussed in this chapter.

[1] For further information on air-conditioning and ventilating systems, consult the *American Society of Heating and Ventilating Engineers' Guide,* published annually (see especially the chapter on Sound Control).

Origin of Noise in Ventilating Systems

The principal sources of noise from a ventilating system are: the motor and fan, the turbulence caused by the flow of air through the air-transmitting system, and sounds of outside origin that enter the system through the walls of a duct, or through a duct outlet in one room, and then are transmitted through the duct to another room.

MOTOR AND FAN NOISE. The motor and fan of a ventilating system usually are the predominating sources of noise. The noise is transmitted (1) through the air, and (2) by solid-borne vibration.

Figure 13.1(*a*) shows the frequency distribution of noise from a ventilating system. All frequencies are present, but there is a prominent peak at 355 cycles. This is the *blade frequency* of the fan, which can be calculated from the equation

$$\text{Blade frequency} = \frac{(\text{rpm})(\text{N.B.})}{60} \text{ cycles/sec} \qquad (13.1)$$

where rpm is the number of revolutions per minute of the fan, and N.B. the number of fan blades. For example, Fig. 13.1(*a*) shows the noise spectrum of a six-bladed blower fan having a speed of 3550 rpm. Hence, the blade frequency is $(3550/60)\cdot 6 =$ 355 cycles. Increasing the fan speed increases the total noise level it produces; for a given fan each doubling of the rpm raises the total noise level approximately 17 db.

NOISE RESULTING FROM AIR FLOW. When air flows through a ventilating system, obstructions of all types (bends, side branches, changes of duct size, grilles) produce eddy currents or other forms of turbulent flow. Noise containing sounds of all frequencies is generated as a result of this turbulence. However, noise arising from turbulence usually contains relatively more high-frequency noise than does motor-fan noise. Hence, streamlining often will contribute effectively to noise reduction. Sometimes the turbulence will set into vibration parts of the system, particularly the walls of unlined ducts, and give to the resulting noise a definite pitch.

Noise that results from the turbulent flow of air increases with

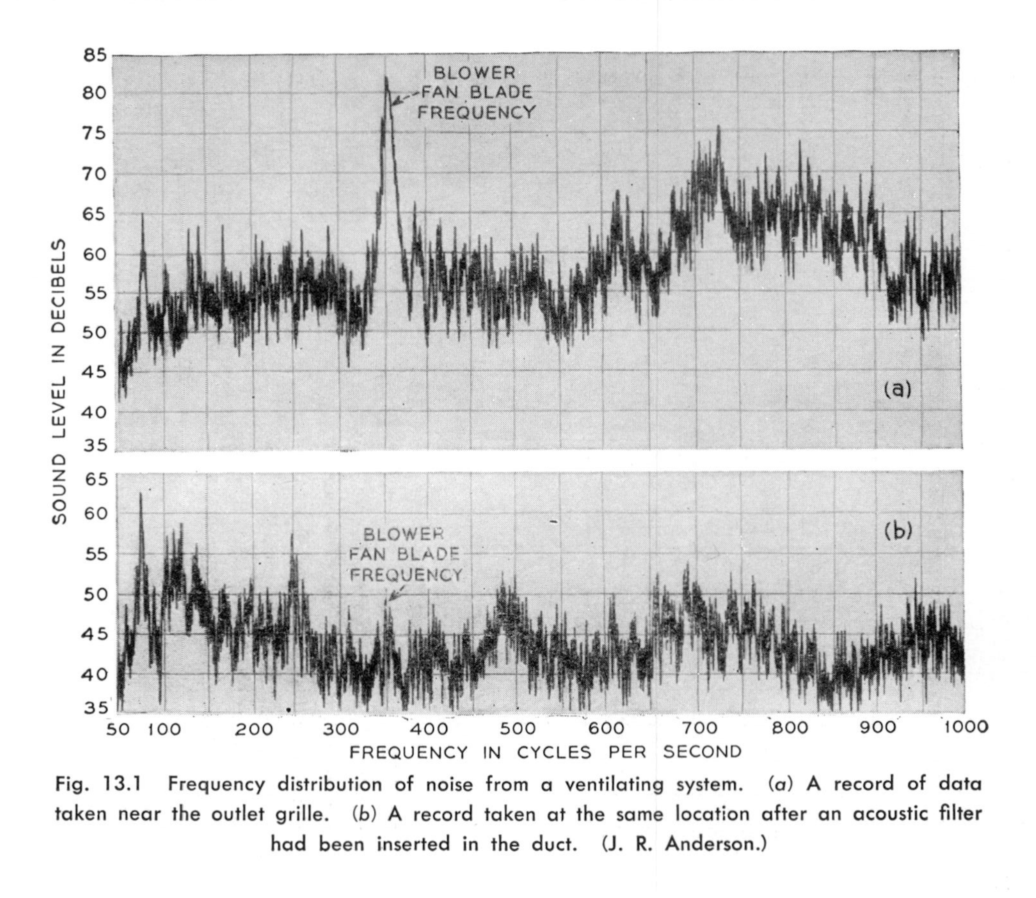

Fig. 13.1 Frequency distribution of noise from a ventilating system. (*a*) A record of data taken near the outlet grille. (*b*) A record taken at the same location after an acoustic filter had been inserted in the duct. (J. R. Anderson.)

increasing velocity of flow. From this standpoint, it is desirable to have the air velocity low; however, this involves relatively larger ducts and hence greater expense. If a certain level of noise is acceptable in a room, acoustical correctives will permit the ventilation system to have a higher velocity. The increase in noise with air flow velocity is illustrated by the curves of grille noise level shown in Fig. 13.2. The measurements were taken at a distance of 6 feet from each of three typical grilles having a face area of 0.5 square foot. These data indicate that grilles that produce a large spread of air by deflectors which offer obstruction to the outward flow produce a somewhat higher noise level than do those having little air resistance.[2] For a grille of a given type, increasing the face area increases the noise level *if the air velocity is constant;* the increase is approximately 3 db for each doubling of the face area. For example, from Fig. 13.2 the noise level for grille *B,* which has an area of 0.5 square foot, is 27 db for a face velocity of 1250 feet per minute. The noise generated by a grille of this type having an area of 1 square foot would be 30 db for the same air velocity of 1250 feet per minute. *If the total amount of air flowing past the grille remains the same,* the noise level decreases rather rapidly as the size of the grille is increased. This is illustrated by the curves of Fig. 13.3, where the noise level is shown as a function

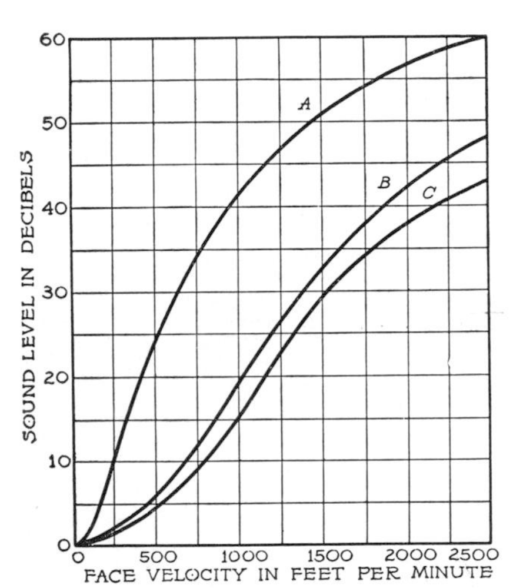

Fig. 13.2 Noise level near grille vs. face velocity of the air stream. Curve A is for a grille providing little spread. Curve *B* is for a grille of the honeycomb type giving a small spread. Curve C is for a grille producing a large spread. (P. H. Geiger.)

[2] P. H. Geiger, *Heating, Piping, Air Conditioning,* 8, 601 (1936). These data were taken with a sound-level meter using the 40-db weighting network.

of the grille area. The volume of air flow in cubic feet per minute is specified for each curve. These data apply to grille *B* of Fig. 13.2, a honeycomb type giving small spread, but they are considered typical for most grilles.

If there is more than one grille in a room, the noise may be computed as if there were but one grille with an area equal to the sum of the areas of the separate grilles. For a given face velocity, the level of the grille noise will be about the same for an exhaust grille as for a supply grille. However, in most rooms the face velocity at the exhaust grille is somewhat lower than it is at the supply grille; therefore, the contribution of the exhaust grille noise to the room noise level can be neglected compared to that of the supply grille. (See p. 212.)

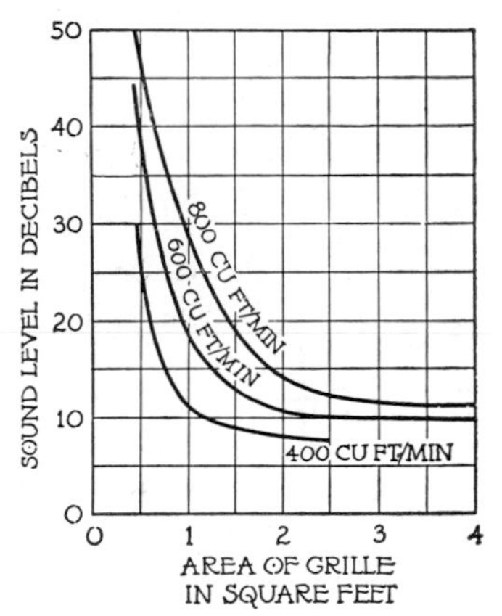

Fig. 13.3 Noise level at a grille *vs.* grille area for a constant volume flow. The volume of flow, in cubic feet per minute, is specified for each curve. (P. H. Geiger.)

It is advisable to limit the air velocity to a value such that the contribution of the grille noise to the general room noise will be negligible. Since noise in a room varies widely with the type and location of the room (see Fig. 10.3), a single figure cannot be given as the value of air velocity which should not be exceeded. This figure depends on the original room noise level. Furthermore, since the magnitude of the grille noise depends on the type and size of grille, a face velocity that is acceptable for one grille may not be tolerable for another.[3] Grille noise will usually not be noticeable if it is

[3] In comparing data of noise level *vs.* face velocity for different grilles, one should consider the location of the microphone with respect to the grille, the frequency-weighting network of the sound-level meter, the face area of the grille, and the total sound absorption in the room when the measurements are made. The greater the sound absorption of the room, the lower will be the measured value of noise level. Usually the 40-db weighting network is the most satisfactory one to use.

5 to 10 db below all other room noise; it is usually negligible when the air velocity is below 1000 feet per minute.

CROSS-TRANSMISSION BETWEEN ROOMS. One of the most frequent complaints against certain ventilating systems is attributable to the communication of noise from one room to another by the duct system. This type of disturbance is called "cross-talk" or "cross-transmission." Sound may enter the grille of a duct in one room, travel down the duct, and be radiated from the outlet of that duct in another room. Noise also may enter the system directly by being transmitted through the duct walls, especially if they are unlined and thin. Methods for suppressing these and other sources of noise in the system are discussed in the following section.

Noise Suppression in Ventilating Systems

In order that the noise output from a ventilating or air-conditioning system will be sufficiently low for comfort, it is advisable to take the following steps:

(1) Select and install the equipment so that the noise is minimized at its source.
(2) Determine the amount of noise reduction that must be supplied by sound control measures in order to reduce the noise from the ventilating system to a level that will not be annoying.
(3) Select the most suitable means for attaining the required noise reduction determined in step (2).

MINIMIZE NOISE AT ITS SOURCE. Two types of noise should be suppressed in a ventilating system—that resulting from solid-borne vibration and that which is air-borne. The principal sources of solid-borne vibrations are the motors and fans. In addition, turbulence in the air stream can cause the duct walls and other parts of the system to rattle. Motors and fans in which noise and vibration are deliberately and effectively suppressed are now manufactured. They are especially desirable, but if their cost is very much higher than others which are less quiet, it is sometimes cheaper to control the noise and vibration by other means. *Proper mounting of the motors and fans, so that they will not communicate vibration to the ducts, walls, or floor, is important.* There should be no direct contact between the building structure and the foundation of the motors and fans. Isolation of the machinery from the floor can be accomplished by methods described in the previous chapter. The blower and exhaust fans should be isolated from the ducts by a flexible sleeve, for example, one fabricated of canvas. It is also helpful to use rubber hose for the piping connections from pumps. The tendency of unlined duct walls to be set into vibration can be reduced by the application to its surface of a material which adds mechanical damping.

The noise level is high in the equipment room of a ventilating system. This fact may not be important if the equipment is in an out-of-the-way location such as the basement. However, where the equipment noise is likely to be a source of annoyance to occupants adjacent to the equipment room, the motors and fans should be selected with regard to their quietness, acoustical treatment of the equipment room may be desirable, and the wall partitions should provide adequate attenuation of noise. In the past few years air-conditioning units for single rooms, small enough to fit into a window, have become popular. Such a unit may not be a source of annoyance in the room itself, but it may be a source of considerable disturbance to neighbors having windows near-by. Before purchasing such an air-conditioning unit, one should listen to it or check its specifications with regard to the noise it generates outside as well as inside the room it air-conditions.

Streamlining of the air-transmitting system, particularly at the elbows, aids in minimizing the noise resulting from air flow. Edges of metal surfaces perpendicular to the air current should be streamlined or rounded. Vanes are especially important in reducing the noise generated at elbows. Such vanes are illustrated in Fig. 13.4, which is a typical detail of an air-supply duct and outlet for a radio studio, where the noise level from the ventilation system must be very low.

The design for a ventilating system should specify the amount and type of sound treatment the system shall contain and where it shall be located. A useful rule-of-thumb used by many contractors for lining ducts (with absorptive material) is that the duct should be lined for a length equal to ten times the average duct "diameter"; accordingly, a duct 10 inches by 20 inches would be lined with a standard duct-liner material [4] for a distance of about 150 inches. If the ducts supply locations where the room noise is lower than average, the duct is lined for a distance equal to about 20 times the average duct diameter. It is advisable to locate the

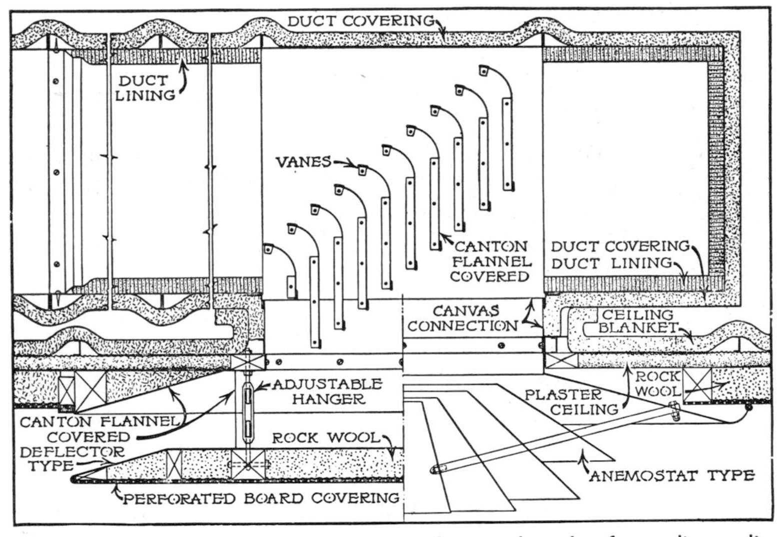

Fig. 13.4 Typical detail of air-supply duct and outlet for radio studio. (Courtesy National Broadcasting Co.)

lining at the duct outlet so that it can aid in suppressing all noises originating within the duct. However, when a number of branch ducts are used to ventilate many rooms, it is economical to place a sound-absorptive plenum chamber immediately following the fan. The cross-sectional area of the plenum should be at least ten times the discharge area of the fan. One advantage of this type of treatment is that absorptive material, 2 inches thick or more, can be used so that a relatively high value of low-frequency absorption can be obtained. Muslin-covered rock-wool blanket (mineral wool or Fiberglas) will be satisfactory for this purpose. The decibel reduction provided by an absorptive plenum is given approximately by

[4] Usually a mineral-wool material 1 inch thick.

$$\text{Reduction} = 10 \log \left(\frac{a}{S_f} \right) \text{ db} \qquad (13.2)$$

where a represents the total units of absorption and S_f is the area of fan discharge. Suppose, for example, that the plenum has a surface area of 120 square feet with an absorption coefficient of 0.5, and that the area of the fan discharge is 2 square feet. Then $a = 120(0.5) = 60$ square-foot-units (sabins); from Eq. (13.2) the decibel reduction is 15. Calculations preferably are made for a frequency close to the blade frequency of the fan, since this frequency is generally the one that determines what the resultant noise level will be.

Computation of Required Noise Reduction. The following steps are necessary in the computation of the noise reduction for a ventilating system required to reduce the noise to an acceptable level.

(1) Estimate the probable noise level in the room. The figures and tables of Chapter 10 are useful for this purpose.

(2) Determine the permissible noise level at the duct outlet. For example, if the noise level from the duct is equal to the level of all other noise in the room, Fig. 10.1 shows that the noise level in the room will be increased approximately 3 db by the addition of the duct noise; similarly, if the noise radiated from the duct is 10 db below the level of all other room noise, Fig. 10.1 shows that this duct noise will raise the room noise level by less than 1 db. A satisfactory working rule is to specify that the noise level in the room resulting from the ventilating system shall be at least 5 db below the acceptable noise level for that room. (See p. 221.)

(3) Estimate the noise level at the source of noise. Data on the noise generated by a fan should be supplied by the manufacturer.

(4) Subtract the permissible noise level of step (2) from the level of step (3). This gives the total attenuation required in the duct system.

(5) Estimate the total attenuation that would exist between the fans and the duct outlets if no acoustical control measures were applied. This can be done by the empirical method of the following section.

(6) The difference between the attenuation of steps (4) and (5) represents the additional noise reduction that must be supplied.

ATTENUATION BY UNLINED DUCTS, ELBOWS, AND GRILLES. Sounds which are propagated through unlined ventilating ducts usually are attenuated very little in the important frequency range, unless the ducts are quite long. The losses are least when the ducts are large and the frequencies are low. For example, for a sheet metal duct 24 inches by 24 inches and 60 feet long, the loss would be approximately only 3 db. At high frequencies, where the wavelength of sound is comparable to the cross-sectional dimension, the attenuation is greater. However, in most calculations the attenuation in unlined ducts can be neglected. If the duct contains elbows, there will be some additional attenuation—approximately 1 to 2 db for each elbow.

Reflection of sound takes place in a duct where there is a transformation of the cross-sectional area. For this reason there is some attenuation of noise at branches and outlets where the size of the duct changes, but the greatest change in area takes place at the grilles. If A represents the total grille area (supply plus return) and a represents the total number of units of absorption in the room, the grille attenuation due to the area transformation is approximately

$$\text{Grille attenuation} = 10 \log \frac{a}{A} \quad \text{db} \tag{13.3}$$

Thus we note that noise is both generated and attenuated at a grille. The attenuation is an important factor in the control of cross-talk between rooms connected by a common air-supply duct. In general, the attenuation can be neglected, but the noise arising from turbulence caused by the passage of air through the grille cannot be neglected.

Acoustical Materials for Duct Lining

An acoustical absorptive lining increases the attenuation of sound which travels down a duct. Furthermore, it increases the transmission loss of the duct walls for sound entering the system by passage through the walls. This may be important if the duct has thin walls and if it passes through a noisy room. The lining may also add considerable mechanical damping to the duct walls and act as a safeguard against resonant vibration of these walls. Furthermore, if the lining has low thermal conductivity, as it usually has, additional thermal insulation on the outside of the duct will not be necessary. However, absorptive lining does slightly increase the resistance to air flow in the duct.

The properties required of acoustical materials for use in ventilating systems are somewhat more exacting than those for use in the treatment of rooms (discussed in Chapter 6). Materials used in ducts should be moisture-resistant and should offer as little resistance to air flow as possible. *It is essential that the lining be fire-resistant; combustible duct linings are prohibited by most building codes.* Hence commercial lining materials are usually mineral in composition. They should be held together by a binder that does not support combustion, that gives the material adequate mechanical strength, and that prevents particles from being carried off by the air stream. Sheets of commercial duct-lining material can be cut conveniently and installed in ducts by bolts or special adhesive. Rock wool covered by perforated metal is another acoustical treatment that has been used to line ducts, especially at very high velocities. For this application the perforated metal surface should be such that it offers little resistance to air flow, and such that the passage of air over it does not produce objectionable noise.

The absorption coefficients of a number of acoustical materials made especially for lining ducts in ventilating systems are given in Tables A.1 and A.3, Appendix 1. They show that the materials are least efficient in absorbing sound at low frequencies. The principal benefit of increasing the thickness of the lining is to increase the absorption coefficient at low frequencies. Since the noise from a blower or exhaust fan produces considerable sound at its blade frequency, which is usually between 100 and 500 cycles, special emphasis should be given the absorption coefficients in this frequency range if such noise is to be suppressed by lining material. In computing the attenuation requirements for lining materials, calculations can be made for a frequency at 1024 cycles if the material is used principally for reducing grille noise or other noise due to turbulence.

The smaller the cross section of a lined duct, the greater is its attenuation in decibels per foot. Hence the division of a large duct into a number of smaller lined ducts greatly increases its total attenuation. (For example, see section *B–B* in Fig. 13.9.) Absorptive "cells" of this "honeycomb" type may be used to advantage where the portion of the duct available for treatment is not sufficiently long to permit the required noise reduction by

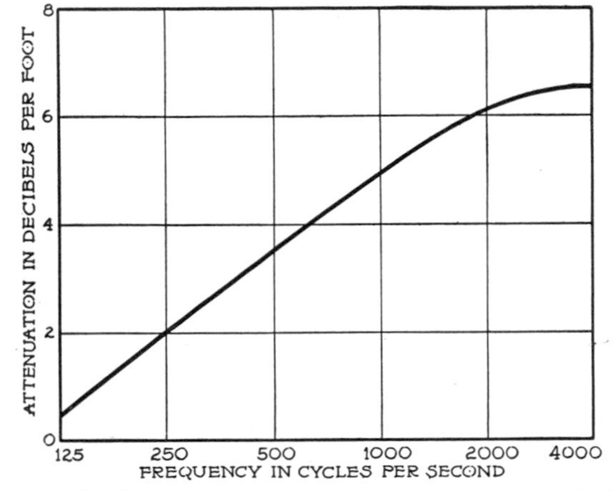

Fig. 13.5 Example illustrating the variation of attenuation with frequency in a lined duct. These data are for a lined duct 30 inches long, between floors in a building. The 48-inch by 66-inch duct cross section is divided into 13 flues, each 48 inches by 4 inches, by means of 1-inch Air-acoustic sheet, 5 inches o.c. (W. A. Jack.)

merely lining the duct walls. This situation occurs where two rooms are connected by a short section of duct. The cross-sectional dimensions of cells should be small—sometimes as small as 2 inches by 4 inches. Where the noise from an existing installation must be reduced, and where the ducts are not readily accessible, one solution is to place such a honeycomb structure of absorptive cells directly behind the grille. Such structures increase the resistance to the flow of air because of the nature of the material and because the structure reduces the area through which the air flows.

Figure 13.5 is an example showing how the attenuation in a lined duct varies with frequency. The decrease in attenuation at low frequencies illustrated here is characteristic of the commercial materials 1 inch thick used for duct lining. They are not efficient absorbers at low frequencies, and, as mentioned previously, the use of thicker materials offers one means of absorbing the prominent noise peak at the fan blade frequency (usually below 512 cycles). If the suppression of this noise peak is to be accomplished by acoustical materials: (1) determine the lowest blade frequency by Eq. (13.1); (2) design the duct to have sufficient attenuation to reduce adequately the noise at the blade frequency. If there is adequate noise reduction at this frequency, it is probable that there will be sufficient attenuation at the higher frequencies.

Calculation of Noise Reduction in Lined Ducts

The attenuation of sound in lined ventilating ducts depends on the characteristics of the duct lining and on the size and shape of the duct. This noise reduction can be expressed in decibels

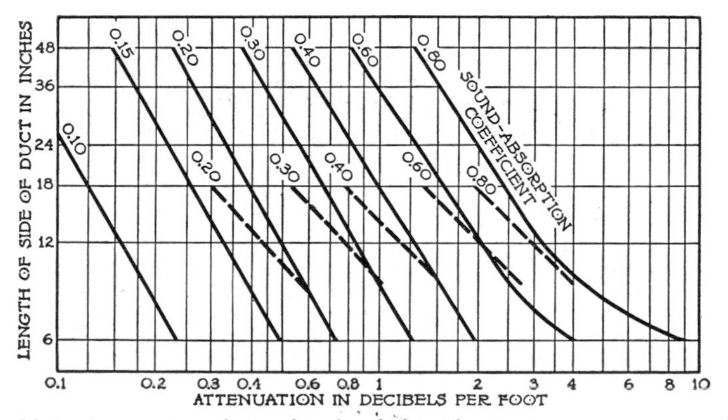

Fig. 13.6 Attenuation of sound in lined ducts having a square cross section. The curves are for various absorption coefficients. (J. S. Parkinson.)

per linear foot of duct. It is possible to compute these losses by means of rigorous methods which involve the acoustical impedance of the material.[5] Although these methods should be employed where the highest standards of noise control are required, the use of a relatively simple formula or of curves compiled from empirical data will yield satisfactory results for most design purposes. Figure 13.6 is a chart which was empirically

[5] See C. T. Molloy, *J. Acoust. Soc. Am.*, **16**, 31 (1944).

derived by Parkinson.[6] It gives, in terms of the absorption coefficient of the lining material and the dimensions of the duct's cross section, the attenuation in ducts having a *square* cross section. The greater the cross section is, the less the attenuation is. As an illustration, suppose that a 20-foot duct has a cross section 18 inches by 18 inches, and that it is lined with a material having an absorption coefficient of 0.40. Then, according to Fig. 13.6, there will be an attenuation of 1 db per foot, hence a total of 20 db for the entire length. The data represented by the solid lines are the *average* results for a number of types of material. Individual materials may depart somewhat from these curves. For example, the broken light lines represent observed data for a 1-inch commercial lining of sheet rock wool. Sabine [7] showed that, for this material, the attenuation in a rectangular duct can be expressed as

$$\text{Attenuation} = 12.6\alpha^{1.4}\frac{P}{A} \quad \text{db/ft} \tag{13.4}$$

where α = absorption coefficient of the lining material
P = perimeter of duct in inches
A = cross-sectional area of duct in square inches.

This formula was found to be accurate, with a possible error of 10 per cent, for ducts having cross-sectional dimensions in the ratio of 1:1 to 2:1, for absorption coefficients between 0.20 and 0.40, and for frequencies between 256 and 2048 cycles. An alternate form of Eq. (13.4) is

$$\text{Attenuation} = K\frac{P}{A} \quad \text{db/ft} \tag{13.5}$$

where K is a number which depends on the absorption coefficient of the lining; it is given by Fig. 13.7. For example, suppose that a duct is 10 feet long, has a rectangular cross section 10 inches by 15 inches, and is lined with a material having an absorption coefficient of 0.60 at 512 cycles. What is the total attenuation of the duct at this frequency? $P = 50$ inches, $A = 150$ square inches, and, from Fig. 13.7, the value of K corresponding to an absorption coefficient of 0.6 is 6.2. Then according to Eq. (13.5)

[6] J. S. Parkinson, *Heating & Ventilating,* **36,** 23 (1939).
[7] H. J. Sabine, *J. Acoust. Soc. Am.,* **12,** 53 (1940).

the attenuation per foot is 2.1 db; for the 10-foot section it is 21 db.

It follows from Eq. (13.5) that for a duct of given cross-sectional area, the greater the ratio of P/A, the greater is the attenuation per linear foot of duct. Thus, a lined rectangular duct 9 inches by 18 inches is more effective in suppressing the transmission of noise through it than is a square duct of equal cross-sectional area and length lined with the same material; however, because the rectangular duct has the greater wall area more lining mate-

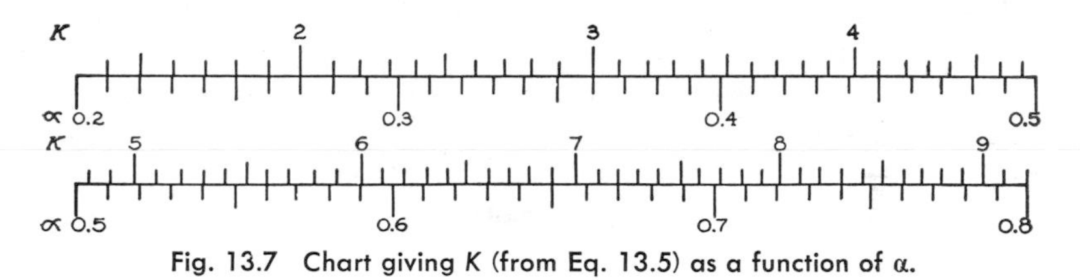

Fig. 13.7 Chart giving K (from Eq. 13.5) as a function of α.

rial is required. As might be surmised, it follows that for rectangular ducts of the same cross-sectional area, the same area of lining material is required to produce a given amount of attenuation regardless of the duct proportions. Within the limits of accuracy of Eq. (13.4), the proportions affect the number of decibels of attenuation per foot or the length of duct to be lined—not the amount of material required.

Acoustical Filters

An acoustical filter in a duct consists of a combination of volumes and masses of air (for example, chambers, short duct sections, perforations) which are combined so as to suppress a range of frequencies of noise transmitted through the duct to the filter. Methods of noise suppression that rely on the use of sound-absorptive materials may not always yield adequate attenuation in systems that supply air to rooms, especially where extreme quiet is required, as in radio, television, and sound-recording studios. The reason is that the absorption coefficient of duct-lining material is relatively small at low frequencies. It is possible to design acoustical filters of the type considered here to eliminate low-frequency noise, such as "blade frequency" noise, which is difficult to suppress by other means. On the other hand, such filters are not effective at high frequencies where their dimensions become comparable with the wavelengths. Hence, a filter in a ventilation system is generally used in conjunction with absorptive material—the filter to attenuate the low frequencies, followed by duct lining to attenuate the high frequencies.

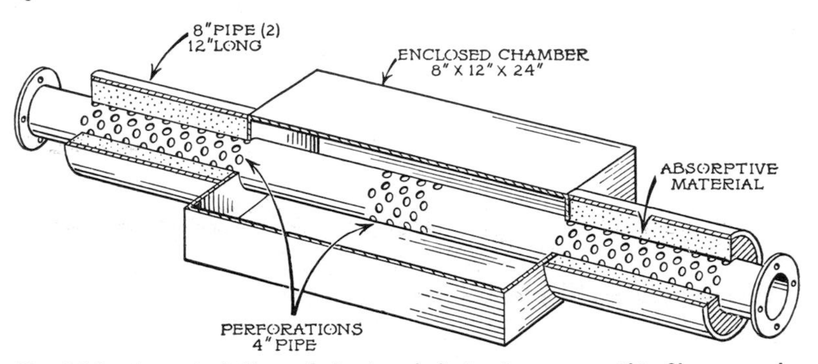

Fig. 13.8 Acoustical filter of the band-elimination type. This filter was designed for a system having a 4-inch duct. (J. R. Anderson.)

An acoustical filter for a ventilating system, it will be seen, is a "tailor-made item" which can be extremely useful if *it is correctly designed.*[8] In ventilating systems, "band-elimination" filters and "low-pass" filters are two of the most useful types. In the first type, the band that is eliminated is centered at the blade frequency of the exhaust or blower fan; higher frequencies are suppressed by sound-absorptive linings, splitters, or cells. Figure 13.8 shows the design characteristics of a band-elimination filter [9] that was installed in a 4-inch circular duct having an Anemostat outlet. Absorptive linings on either side of the filter are an aid

[8] For useful design information see W. P. Mason, *Electro-Mechanical Transducers and Wave Filters,* 2nd ed., Section 4.51, D. Van Nostrand (1948); see also P. M. Morse, *Vibration and Sound,* 2nd ed., Section 23, McGraw-Hill Book Co. (1948).

[9] There is a direct analogy between electrical and acoustical filters. The electrical analogue of the band-elimination acoustical filter is an inductance in series with condenser, the combination being across the transmission line. The magnitude of the acoustical "inductance" is determined by the effective mass of the air in the perforations, whereas the size of the acoustical "capacitance" is determined by the volume of enclosed air. As in the electrical case, an acoustical filter must be properly terminated. If improperly terminated, it may actually increase the noise level at the duct outlet.

in properly terminating the filter, and they also suppress high-frequency noise generated by the passage of air through the filter itself. The filter section consists of a perforated-metal solid pipe surrounded by a rectangular chamber—an enclosed volume. The walls of the outer pipe and chamber are highly damped to

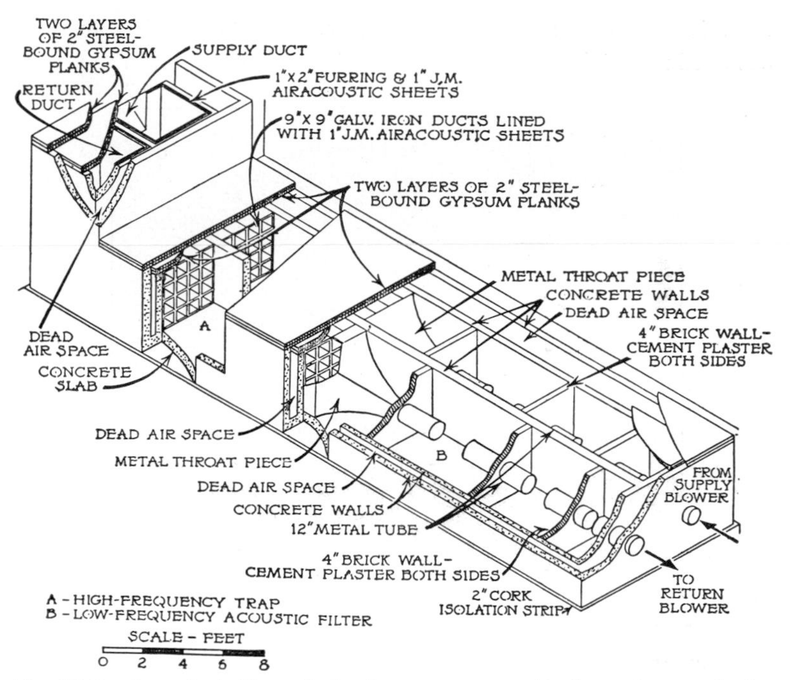

Fig. 13.9 Acoustical filter of the low-pass type. Air from the supply fan passes through the filter to an absorptive cell structure and then on to a room. (R. L. Hanson.)

prevent radiation of sound by the filter to its surroundings. Comparing the measurements shown in Figs. 13.1(*a*) and (*b*) shows that the filter of Fig. 13.8 is designed to suppress the blade frequency of the blower fan (355 cycles), and that it functions very well. The upper noise spectrum is for the system without the filter; the lower spectrum shows the effect of the filter in the system. The peak at 355 cycles has been eliminated, and the noise level in this frequency region has been reduced about 30 db.

Figure 13.9 shows the design of a "low-pass" acoustical filter in the air-conditioning system for a large room at the Bell Telephone Laboratories in which extremely quiet conditions are required. One section of this filter was designed to suppress noise at 535 cycles, the blade frequency of the blower fan. A filter of large dimensions is required to suppress noise of relatively low frequency. A similar filter is used in the exhaust duct. Each filter consists of a series of chambers connected by short sections of flared metal ducts; see Fig. 13.10. The flare reduces the turbu-

Fig. 13.10 The interior of the filter of Fig. 13.9 with the upper surface of the chambers removed. (Courtesy Bell Telephone Laboratories.)

lence and the associated noise generated by the air stream. The acoustical filter suppresses low-frequency noise from the fan. High-frequency noise is attenuated by the absorptive cells joined to the filter by a flared duct which acts as an "acoustic transformer."

Elimination of Cross-Talk

Good sound insulation between two adjacent rooms can be entirely ruined by "cross-talk" via a duct connecting the two rooms. Even if absorptive lining is not needed to suppress motor and fan noise, such lining, or even a honeycomb structure of absorptive cells, may be required in the connecting duct, because the attenuation losses through this duct should be at least as great as the transmission loss of the wall separating the two rooms. For example, if the natural duct attenuation is 25 db, and the transmission loss of the wall is 40 db, the duct lining or cells should supply the additional 15 db. If the connecting duct between the rooms is so short that lining it alone will not provide the required additional attenuation, appropriate absorptive cells should be installed. If an unlined duct passes through a room in which the noise level is high, the noise may be transmitted through the duct walls. Then once it is in the duct system, this noise is a source of cross-talk to other rooms. A plastic coating which adds both mass and mechanical damping is beneficial for suppressing this disturbance.

By the use of procedures described in this chapter and of the measures for the control of vibration and solid-borne noise described in Chapter 12, it is possible to design ventilating or air-conditioning systems to supply air which is noise-controlled as well as temperature- and humidity-controlled.

14 · Sound-Amplification Systems

Practical means for increasing the sound level in the rear portions of a room by proper acoustical design were considered in Chapter 9. It is important to do everything possible to provide the room with the most desirable shape, reflective characteristics, optimum reverberation, and noise insulation. However, even when this has been done the speech level in large auditoriums is usually too low for satisfactory audibility. If high speech intelligibility is desired, Fig. 9.2 indicates that a sound-amplification system should be used if the size of the auditorium exceeds 50,000 cubic feet. This recommended upper limit is not a critical value that must be precisely adhered to (it depends, to a large extent, on the noise level in the room), but to exceed this volume without the assistance of sound amplification is a risky venture. The need for the amplification of speech in legitimate theaters, especially those having large volumes, does not seem to be appreciated by most architects and theater owners. The average person radiates only about 50 microwatts of acoustical power when speaking in a room with a volume of 100,000 cubic feet. From Eq. (8.2) it follows that he would produce a sound level of about 57 db in a room of this size. The professional actor, contrary to theatrical and public opinion, has no special "acoustical power house" with which to "throw his voice." He can be taught to hold up his head, open his mouth, face his audience, emphasize the consonants, but he usually cannot generate more than 100 microwatts of speech power over a long period of time. Although he can and frequently does raise his voice, his average speech power at times may decline to one fiftieth of this value or less, which does not provide enough acoustical power for good hearing conditions in the more distant seats. A sound-amplification system may be necessary even in smaller rooms where the average noise level is greater than 40 db. The effect of noise in reducing the intelligibility of speech is illustrated in Fig. 9.1.

Where high standards of musical performance are desired, as in theaters and auditoriums, it is better to dispense with amplification than to employ a sound-reinforcement system of poor or even moderately good quality. A sound system for music requires unusually good equipment, proper placement of microphones and loudspeakers, and operation of the equipment by a skillful technician who appreciates the art of music and the temperament of musicians. A complete discussion of the problems associated with high-quality sound-amplification systems for music is beyond the scope of this book; therefore the discussion in this chapter will be limited to systems used primarily for the reinforcement of speech. We shall discuss the features with which the architect should be familiar in order to specify the characteristics for sound-amplification systems in the buildings he designs.

Uses of Sound Amplification

Among the purposes for which sound-amplification systems are useful are: [1]

(1) *To increase the sound level in large rooms and open-air theaters.*

(2) *To provide paging and announcing facilities.* Many industrial buildings, offices, hotels, schools, railroad stations, and other buildings require sound-amplification systems for paging, announcements, etc. In spaces where such systems are used, the room noise should be kept as low as possible, and the reverberation time should approximate the optimum value within reasonable limits. The loudspeakers should be operated at levels just

[1] A reference book covering many aspects of sound-amplification systems is *The Architect's Manual of Engineered Sound Systems,* Radio Corporation of America (1947); also see *Radio Manufacturers Association Engineering Bulletin No. 39* (1949).

barely loud enough to permit distinct hearing. Several loudspeakers, properly distributed and operated at a low level, may be better for some installations than one or two operated at a high level. The disturbing or annoying effects of paging and announcing systems can be reduced by the use of an automatic volume control which prevents an inexperienced or loud-mouthed announcer from blasting the listener out of his chair. In a room with *excessive reverberation,* the intelligibility of amplified speech can be increased by suitable suppression of the low frequencies. At frequencies below about 800 cycles, the amplification should drop off at a rate of about 6 db or more per octave, depending on the reverberation characteristics of the room.

(3) *To provide radio and recorded programs.* Such programs are used in schools, factories, hospitals, offices, hotels, restaurants, recreational rooms, and elsewhere for entertainment, instruction, and enhancement of morale, and for therapeutic purposes.

(4) *To increase sound level on the stages of theaters and auditoriums.* It is not unusual for singers to complain that the orchestra cannot be heard well on the stage. Anyone who has been seated on a stage ten feet or more behind a speaker, in an auditorium equipped with a sound-amplification system, knows how difficult it is to hear in such a location—loudspeakers are placed and directed to "cover" the audience area, but not the stage area. The need for a higher sound level on the stage usually cannot be met by increasing the acoustical power from the loudspeakers in their conventional locations, or by directing them toward the stage floor, as such procedures introduce possibilities of "excessive feedback" that may give rise to howling from the loudspeakers. One possible solution is to use directional high-frequency loudspeakers on the stage, directed toward those stage areas where higher levels are required, but not toward the microphones.

(5) *For motion picture theaters.* See p. 318.

Single-Channel Sound-Amplification Systems

Figure 14.1 is a simplified block diagram showing the essential parts of a single-channel sound-amplification system. It consists of one or more microphones, all connected to a *single* amplifier channel, which in turn operates one or more loudspeakers. The microphones are located at positions near the actual sources of sound. A good sound-amplification system is characterized by uniform and adequate coverage, low noise level, and negligible distortion. A system with these characteristics properly maintained and operated, in a large room with no acoustical defects, can amplify speech so naturally that the audience will hear perfectly almost every word without being conscious that the speech has been amplified by artificial means.

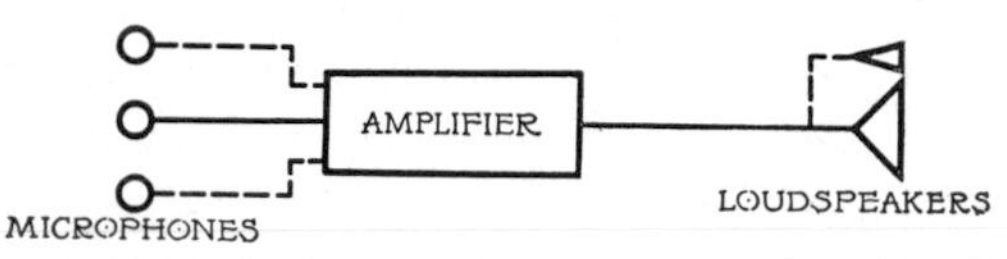

Fig. 14.1 Simplified block diagram showing the essential parts of a single-channel sound-amplification system.

MICROPHONES. The primary characteristics of microphones that determine their suitability for use in a sound-amplification system are ruggedness, size, "impedance," sensitivity, directivity, and frequency response. The sensitivity should be as high as possible, other characteristics being equal, to reduce the gain requirements of the associated amplifier. In most installations a certain amount of directivity is desirable to reduce the pick-up of extraneous sounds and noises. A frequently used microphone directional characteristic is the cardioidal pattern shown in Fig. 14.2(*b*), which indicates a pick-up angle of about 120° in front of the microphone and relatively little pick-up at other angles. Such a microphone is especially desirable for stage use, as it discriminates against noise that originates in the audience behind the microphone, it reduces the effects of reverberation by decreasing the ratio of reverberant sound to direct sound picked up, and it reduces the pick-up of amplified sound from the loudspeakers, thereby lessening the danger of "feedback" howling. Finally, the microphone should respond nearly equally well to all the useful audio frequencies, so that the electrical waves it generates are faithful facsimiles of the sound waves.

AMPLIFIER AND CONTROL CIRCUITS. The primary factors to be considered in selecting the amplifier for a sound-amplification system are power output, gain, frequency response, freedom from hum and noise, low distortion, and mechanical construction. The acoustical power required for the satisfactory *reproduction*

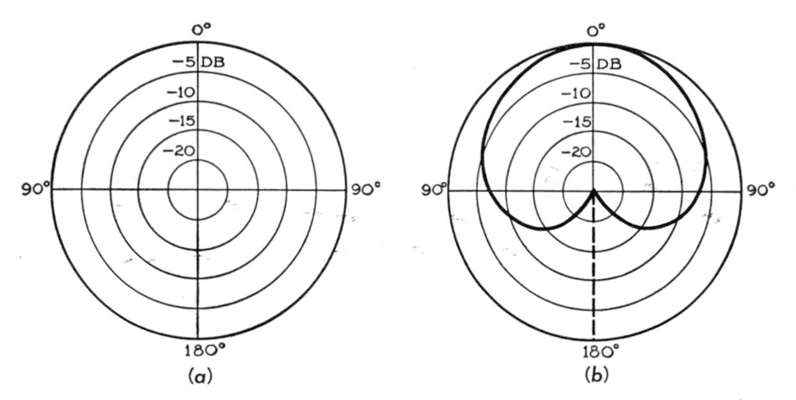

Fig. 14.2 Directivity patterns showing how the sensitivity of a microphone varies with angular direction in a horizontal plane through the microphone. (*a*) A non-directional characteristic, indicating equal sensitivity in all directions. (*b*) The cardioidal characteristic of a Western Electric 639A microphone. (Courtesy Western Electric Co.)

of speech in rooms with volumes of 50,000 to 500,000 cubic feet is approximately equal to

$$\text{Required acoustical power} = \frac{V}{10{,}000{,}000} \text{ watts} \qquad (14.1)$$

where V is the volume in cubic feet. The acoustical power required for the reproduction of music usually requires 30 to 300 times that value given by Eq. (14.1). The *undistorted* electrical power output capacity of the amplifier exceeds this by an amount determined by the efficiency of the loudspeakers. The amplifier must have sufficient gain (amplification) to yield this required power output when used with the particular microphones selected. Thus a less sensitive microphone requires an amplifier of higher gain to achieve the same power output to the loudspeakers. Ordinarily an amplifier with somewhat more than sufficient gain is selected, and a gain (volume) control provides the possibility of compensation for different microphones as well as many other variable factors. For faithful, undistorted reproduction, the amplifier should amplify all the useful audio frequencies nearly equally. In some cases, however, a nonuniform frequency response is required. For example, suppression of the low frequencies is desirable for the reinforcement of speech in rooms that are excessively reverberant at low frequencies, as it increases the intelligibility of the speech and reduces the possibility of acoustic feedback (howling). Many amplifiers provide equalizers (tone controls) for selectively altering the low-frequency or high-frequency response characteristics of the system. Finally, the amplifier should not produce an objectionable level of residual hum or other noise. All amplifiers yield a small amount of background noise, but with proper design this can be reduced to a negligible level.

The amplifier control circuits required depend entirely on the particular installation and its anticipated uses; they may include such elements as power, microphone, and loudspeaker switching circuits, gain controls, volume indicators, circuits to mix the outputs of several microphones, and tone equalizers. In small installations these elements may be an integral part of the amplifier itself. In larger auditoriums the control circuits may be located remotely from the rest of the sound-amplification equipment, preferably at an accessible monitoring location in the main seating area.

LOUDSPEAKERS. The quality of performance of a sound-amplification system is more likely to be limited by its loudspeakers than by any of its other components because of cost considerations. Too frequently a sound system has good microphones, good amplifiers, but loudspeakers of poor quality. *A sound system is no better than its loudspeakers.* The three major factors to be considered in selecting loudspeakers are power-handling capacity, frequency response, and spatial distribution of the sound produced. The loudspeakers generally used in sound-amplification systems can be classified as (1) the *horn type,* in which a speaker "driver" unit is terminated with a horn, (2) the *direct radiator type,* usually mounted in a baffle or cabinet, and (3) a combination of (1) and (2). Figure 14.3 is a photograph of a combined unit. The low-frequency direct radiator is shown enclosed in the cabinet; the high-frequency horn is mounted on the cabinet.

The acoustical power required for the satisfactory reproduction

Fig. 14.3 Photograph of a low-frequency loudspeaker unit mounted in a cabinet and a high-frequency horn-type loudspeaker mounted on the cabinet. (Jensen Mfg. Co.)

of sound in rooms is given in Eq. (14.1). Since the efficiency of most loudspeakers in converting electrical energy into acoustical energy is very small, the electrical power supplied to the speakers must be many times these values.

A common defect in many loudspeakers is an excessive response at one or more bands of frequencies and a deficient response at others; serious distortion is introduced by such a condition. Many loudspeakers also introduce spurious frequencies in the reproduced sound. Another undesirable characteristic of most loudspeakers is the tendency to concentrate the high frequencies in a cone along the axis of the radiator. The quality of reproduced sound from such loudspeakers deteriorates appreciably for listeners outside the beam.

In order to have high quality reproduction of sound, it is necessary to provide complete coverage of the entire seating area of the auditorium. In large rooms, this requires the use of groups of loudspeakers designed and arranged to radiate the high frequencies uniformly to all sections of the seating area where amplification is needed. The angle of divergence between the axes of two adjacent horns should be such as to avoid noticeable "dips" or "bumps" in the sound field produced by the horns.

In an auditorium, a single location for the loudspeakers (of a single-channel system) is usually preferable to two or more locations. It is highly desirable to *preserve the illusion that the sound comes from its original source,* and *not from the loudspeaker to the right or the one to the left.* A single location for the speakers, as near the source as possible, helps to preserve this illusion, though such a location may increase the possibility of feedback and may add to the difficulty of obtaining complete coverage for the entire seating area, especially in wide rooms. In order to obtain a good coverage of the entire seating area the loudspeakers are usually elevated. Fortunately one's ability to localize sounds in the vertical direction is rather poor. Proscenium openings present difficulties in working out the most favorable location for loudspeakers, as the best height is usually 10 to 15 feet above the stage or platform floor. When a single loud speaker or a single group of speakers is used, it should be centrally located, just above the proscenium opening. From this standpoint it is desirable to keep the proscenium opening low. In factories, restaurants, and large reverberant churches, the use of many low-level loudspeakers properly distributed over the listening area may be preferable to placing them all in a single location.

Stereophonic Sound Systems

Suppose that an actor is talking on the stage of an auditorium. A listener in the audience can close his eyes and determine, with reasonable accuracy, the position of the actor on the stage. This is called *lateral* or *angular* localization. Speech of a performer on the right side of the stage seems to come from the right. On the other hand, the *vertical localization* of sound sources is rather inaccurate. It is fortunate from the standpoint of amplification systems that localization is so poor in the vertical direction. As a result, fewer loudspeakers are needed to preserve the spatial distribution of the true sources, and they need not be placed in the horizontal plane of these sources. The loudspeakers can usually be somewhat higher than the stage and yet give the audience the illusion that the amplified sound is coming from its usual position on the stage.

When sound originating at several locations on a stage is amplified by a single-channel amplification system, the spatial character of the original sound is almost entirely lost. The listener usually tends to identify the loudspeaker instead of the original source as the source of the sound. For this reason, when such a system is used, a conflict arises between the aural and visual senses; for example, in a legitimate theater the eyes see an actor move across the stage, but the ears hear his voice coming from a fixed location. It is probable that this conflict between what the eye sees and what the ear hears is a prime factor among those responsible for the lack of greater acceptance of sound amplification in the legitimate theater in spite of the acute need for a higher sound level. Even when a large auditorium has been well designed, there is need for amplification—decibels cannot be "squeezed" out of a stage whisper.

The sounds that reach each of the two ears of a listener in an auditorium are different. These differences enable the listener to localize the direction of the sound source. Steinberg and Snow [2] have shown that angular localization depends largely on the loudness difference of the sound reaching the two ears. Since this difference is greater at the higher frequencies than at the lower ones, high-frequency sounds are much easier to locate. Because the loudness difference at the two ears depends on the frequency of the sound, the *quality* of a complex sound is not the same at the two ears, and this difference aids in auditory localiza-

[2] J. C. Steinberg and W. B. Snow, *Bell System Tech. J.*, **13**, 245 (1934).

tion. Difference of arrival times at the two ears also plays a vital part in auditory localization.[3]

In order for sound to be more perfectly preserved in reproduction, its spatial character should be maintained. Amplified sounds should create the illusion of coming from their true sources. Although it would require a very large number of microphones, amplifiers, and loudspeakers to do a perfect job of such sound reinforcement,[4] good auditory perspective can be obtained if as few as three single channels are employed. This requires the use of multiple-channel systems to give the listeners' two ears the necessary differences of sound quality and arrival time. A system of this type, which reproduces sound so that it tends to maintain its spatial character, is called a *stereophonic system.*

Before a stereophonic sound-amplification system is planned, technical advice should be sought to obtain the best possible layout of equipment. Figure 14.4 gives simplified block diagrams showing two practical arrangements. Of the two, Fig. 14.4(*a*) is the more desirable system. All components of the system should be of high quality. The microphones may be directional so they will discriminate against audience noise in favor of sound on the stage and give a higher ratio of direct to reverberant sound. However, the angle of response toward the stage should be broad, as for example in the cardioidal pattern of Fig. 14.2(*b*). If the microphones are highly directional, the reproduced sound may appear to *jump* from one loudspeaker to another as the performer moves across the stage. A high ratio of direct to reverberant sound is desirable because auditory perspective depends, to a large extent, on the relative loudness of the *direct* sound that strikes the microphones. For this reason, microphones in stereophonic systems are sometimes placed nearer the action than they would be in single-channel systems. Each loudspeaker should provide coverage for the entire audience and should be mounted at approximately the same height, about 10 to 15 feet above the stage; the side loudspeakers should be

[3] K. deBoer, *Philips Tech. Rev.*, **5**, 107 (1940); **8**, 51 (1946).

[4] See "Auditory Perspective Basic Requirements," H. Fletcher, *Elec. Engr.*, **9**, 53 (1934).

close to the edge of the proscenium. Reproduction from a system using a center "phantom" channel, as in Fig. 14.4(b), is not quite as realistic as that from a system employing three separate

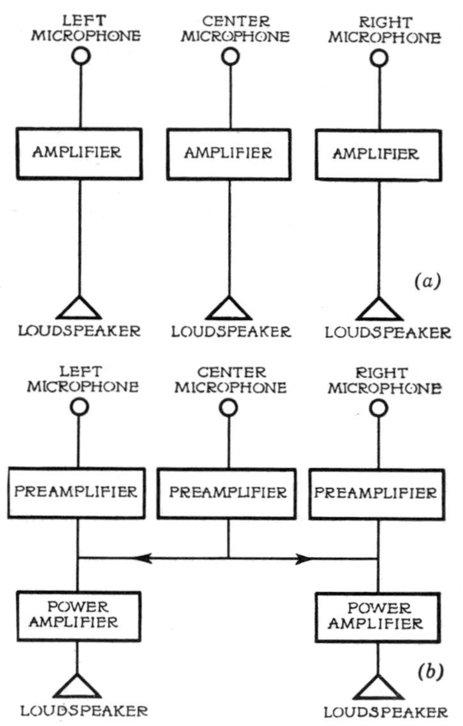

Fig. 14.4 Simplified block diagram showing two possible arrangements for components of a stereophonic sound-amplification system. (*a*) Three single-channel systems. (*b*) Two single-channel systems plus one "phantom" center channel. The electric currents from the center channel move only in the directions indicated.

channels, as in Fig. 14.4(a). However, a phantom channel system is very useful in an installation where a loudspeaker cannot be mounted conveniently above the center of the stage. Also, it is less expensive than the three-channel system. A commercial installation utilizing a phanton center channel is shown in Fig. 14.5.[5] This stereophonic system provides acoustical cover-

[5] A. W. Colledge, *Western Electric Oscillator,* **4**, 11 (1946).

Fig. 14.5 A stereophonic sound-system installation at Pitt Auditorium, Pittsburgh. The squares indicate the positions of the loudspeakers; the circles indicate the positions of the microphones. (Courtesy Western Electric Co.)

age for 8000 persons in an open-air theater. It illustrates the practicability of providing sound amplification to supply the required additional acoustical power for satisfactory listening conditions in a legitimate theater while maintaining the illusion of the spatial distribution of the original sound sources.

15 · Auditoriums

In the preceding chapters, we have discussed the basic principles and general procedures of design that determine the acoustical properties of all types of buildings. In this chapter and those that follow, we shall show how these principles and procedures can be applied to obtain the best possible acoustics in theaters, auditoriums, school buildings, churches, civic buildings, offices, factories, hospitals, residences, and radio studios. The present chapter describes practical aspects of designing that lead to good acoustics in legitimate, motion picture and "little" theaters, and in school and civic auditoriums.[1]

Historical Development of the Auditorium

Although the early Egyptians probably used the courts of their temples as "auditoriums" where ceremonial services were seen and heard by large audiences, auditoriums as we know them evolved from classical Greek and Roman open-air theaters. The first forms of enclosed theaters bear a close resemblance to the Roman type. Thus, the Theater of the Olympian Academy, at

[1] The occasions for designing very large concert halls and opera houses occur so rarely in the practice of the general architect, and the acoustical problems are so specialized, that it does not seem appropriate to include a detailed discussion of these problems in this book. Aspects of the acoustics of concert halls and opera houses, and some examples of good design, will be found in V. O. Knudsen, *Architectural Acoustics,* John Wiley & Sons (1932), pp. 540–555.

Vicenza, Italy (sixteenth century), was essentially of the classical Roman design, except that a roof and side walls were added. It was soon learned that the reflection of sound from the scenes could contribute to conditions for good hearing. The doorway through the rear wall of the Roman stage developed into the proscenium opening. Then the walls and splays of the proscenium were utilized to reflect sound to the audience. The seating area was elongated into a U-shape which was well adapted to accommodate multiple balconies. These provided the means for "lining" the walls of the theater with an audience, which is a highly absorptive treatment; hence, there was little danger of excessive reverberation, and serious echoes from the walls were improbable. In the San Carlo Theater of Naples, in the Drury Lane of London, and in the Burgtheater of Vienna—theaters with three or four balconies—the audiences were seated relatively near the stage, in quiet, and not excessively reverberant, auditoriums. Many such theaters provided an almost ideal environment for listening. When the actors practiced the art of elocution, as they usually did in those times, the theater was a place worthy of the name "auditorium." During the eighteenth century, the introduction of domes, the elimination of multiple balconies, and the use of concrete and hard wall plaster did much injury to the acoustics of theaters. Acoustical difficulties multiplied but were ignored. Pleasing the eye and the fancy of the public took precedence over providing good acoustics.

In the nineteenth century scientists began to appreciate the nature of the acoustical problems in auditoriums, but it was not until the advent of the present century that physicists and engineers undertook systematic researches in room acoustics. The principles of acoustics have now become well established; sound has been united with light on the cinema screen. Good acoustics has become a prime requirement in all auditoriums. Community and college theaters are in vogue, and their number is increasing at an extraordinarily high rate. According to a recent report there is a prospect of 4,500,000 students in the colleges of the United States in 1965, compared with 2,500,000 in 1949 and 1,500,000 in 1940. Many colleges will want a little theater as well as a larger auditorium, and so will many high schools. Similarly, civic auditoriums, forums, and recreational halls are objectives of nearly all communities. The public expects and deserves good acoustics in all these buildings. Architects can undertake these tasks of design with confidence, if they will follow the procedures exemplified in this chapter.

Planning the Auditorium

In working out the acoustical plans for an auditorium, the architect should take the steps enumerated in the first part of Chapter 9. Specifically, these four steps are necessary:

(1) Examine the site with respect to *noise*. The information gained from a study of existing and prospective sources of noise in the neighborhood of the site is necessary to determine the nature of sound insulation that must be incorporated in the building to prevent the noise level in the finished auditorium from exceeding the maximum tolerable level—30 to 40 db. It is also necessary to reduce noises originating within the building so that the composite noise level from all sources does not exceed the appropriate tolerable level.

(2) Limit the *size* of the auditorium.

(3) Design the *shape* of the auditorium in accordance with principles described on pages 181 to 191, so that a plentiful flow of direct and beneficially reflected sound will reach all auditors, and so that the audience area will be free from echoes, flutters, and sound foci.

(4) Provide the *optimum reverberation* in all parts of the auditorium throughout the range of frequencies that are important for speech and music. The curves of Figs. 9.11 and 9.12 should be used for determining the optimum reverberation time *vs.* frequency characteristic.

The Little Theater

In the little theater the architect has an opportunity to design a structure that will embody the highest attainable standards of acoustics. If the seating capacity is limited to 300, the volume of the auditorium should not exceed 50,000 cubic feet. All seats are located on one floor, the rear portion of which should have a steep slope (see p. 184) so that auditors will have good sight lines and good sound lines in all parts of the auditorium.

The ceiling should not be more than 20 feet high and should be left smooth. A material with a highly reflective finish will serve to direct the sound toward the rear seats. Diverging walls are desirable but not so necessary as in larger theaters. The lower 6 to 8 feet of the side walls should be of reflective material. The front portions of these walls should not be pierced with boxes.

The rear wall should *not be concave.* If this wall is a probable source of echoes or interfering reflections (as it may be if it is far enough from the stage), it should be treated with highly absorptive material and broken with deeply inset doors, hangings, or other ornaments that will minimize the effects of long-delayed reflections and will facilitate proper diffusion. If additional acoustical treatment is required to impart to the room the optimum reverberation characteristic, it should be applied to the upper portion of the side walls in non-uniform strips, panels, or patches.

The chairs should be upholstered with absorptive cloth, such as mohair, over deep, porous padding. The absorption of each chair should be 3 to 4 square-foot-units (sabins) at all frequencies above 512 cycles, and 2 to 3 sabins at frequencies of 128 and 256 cycles. The reverberation time of the theater will then be nearly independent of the size of the audience. Even during rehearsals the reverberation will be close to the optimum value.

The benefits associated with the small volume (less than 50,000 cubic feet for the audience space) should not be nullified by making the stage recess so large that the sound is dissipated before it reaches the seating area. The volume of the stage should be reduced to a minimum consistent with other requirements. Stage settings with rear, side, and overhead reflective surfaces should be designed to confine the sound to a small volume and reflect it to the audience. The use of plywood flats or heavily painted and back-painted canvas flats is advantageous for the ceiling as well as for the side and rear walls of the stage set. Designers of stage sets should be instructed to recognize these pertinent requirements for good acoustics, which are especially necessary when the stage is large.

The stage floor should be elevated as much as possible, but it should provide also good sight lines from all seats; this usually will allow an elevation of about 42 inches above the front level portion of the main floor. Orchestra pits should be avoided whenever possible; if indispensable, it is advisable that they be covered with a sound-reflective apron (plywood or heavily painted canvas) when not in use.

The optimum times of reverberation for the auditorium in a little theater with a volume of about 40,000 cubic feet (the volume of stage not included) are approximately 1.5 seconds at 128 cycles, and 1.0 second at 512 to 4096 cycles. These values are a compromise between those given by the curves for "speech" and "average music" in Fig. 9.11.

The exclusion of both outside and indoor noise should receive study whether the site is quiet or noisy. The average level of noise in the unoccupied auditorium should not exceed 35 db, and if the highest standards of acoustics are required this level should be reduced to 30 db. This reduction will necessitate considerations of noise conditions at the site and calculations of sound insulation such as have been described in Chapter 11. Suitable wall and ceiling structures and entrance and exit doors must be chosen to provide adequate protection against outside noise. Corridors, promenades, lobbies, and vestibules should be interposed between the auditorium and probable sources of outside or indoor noise. These interposed spaces should be treated with highly absorptive material, not only to reduce outside noise but also to exclude the possibility of feedback of reverberant sound from these "coupled spaces" to the auditorium. About one half of the wall and ceiling surfaces should be covered with highly absorbent materials, and the floors should be carpeted over heavy felt pads.[2] The combined effects of these noise-reduction measures should reduce the average noise level in the theater to 35 db. Thus, if the outside noise transmitted into the auditorium has a level of 32 db, the noise from inside origin—such as that from the ventilating and other equipment, and from diverse activities in the building—must not have a level of more than 32 db so that the level of the combined noise, from outside and inside, will not exceed 35 db. (See Fig. 10.1, which shows that, when two "ran-

[2] Vestibules and similar spaces, with highly absorptive interior boundaries, are sometimes referred to as *sound locks*. When properly designed, they can be very useful for sound-insulation purposes.

dom" noises of the same level are added, the combined noise level is 3 db greater than the level of either component.) A noise level of 35 db will not be noticed by an audience except during intervals of unusual dramatic suspense; for example, when the audience listens in rapt silence to a faint whisper.

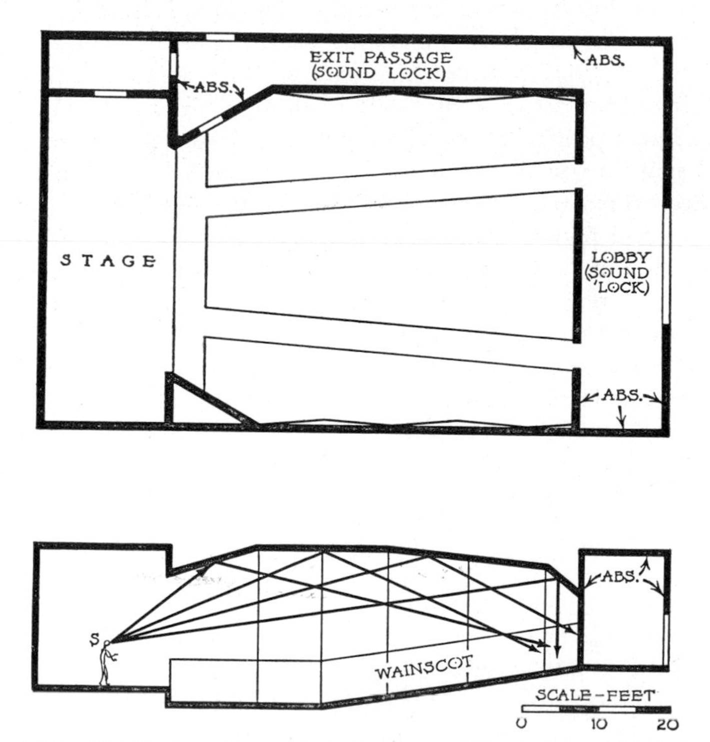

Fig. 15.1 Sketch of a plan and section for a little theater which is based on the requirements for ideal acoustics.

A sketch of a plan and section for a little theater, based on a study of the requirements for ideal acoustics, is shown in Fig. 15.1. The auditorium is isolated by two walls on the sides that are adjacent to streets. These walls, in combination with the promenade and lobby which they enclose (and which, with heavily carpeted floors, act as sound locks), provide an average transmission loss of at least 60 db. This amount of insulation is sufficient to exclude all ordinary sounds, even in a noisy metropolitan locality. A transmission loss of not less than 60 db is required for the ceiling, especially if air traffic is a present or potential source of noise. If the theater is located on a quiet site, it is unnecessary to provide this much insulation for walls or ceiling. Where the site is subject to excessive earth vibrations, such as result from near-by bus or trolley lines, it is advisable not only to provide double walls, but also to isolate the inner walls of the auditorium from the earth. The splayed walls and ceiling of the proscenium, the flat ceiling of the auditorium, and the lower portion of the side walls are designed to reflect useful sound upon the audience and accordingly are finished with reflective materials (suspended plaster ceiling, furred-out plaster walls, and plywood wainscot applied to randomly spaced wood strips). The aisles are covered with cork carpet in order to reduce the noise of footfalls, and the floor under the seats is covered with linoleum to reduce the noise of scuffing feet.

If the theater has a seating capacity of 270, a volume V of 40,000 cubic feet, and a total interior surface area S of 7500 square feet, then, by the use of Eq. (8.12), the total absorption required to give the optimum reverberation time is calculated as shown in step A of Table 15.1. [Equation (8.17) should be used for larger auditoriums or for higher frequencies than those considered here.] We shall now determine the amount of additional absorptive material, if any, that should be installed if the room is to have the optimum reverberation time. This is done by computing the amount of absorption in the room without any special absorptive treatment; this value is then subtracted from the total required absorption of step A, the difference being the addition absorption that should be furnished. We shall carry out these calculations by first assuming that unupholstered chairs are used and then comparing the results with similar computations based on the assumptions that upholstered chairs are used.

In Table 15.1, the absorption furnished by the wall surfaces is tabulated in step B. If we add these values to those of step C, which gives the absorption supplied by chairs and audience, we find that the total absorption in the room, in the absence of any absorptive treatment, is: 862 square-foot-units (sabins) at 128 cycles; 1130 sabins at 512 cycles; and 1265 sabins at 2048. Hence, subtracting these values of absorption from corresponding values in step A, we find that the following amounts must be added to provide the optimum reverberation time at these frequencies: (1200 − 862) = 338 sabins at 128 cycles; (1725 − 1130) = 595 sabins at 512 cycles; and (1725 − 1265) = 460 sabins at 2048 cycles.

TABLE 15.1

A. Required Absorption

	128 Cycles	512 Cycles	2048 Cycles
Optimum reverberation time in seconds	1.5	1.0	1.0
$-2.3 \log_{10} (1 - \bar{\alpha})$	0.174	0.261	0.261
$\bar{\alpha}$	0.160	0.230	0.230
Total square-foot-units of absorption required = $S\bar{\alpha}$	1200	1725	1725

B. Absorption Furnished by Wall Surfaces

Absorptive Material	128 Cycles		512 Cycles		2048 Cycles	
	Abs. coef.	Abs., in sq-ft-units	Abs. coef.	Abs., in sq-ft-units	Abs. coef.	Abs., in sq-ft-units
Cork carpet, 380 sq ft	0.04	14	0.05	19	0.05	19
Linoleum floor, 2000 sq ft	0.04	80	0.04	80	0.04	80
Ceiling, 2000 sq ft	0.05	100	0.06	120	0.06	120
Wood wainscot, 1060 sq ft	0.06	64	0.06	64	0.06	64
Proscenium opening, 450 sq ft	0.30	135	0.40	180	0.50	225
Stage wall, 430 sq ft	0.05	21	0.06	26	0.06	26
Rear wall, upper side walls 1230 sq ft	0.05	61	0.06	74	0.06	74
Total absorption from above required materials	475		563		608	

TABLE 15.1 (*Continued*)

C. Absorption Furnished by Unupholstered Chairs and the Audience

	128 Cycles	512 Cycles	2048 Cycles
	Abs., in sq-ft-units	Abs., in sq-ft-units	Abs., in sq-ft-units
90 unupholstered chairs	27 (0.3 per chair)	27 (0.3 per chair)	27 (0.3 per chair)
180 auditors in unupholstered chairs	360 (2.0 per person)	540 (3.0 per person)	630 (3.5 per chair)
Total absorption by chairs and audience	387	567	657

D. Absorption Furnished by Upholstered Chairs and the Audience

	128 Cycles	512 Cycles	2048 Cycles
	Abs., in sq-ft-units	Abs., in sq-ft-units	Abs., in sq-ft-units
90 upholstered chairs	180 (2.0 per chair)	270 (3.0 per chair)	270 (3.0 per chair)
180 auditors in upholstered chairs	540 (3.0 per person)	810 (4.5 per person)	900 (5.0 per person)
Total absorption by chairs and audience	720	1080	1170

It is possible to select from the tables of Appendix 1 material which, if properly mounted, would furnish this required amount of additional absorption. Even then, however, desirable acoustical conditions would not prevail, for there would be a large difference between the reverberation times for the relatively empty theater and the full theater. This situation can be remedied by replacing the chairs with upholstered ones. Suppose they have the absorption listed in step D. Then, adding the absorption in D to the absorption supplied by the wall materials of step B as above, we find that the following amounts of absorption should be added: (1200 − 1195) = 5 sabins at 128 cycles; (1725 − 1643) = 82 sabins at 512 cycles; and (1725 − 1778) = −53 sabins (that is, subtract 53 sabins) at 2048 cycles; these values are negligible.

Hence, the analysis in Table 15.1 leads to the conclusion that no special absorptive materials are needed in this theater in order to provide the optimum reverberation times if upholstered chairs are used; the total absorption furnished by the audience, chairs, and the indicated materials for the walls, floor, and ceiling does not differ more than 5 per cent from total required absorption. This desirable outcome follows from the choice of a small volume per seat for the theater (148 cubic feet) and proper furnishings, including the highly absorptive upholstered chairs. If a much larger volume per seat had been used, as is customary in the design of many theaters, it would have been necessary to add special absorbents to the walls or to the ceiling or to both. This should be avoided if possible because an increase of absorption would result in an unavoidable decrease in the average sound level of speech in the theater [see Eq. (8.6)]. Furthermore, when most of the required absorption is furnished by upholstered chairs, as it is in the theater here considered, the reverberation time is relatively independent of the size of the audience—a highly desirable condition.

It is necessary to control the reverberation on the stage. Reverberation times of the same values as those specified for the auditorium will be satisfactory. This condition is usually closely approximated if the stage is not excessively large and is equipped with a full set of flies and with side, rear, and overhead hangings of velours or monk's cloth. The hangings should be hung with deep folds, at least 100 per cent being allowed for gathers. This will contribute to the absorption at low frequencies. The reverberation times should be calculated by the procedure followed in Table 15.1. If additional absorption is required, it can be furnished by treating the upper walls or ceiling of the stage with a suitable type and amount of fiberboards or other absorptive material.

A small, enclosed set, made of reflective material (such as

pressed fiberboard, plywood, painted canvas) and placed forward on the stage, affords the best means of projecting sound to the audience, and its use should be encouraged whenever feasible. Some tests conducted in one rather large theater show the importance of having the action take place on the front of the stage, and consequently of having an enclosed set to direct the speech to the audience. With a speaker on the front part of the stage, the syllable articulation of speech was 85 per cent in the balcony. When the speaker moved to the rear part of the stage (with an open setting on the stage) the articulation was reduced to 60 per cent.

The intelligibility of speech in the little theater described in this section will be excellent. In general, it will not be necessary for the actors to raise their voices. They can therefore give their entire thought and feeling to the dramatic expression that will best portray the lines of the play, and they can act with the assurance that every word will be heard by the audience.

The Legitimate Theater

In this section, consideration will be given to legitimate theaters that are larger than the one described in the preceding section. Although the same general principles of design apply here with equal relevance, there is one important point of difference. In legitimate theaters, because of their larger size, speech is at a lower sound level than it is in little theaters. In fact, it frequently is not loud enough for good audition.

Therefore it is of the utmost importance to design the shape of the auditorium so that it will provide the audience with the greatest possible amount of direct and of beneficially reflected sound. The divergence of the side walls, the slope of the overhead proscenium splay, and the slope of the main ceiling of the auditorium should be carefully designed to reinforce the sound propagated to the audience, and some preferential reflection of sound should be provided for the rear seats under and in the balcony.

It is good acoustical design to keep the balcony overhang (depth) less than twice the height of the balcony opening, and to keep the balcony soffit reflective and inclined downward toward the rear wall. Heavily upholstered chairs, carpets on the aisles, and such absorptive treatment of the rear wall as is required to prevent objectional reflections ordinarily provide satisfactory reverberation characteristics in the balcony recess. When the ratio of depth to opening height does not exceed 2, this space can be regarded as an integral part of the auditorium, and it then is not necessary to make separate calculations of reverberation in the two spaces—the main part of the auditorium and the balcony recess. In routine calculations of reverberation time, it is customary to regard these two spaces as one single volume and the stage recess as another. It is important that the stage have approximately the same reverberation characteristic as the auditorium. For such computations, the coefficients of absorption in Table 15.2 are applicable to the stage opening. If

TABLE 15.2

Absorption Coefficients for Stage Openings

Frequency in cycles	128	512	2048
Absorption coefficient	0.30	0.40	0.50

a theater is to be used for musical as well as dramatic productions, the reverberation characteristic should be based upon the requirements for both speech and music, and the absorptive materials should be carefully located to favor a uniform average rate of decay in all parts of the theater.

Calculations such as have been described in Chapter 11 should be made to determine the insulation required against outside noise. Noise of inside origin, such as that from ventilating equipment, should be suppressed.

Theaters which are circular or elliptical in plan, or theaters with domed or cylindrical ceilings, are *especially likely to give difficulty,* although when the curved surfaces are well broken or coffered, or treated with a material having a coefficient of absorption in excess of 0.70, the difficulty due to the curved surfaces may be largely overcome. For example, consider the acoustical treatment that was applied to one theater having a circular plan with the stage opening forming a part of the cylindrical walls. If the walls of this theater had been finished with reflective materials there would have been a very pronounced focusing of sound at a point about half way between the rear and the center of the auditorium. To overcome this anticipated defect, two layers of 1½-inch fibrous blanket, separated by a 2-inch air space and covered with a perforated membrane, were applied to the side and rear cylindrical walls in the form of a band extending in length approximately three fourths of the way around the interior, and in height from about 3 feet up to 16 feet from the floor. This expedient gave satisfactory results. The front portions of the side walls were finished with hard material to give helpful reflection toward the central and rear parts of the theater. Another method of handling the problem of large concave surfaces is to divide them into secondary convex surfaces as in Figs. 9.7 and 9.8. Although it is possible to treat concave surfaces in such a way as to overcome most of the acoustical defects they produce, it is preferable to use forms that are free from undesirable curvatures. This does not mean that concave surfaces are always to be avoided, but if they are employed without careful consideration they may lead to disastrous results.

Figure 15.2 shows an acoustical study of a longitudinal section of a theater in which the ceiling surfaces have been designed to reinforce sound by reflection. The overhang of the balcony is short, and the opening under the balcony is high; therefore adequate sound will reach the rear seats under the soffit. These seats, which are usually the poorest ones in most theaters, are further benefited in this design by the reflections of sound from both the splayed walls and the ceiling of the proscenium. The main part of the ceiling has a gently rising slope which provides the most favorable reflection of sound. Heavily upholstered chairs are used throughout, and the aisles are carpeted over a ½-inch carpet pad. Most of the absorption required to provide the optimum reverberation is applied to the rear wall, under and above the balcony, to prevent echoes and interfering reflections from these surfaces. A 2-inch or 3-inch mineral-wool blanket covered with perforated plywood, or similar facing, is suitable here. The highly absorptive material should not extend below the height of the heads of the audience. Below this level the rear wall is paneled wainscot, which with the similar side wall wainscot provides much of the required low-frequency absorption. Calculations similar to those described in Table 15.1 [using Eq.

(8.17) for the computation of reverberation times] should be made to determine the kind and amount of additional absorptive material, if any, required to give the optimum reverberation characteristic. The directions and procedures outlined in the preceding section for sound insulation and for other acoustical aspects of the little theater also apply to the larger legitimate theaters.

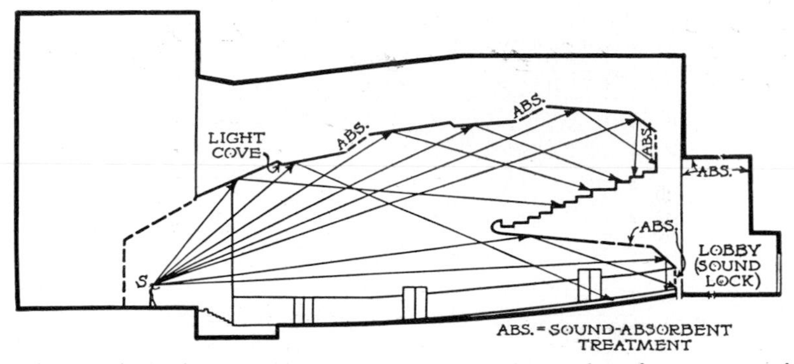

Fig. 15.2 Longitudinal section of a legitimate theater based on acoustical requirements.

An enclosed set, like that indicated by the dotted lines of Fig. 15.2, should be used whenever possible. The set reflects toward the audience a large and much-needed amount of sound which originates on the stage, and which would otherwise reverberate and be lost by absorption in the upper part of the stage. Sets with parallel side walls may cause flutter echoes. On the other hand, the side walls should not diverge too rapidly. If they splay outward too much the performers on a large stage may have difficulty in hearing each other.

Motion Picture Theaters

The general considerations of shape, discussed in detail in Chapter 9, apply to motion picture theaters. Furthermore, certain admonitions given in that chapter are especially pertinent here. For example, concave rear walls, parallel side walls, parallel ceiling and floor, and surfaces that give long-delayed reflections in the seating area should be avoided. Long, narrow theaters often have very poor acoustics; they are likely to require so much acoustical power from the sound system, in order to give adequate sound level in the rear seats, that the loudness will be excessive in the front and central seats.

Because sound is reproduced in motion picture theaters by means of electro-acoustical equipment that can furnish adequate sound levels in all parts of even very large theaters, the acoustical design of the cinema is not so dependent on beneficial reflections

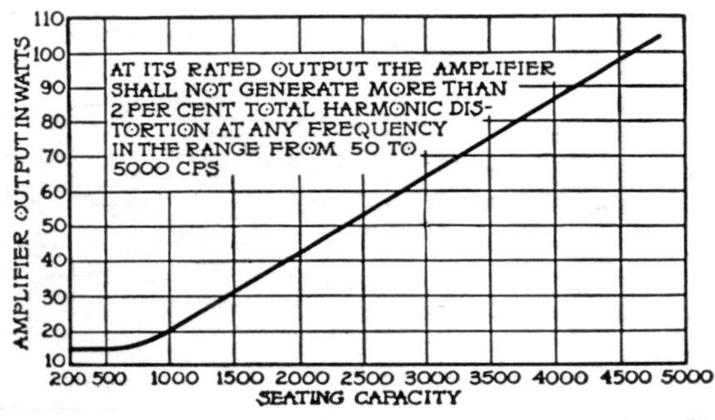

Fig. 15.3 Minimum requirements for the amplifier power-handling capacity in the sound system of a motion picture theater. (Motion Picture Research Council.)

from the walls, proscenium, splays, and ceiling as is the design of the legitimate theater. The average sound level of speech in the cinema is usually about 65 db for dialogue, which is 10 to 15 db higher than the average unamplified speech level in the legitimate theater. The acoustical power required to maintain this level depends on the size of the theater. The minimum power requirements of the amplifier that actuates the loudspeakers is given in Fig. 15.3. The amplifier rating is shown as a function of the seating capacity of the house.[3] For example, a theater seating 1000 should be equipped with an amplifier having a power output of at least 20 watts.

Lengths greater than about 150 feet should be avoided in

[3] A volume per seat of 125 cubic feet is assumed. Even if the volume per seat is somewhat higher, the values given by the chart can be used, since these data include a safety factor which takes this into account. It also allows for differences in loudspeaker efficiency.

order to prevent a noticeable delay in the arrival of the sound to persons in the rear of the theater. It requires about 1/7 second for sound to travel 150 feet. The lack of synchronism between sight and sound becomes quite annoying when the difference exceeds about 1/7 second. Since the length of the theater may be as great as double the width, it is necessary to design the side walls, floor, and ceiling so as to minimize the attenuation of the sound transmitted toward the rear seats. Sound which is propagated over an absorptive surface, such as an audience or an acoustically treated ceiling, is greatly attenuated. Hence the floor should rise steeply toward the rear, the loudspeakers and screen should be well elevated, and the ceiling and side walls should neither be highly absorptive nor obstruct unduly the flow of sound from front to rear. Splays and other functional deviations in the wall and ceiling contours can be used to give the proper diffusion without hindering the efficient transmission of sound to the rear of the auditorium.

The Motion Picture Research Council recommends, for proper viewing and listening conditions, that the first row of seats be at least 20 feet from the screen—for screen widths not greater than 16 feet. For wider screens, the first row of seats should be back an additional 15 inches for each foot of screen width over 16 feet.

If there is a balcony, its depth should not be more than three times the height of the balcony opening. A relatively deeper overhang can be tolerated here than for a legitimate theater, since the average speech levels in a cinema are somewhat higher. The balcony soffit should slope downward toward the rear, and should not be absorptive. See p. 189 for other features of design.

A volume per seat of 125 to 150 cubic feet is a good figure to use in determining the optimum volume, the lower value being preferable. The design of a house with a low volume per seat has several advantages over designs with the usual larger values, acoustically and otherwise. The building cost is reduced; the costs of the correspondingly smaller air-conditioning equipment (and, to a much lesser extent, of the sound-amplification system) are likewise reduced; and the optimum reverberation can be obtained with the use of little or no special sound absorbents added to the walls and ceiling, if thick carpets are used on the aisles, and heavily upholstered chairs are installed. The optimum reverberation times for motion picture theaters can be determined from Figs. 9.11 and 9.12. After calculation of the total square-foot-units (sabins) of absorption supplied by the upholstered chairs, audience, carpeting, and the walls and ceiling, the total required number of sabins is determined as in Table 15.1. Then the number of sabins of absorption that must be added is obtained by subtracting from the total number of required units the units of absorption furnished by chairs, audience (assume a two-thirds capacity audience), and all the boundaries of the auditorium. Absorptive material should be applied to the rear wall to eliminate "slap-back." Additional absorptive material may be applied to the side walls in accordance with the general principles and recommendations of Chapter 9.

Treatment of the walls behind the screen with highly absorptive material prevents sound radiated from the back of the loudspeakers from being reflected to the audience. It also suppresses acoustical resonances that occur on some stages. Mineral-wool blankets have been used in many theaters to treat this area. The surface of the backstage acoustical treatment should be very dark, preferably black, in order to avoid light reflection from it. As indicated in Chapter 6, the absorption characteristics of an acoustical material can be enhanced, especially at low frequencies, by furring it out from the wall. If a blanket consisting of glass wool is used, it should be at least 2 or 3 inches thick and have a density of about 4 pounds per cubic foot. The floor between the screen and the first row of seats also should be highly absorptive, in order to prevent sound from reaching the audience in the front seats by reflection from this area. Such reflections contribute to the loss of "intimacy"; that is, the loss of feeling that the sound is actually coming from the screen. They may be suppressed by covering the stage floor with heavy carpets over 1-inch Ozite or similar absorptive pad.

In many respects the acoustical problems of motion picture and legitimate theaters are similar. Both should be properly insulated against noise according to the principles of Chapters 10 through 13. In general, a slightly greater noise level can be tolerated in motion picture theaters than in legitimate theaters because of the higher speech level. The average "film (background) noise" level is about 35 db, whereas the average audience noise level in a cinema is about 40 to 45 db.[4] Since the projection booth is a potential source of noise, all available interior surfaces should be heavily treated with fireproof acoustical material, such as a 2- to 3-inch mineral-wool blanket covered with perforated Transite. Double panes of glass of different thicknesses should be employed in the portholes. The windows should fit tightly in their frames so that there are no threshold cracks. It also is helpful to cover with absorptive material the peripheral surfaces separating the double windows. The wall between the projection room and the auditorium should have a transmission loss of not less than 35 db at 128 cycles and not less than 45 db at 512 to 2048 cycles.

School Auditoriums

The school auditorium usually serves a wide range of functions. It is used as an assembly room, large classroom, theater, cinema, concert hall, community auditorium, and it houses a host of other activities. The elements of design given in Chapter 9, regarding shape, size, reverberation, and diffusion, are applicable here. Furthermore, the principles and practice of noise control as described in Chapters 10 through 13 should be followed scrupulously.

In regard to theatrical uses, most school administrators and instructors of drama expect the *impossible* when they produce stage plays in a large auditorium without the benefit of a high-quality sound-amplification system. Auditoriums which are to be used without sound amplification, even if only occasionally, should not have volumes in excess of the following: for elementary schools, about 40,000 cubic feet; high schools, about 50,000 cubic feet; colleges and universities, about 60,000 cubic feet. (These volumes include the volume of the recess under the balcony but not the volume of the stage recess.) A great deal of dissatisfaction will be eliminated by avoiding the design of larger auditoriums for schools. If for any reason it should become necessary to construct a larger auditorium, provision should be made for sound reinforcement for speakers with weak voices, for occasional musical programs, and for all theatrical programs. Nothing less

[4] W. A. Mueller, *J. Soc. Motion Picture Engrs.*, **35**, 48 (1940).

than a stereophonic sound system will be entirely satisfactory for theatrical purposes. (See Chapter 14.)

The auditorium should be located in a quiet section of the campus. If it forms a part of another building, it should be thoroughly insulated from the remainder of the building. There should be two sets of tightly fitting doors between the auditorium and adjacent corridors or the outdoors. If a high degree of insulation is required, it will be helpful to dispense with windows. With the increase in airplane traffic it has become increasingly necessary to eliminate windows; with the good air-conditioning systems available, they are no longer a necessity. Any noise from the ventilating or other mechanical equipment should be adequately suppressed. The floor should be covered with linoleum or some other soft covering. The chairs should be heavily upholstered, of a rigid, substantial construction, and securely fastened to the floor so that there will be no creaking or squeaking.

It is necessary to make a compromise between the optimum acoustical properties for speech and for music in order that the school auditorium may best serve its diverse uses. The optimum reverberation times for school auditoriums are given in Figs. 9.11 and 9.12. The exact calculation of reverberation involves a three-space problem: the stage recess, the main part of the auditorium, and the recess under the balcony. However, if the stage has an enclosed set and if the balcony recess it not too deep, the calculation of reverberation time reduces to a one-space problem. In order to make this simplification, it is assumed that each of these three spaces contains an appropriate amount of absorption to permit a uniform average rate of growth or decay of sound in all parts of the auditorium. The complete set of hangings required for the stage setting ordinarily will supply a sufficient amount of absorption for the stage recess. In fact, a full set of stage hangings may make the stage *too dead* for musical settings. For this reason it is advisable to provide an enclosed wood veneer or heavily painted canvas set for musical programs, such as is shown by the dotted lines in Fig. 15.2. If upholstered chairs are provided and the aisles of the floor are carpeted, the recess under the balcony ordinarily will not require

Fig. 15.4 Photograph of a side wall of the Arnold Auditorium, Bell Telephone Laboratories, showing non-uniform absorptive treatment and splays which insure proper diffusion. (Courtesy Bell Telephone Laboratories.)

additional absorptive treatment of its side walls in order to provide a suitable reverberation time. If additional absorptive material is required, it should be applied to the side walls in strips, panels, or patches to give added diffusion; if none is required, it may be necessary to introduce splays, or other means of insuring proper diffusion. An example of good side wall design is shown in Fig. 15.4.

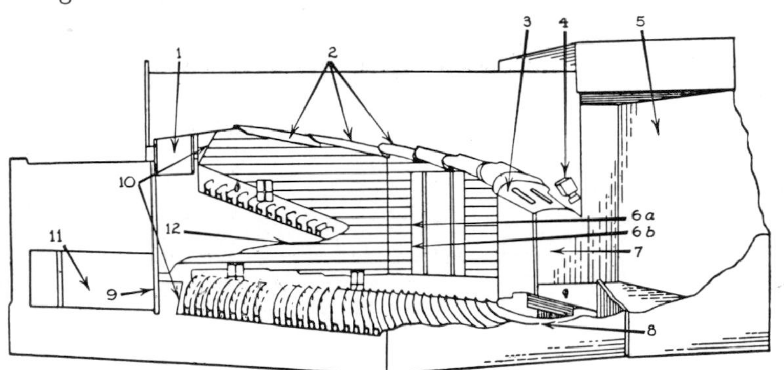

Scale: 3/16" = 1' 0"

1. Acoustically treated projection booth with sound-amplification equipment and controls for sound monitoring.
2. Ceiling planes reflect sound to all parts of the auditorium.
3. Three-channel public address system to reproduce stage sound in "auditory perspective."
4. High-fidelity speakers: bass-compensated dynamic speaker for low tones; high-frequency directional horns for high tones.
5. Backstage treated with acoustical plaster to reduce "stage echoes."
6. Acoustical treatment on walls: overall distribution in alternate bands of (*a*) acoustic tile and (*b*) hardwall plaster.
7. Proscenium splays; horn-like shape of stage opening projects sound to audience.
8. Upholstered seats; absorption value of each seat equivalent to that of a person's clothing.
9. Double doors to foyer insulate against external noises.
10. Slanting rear walls on main floor and balcony reflect sound down toward rear seats.
11. Acoustically treated foyer to reduce external noises.
12. Streamlined balcony improves flow of sound to rear seats.

Fig. 15.5 Section of a high school auditorium that incorporates essential characteristics of good acoustical design.

Figure 15.5 shows a section of a high school auditorium that incorporates the essential characteristics of good acoustical design. The following features were given careful consideration during the design and construction of the auditorium: the floor plan, the elevation of seats, the diverging proscenium splays, the functional ceiling, the shape and dimensions of the balcony recess, the control of reverberation and diffusion by alternate horizontal strips of absorptive tile and reflective plaster for the side walls, the upholstery of the chairs, the stage furnishings (including an enclosed reflective stage set for musical programs), the planned insulation against outside noise, the control of inside noise, and the sound-amplification system.

Civic Auditoriums

Many school auditoriums, especially in small towns, are used for community purposes—town meetings, debates, concerts, and a variety of other gatherings. But as a town grows to a city, there develops a need for a separate civic auditorium to serve the above purposes and a number of others, such as dances, bazaars, conventions, and activities that require (1) a level hardwood floor and (2) readily removable chairs. The present section is concerned with the latter type of auditorium. The two features just mentioned introduce acoustical problems that do not occur in the usual auditorium. The level floor, especially if it extends more than about 50 feet from the stage, requires the stage to be as high as sight lines will allow. The portable chairs generally will not be upholstered, or they will have only thin pads of soft and absorptive material on the seats and backs. These chairs furnish much less absorption than do fully upholstered ones. (The fixed chairs should be heavily upholstered, of the type previously advocated in this chapter for theaters and school auditoriums.) It usually will be necessary to compensate for the lack of absorption in the chairs by the introduction of absorptive strips, panels, or patches on the walls and ceiling in such amounts as will provide the optimum reverberation and good diffusion. The optimum reverberation characteristic will be provided in most cases if the curve which applies to school auditoriums in Fig. 9.11 is used. Since civic auditoriums are often used for small audiences and since the chairs are often unupholstered, it is advisable to provide the optimum reverberation for one half of capacity audience. If the room has a volume of more than about 50,000 cubic feet, a sound-amplification system is necessary. The loudspeakers should be located somewhat higher than they would be in an auditorium with a sloping floor.

All the acoustical problems considered in the previous sections of this chapter, namely those relating to theaters, cinemas, and school auditoriums, are likely to arise in planning the acoustics of civic auditoriums, and it is recommended that these sections be carefully reviewed. The acoustical problems relating to the design of a municipal auditorium become increasingly complex and difficult as the size of the auditorium increases. Echoes and interfering reflections become much more probable, and therefore appropriate plans should be worked out at the very start of the design to avoid these defects and to insure a good distribution of sound to all seats. Acoustical studies should be made of all feasible shapes and seating arrangements, supplemented by model testing as described in Chapter 9, if there are any uncertainties in determining the best plans.

Existing municipal auditoriums are notoriously defective in regard to acoustics. Until recently, most of them were excessively reverberant. Many of these have been "corrected" by "acoustical treatment," that is, by adding acoustical tile or plaster to the entire ceiling, or to almost every available surface that can be so treated. This has given quite satisfactory results in many cases. However, in some instances, the auditoriums have been *over-treated,* or mistreated by placing the absorptive material in the wrong places, and the correction of other serious defects has been overlooked. For example, a study was made recently of a municipal auditorium seating 1800 persons. At the time of construction, acoustical plaster was applied to all side walls, the rear wall was treated with a thin acoustical tile, and the entire ceiling was treated with thin fiberboard applied directly to concrete ceiling slabs. The stage was furnished with overhead, side, and rear hangings of unlined velours. As a result, the auditorium was somewhat overtreated at medium and high frequencies, although it was too reverberant at low frequencies. Complaints were made about the acoustics, most of them by the artists who sang or played solo instruments in this auditorium. They averred that their voices were "smothered," that they felt unable to "project" sound out into the audience. The management, heeding these complaints, added still more absorptive treatment—a burlap type of material festooned between the exposed trusses—and made the situation worse. Instead, the following correctives should have been supplied: (1) a stage setting of ¼-inch plywood with overhead and side splays; (2) a suspended ceiling designed in accordance with the principles of Chapter 9 to give beneficial reflections and made of a material that would add considerable absorption at the low frequencies and diminish the absorption at the high frequencies; (3) replacement of the thin absorptive tile on the rear wall with a material that is highly absorptive at low as well as high frequencies; and (4) a high-quality sound-amplification system. These correctives would not provide ideal acoustics in this auditorium, but they would eliminate the objectionable defects and make the building quite satisfactory for its many functions.

16 · School Buildings

The school was established to promote learning, which is acquired largely by word of mouth and by listening. Therefore, acoustics is one of the most important physical properties that determine how well the school building can serve its primary function. Thus, the exclusion of noise and the reduction of reverberation are indispensable in adapting classrooms to the function of oral instruction.

The acoustical design of school buildings requires consideration of the following: (1) the selection of the site; (2) the location on the site of buildings, playgrounds, and planting, and the arrangement of rooms within the buildings; (3) the sound-insulation requirements of the building; and (4) the acoustical planning of classrooms, laboratories, offices, corridors, and all other rooms in which a quiet environment or good acoustics is needed. The general procedures for handling these problems are discussed in this chapter.

All the building requirements for the school plant must share in determining the most feasible design, but the nature of the above considerations indicates that good acoustics requires deliberate and early planning. It cannot be postponed until after the other features have been determined, as has been done so often in the past. In general, good acoustics will require much more than the treatment of the ceilings of certain rooms and corridors with absorptive material.

School Sites

School boards should be encouraged to consult their architects regarding the selection of school sites; and architects should be prepared to advise school boards on such matters. The acoustical relevancies affecting the choice have been discussed in Chapter 10 (see pp. 218 to 222). Many existing schools are so disturbed by traffic noise that the classroom efficiency is greatly reduced. Noise is not only distracting to effective thinking, but it also interferes with audition to the extent that the hearing of speech is made difficult and at times impossible.

Proximity of the site to the homes of the students and to transportation facilities is an important consideration, but it does not justify a location near the intersection of main traffic arteries. Such a site may be convenient, but it will entail costly sound insulation in the buildings if the noise is to be reduced to acceptable levels; or, if appropriate means for providing the required sound insulation is disregarded, as it frequently is, noiseful rooms will result. A quiet site, even if more costly than a noisy one, may prove to be the more economical one in the long run, and it will add immeasurably to the utility and attractiveness of the school. The choice of a location adjacent to quiet streets, or even of one removed only a few hundred feet from noisy streets, can help appreciably to insure freedom from disturbing traffic noise.

It is not sufficient merely to select a quiet site. In addition, zoning ordinances should be established to require trucking routes, main traffic arteries, trolley lines, and airplane routes to be located sufficiently far from the school. The noise level at the site should be below 70 db. In order to insure this condition, it is necessary to eliminate honking of automobile horns within about 400 feet of the site, to keep motorcycles at least 300 feet away, trucks about 200 feet away, and private automobiles 100 feet away. These estimated distances are based on the assumption that the noise travels through the open air along unobstructed paths; they can be approximately halved if buildings, embankments, trees, or other obstacles are located between the sources of noise and the school site. For all the above-named sources, such obstacles frequently exist. This is not the case for air traffic. There are no obstacles to shield school buildings from the noise of passing airplanes, and, furthermore, airplanes generate much more noise than motor vehicles do. In fact, planes are so noisy that they must not be allowed to pass within 4000 feet of the school site because ambient noise level should be kept below 70 db. These data are based on the average noise levels of two- and four-motor propeller-type airplanes. If airplanes are made quieter than they now are—now a possibility with jet propulsion—the 4000-foot distance can be diminished.

Most school sites have been selected in advance, and many are partly or largely developed. Whether in quiet or noisy locations, they should be noise-surveyed prior to the designing of additional buildings. If the site is a quiet one, no extraordinary precautions need be taken for the insulation of outside noise. On the other hand, if the noise level is high, the survey should be made for the purpose of ascertaining the noise-insulation factor [see Eq. (11.3)] that will be required to reduce the noise from the outside to the acceptable levels: about 40 db for classrooms and lecture rooms; about 40 to 45 db for libraries; etc. For example, if the noise survey indicates that the prevalent noise at the site (during the times of day, week, and year the school will be used) will be 70 db, with a standard deviation of 5 db, the required noise-insulation factor must be not less than 40 db if it is to reduce this noise to 40 db, the acceptable level in classrooms. This provides a 5-db margin of safety, which allows for a reasonable increase of traffic noise—a probable eventuality for many sites, especially in growing communities.

Layout of School Buildings

In Chapter 10 it was emphasized that a significant part of the noise-insulation factor required for certain buildings can be obtained by: (1) well-planned grading and landscaping of the site, and (2) laying out the buildings and arranging the rooms within each building so that noise-insulation requirements will be reduced to the minimum. The two items together may reduce noise by more than 10 db. The auditorium, classrooms, lecture halls, music rooms, library, and all other rooms used for speech, music, or study should be located on the quieter sections of the site, well removed from noisy streets and playgrounds. These rooms should not have windows that open onto streets, playgrounds, or other existing or potential sources of noise; nor should they be near noisy machine or wood shops, band rooms,

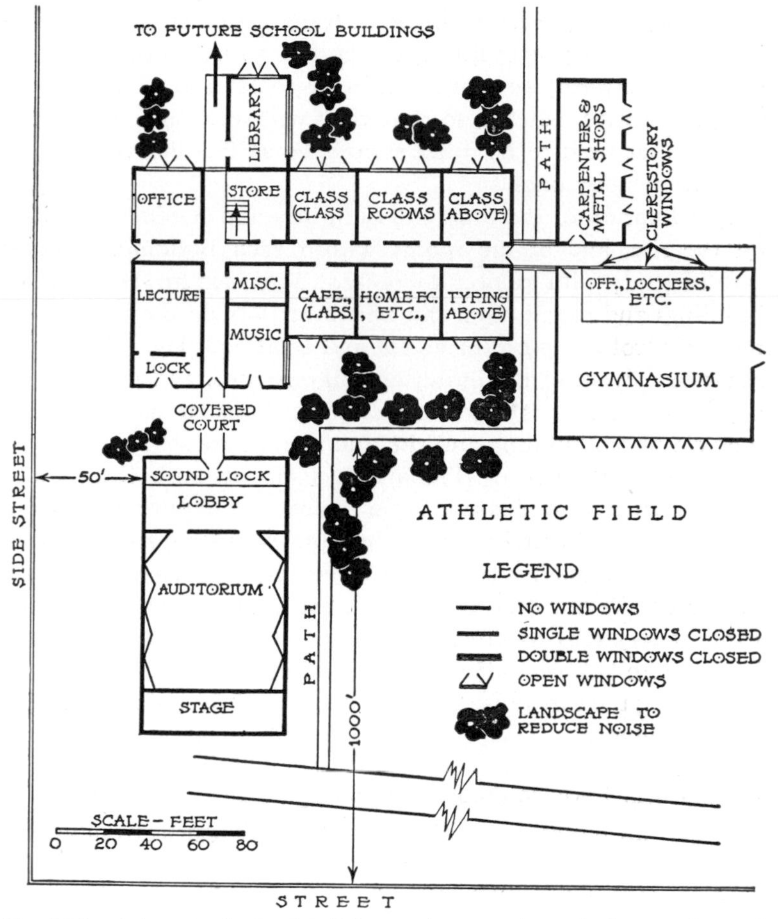

Fig. 16.1 A layout of school buildings, playgrounds, and planting that will minimize disturbances from noise.

gymnasiums, etc. Much of the disturbing noise is transmitted from one room to another *via* open windows. Such transmission is especially facilitated when the windows of one room look upon the windows of another near-by room, as in enclosed or partially enclosed courts, or when the separate buildings are near each other. It is much better to avoid windows that open where such noise hazards exist. If the windows do or must open, intervening trees and shrubs will provide a little noise reduction.

Another common means of noise transmission in school buildings is through poorly fitting doors and along corridors. The corridors between the quiet and noisy rooms should be treated with absorptive material so that they will serve as sound locks. Rooms that need to be quiet should be as far removed as possible from noisy ones.

The location of the different buildings and outdoor activities on the site, and of the rooms within each building, deserves careful study with respect to noise. The best arrangement of buildings and rooms often can be worked out in the interests of noise control as well as of access and lighting. Figure 16.1 is a suggested layout of the buildings, playgrounds, and planting on a school site that is based on considerations of noise control. The classroom building and the auditorium are set back at least 50 feet from the street. The arrangement of the buildings and rooms is such as to separate those rooms that require quiet from centers of noisy activity. Double-glazed windows are indicated where the sound-insulation requirements are most stringent. Minor details are considered, such as: the arrangement of the window locations to provide the longest possible path for airborne noise between adjacent rooms, the staggering of the doorways along the corridors, and the planting of dense-foliaged trees and shrubbery.

Rooms that Require Acoustical Designing

Nearly all rooms in a school building will benefit from appropriate acoustical design and treatment, and the resultant improvement usually will justify the expense. In general, careful planning is required for the following: auditorium, classrooms, lecture halls, language rooms, music rooms, gymnasiums, cafeteria, wood and metal work shops, laboratories, offices, all corridors, stairways, vestibules, and passageways that are likely to transmit sound from one part of the building to another. Each type of room should be studied and planned individually. Then it usually will be found possible to meet all sound-insulation requirements by adopting not more than two or three wall and ceiling-floor systems (occasionally a single type will suffice), and to meet all sound-absorption requirements with the use of only one or two types of acoustical material.

The acoustical design of school auditoriums has already been considered in Chapter 15. The other rooms in the buildings will be the subject of the remainder of this chapter.

Classrooms

There are no severe restrictions about the shape and size of classrooms, although long, narrow rooms and excessively large rooms should be avoided. Rooms having dimensions of 25 feet in width by 30 feet in length by 12 feet in height will be found satisfactory for classes of not more than about 40 pupils. Approximately the same proportion of width to length is desirable for either smaller or larger rooms. The acceptable noise level in school rooms (unoccupied) is about 40 db for ordinary classrooms and as low as 35 db for language rooms, music rooms, and other classrooms in which a quiet environment is especially desirable. The noise-insulation factor between adjacent rooms should be not less than 40 db for speech rooms and not less than 45 db for music rooms. (See Chapters 11 through 13 for specific procedures of noise control.)

The amount of absorptive material that must be added to each classroom in order to provide the optimum reverberation characteristic will depend on the room's size, purpose, seating capacity, and the age of the pupils. If the room is to be used by children under 6 years of age, the amount of sound absorption due to the pupils will be relatively small; consequently it will be necessary to use a correspondingly greater amount of absorptive material for the walls and ceiling than would be required in a room for adults.

The manner of determining the proper absorptive treatment of a typical classroom is suggested in the following example: Suppose that a room is 27 feet by 32 feet by 12 feet and has 40 desks. Choose from the tables on sound insulation the type of wall construction that will provide an average transmission loss (T.L.) of not less than 40 db; for example, wood studs with lath and lime plaster on both sides with the plaster applied to a thickness of at least ½ inch, or 4-inch porous clay tile plastered on both sides. If forced ventilation is provided for the recitation rooms, the transmission of sound from one room to another by way of ventilation ducts should be suppressed. Double-glazed windows that fit tightly in their frames should be used. If artificial ventilation is not provided for the rooms, the windows are likely to be open. Then there is no advantage in providing a high degree of insulation for the walls since the limiting factor in the transmission of sound from room to room will be by way of the windows, and the over-all T.L. between adjacent rooms will not be more than about 25 to 30 db. This is inadequate insulation, but nothing would be gained by providing walls that furnish a T.L. of more than about 35 db.

The most important single factor in the acoustics of the classroom is the control of reverberation. The optimum reverberation time in a small recitation room is 0.75 second at frequencies of 512 to 2048 cycles, and about 1.0 second at 128 cycles. (See optimum reverberation time curve for *speech,* Fig. 9.11.) The amount of absorptive material required to provide the above reverberation times can be determined readily by means of Eq. (8.12). A recommended procedure for calculating the required absorption is given in Table 16.1. The volume of the room is 10,368 cubic feet, and the interior surface is 3144 square feet. From the nomogram of Fig. 8.11, the absorption required to give the optimum reverberation times is 510 square-foot-units (sabins) at 128 cycles, and 680 sabins at 512 to 2048 cycles. These are entered as item 2 in the table. Suppose that the room accommodates, on the average, a class of 32 high school students. The absorption for each person, including his desk, is 2.6 sabins at 128 cycles, 4.2 sabins at 512 cycles, and 4.6 sabins at 2048 cycles. The absorption of the 32 pupils at their desks is entered as item 3. Each of the eight unoccupied wood desks has an estimated absorption of 0.4 sabin at the three frequencies. The absorption of the furniture has been estimated and is entered in the table. The floor is to be covered with asphalt tile having an absorption coefficient of 0.05. The absorption of windows, blackboards, and doors also is listed in the table. The total absorption in sabins from items 3 to 9 is 155, 211, and 224 at 128, 512, and 2048 cycles, respectively. The walls and ceiling must

TABLE 16.1

CALCULATION OF REVERBERATION TIMES FOR A CLASSROOM 27 FEET BY 32 FEET BY 12 FEET; VOLUME, 10,368 CUBIC FEET; TOTAL SURFACE AREA, 3144 SQUARE FEET

Item		128 Cycles		512 Cycles		2048 Cycles	
1	Optimum reverberation time in seconds	1.0		0.75		0.75	
2	Total square-foot-units of absorption required = $S\bar{\alpha}$	510		680		680	
	Absorptive Material in the Classroom	Absorption Coefficient	Absorption in Square Feet	Absorption Coefficient	Absorption in Square Feet	Absorption Coefficient	Absorption in Square Feet
3	32 pupils, seated at desks	2.6 units per pupil	83	4.2 units per pupil	134	4.6 units per pupil	147
4	8 unoccupied desks	0.4 unit per desk	3	0.4 unit per desk	3	0.4 unit per desk	3
5	Miscellaneous furniture		10		15		15
6	Floor, 864 sq ft	0.05	43	0.05	43	0.05	43
7	Glass windows, 140 sq ft	0.03	4	0.03	4	0.03	4
8	Blackboard, 280 sq ft	0.03	8	0.03	8	0.03	8
9	Wood doors, 60 sq ft	0.07	4	0.07	4	0.07	4
10	Absorption from items 3 to 9	155		211		224	
11	Additional required absorption, item 2 minus item 10	355		469		456	

furnish the remainder of the required absorption, namely, 355 sabins at 128 cycles, 469 sabins at 512 cycles, and 456 sabins at 2048 cycles. There are several possible methods of treatment of these surfaces that will provide the required absorption. For example, the entire surface of 1800 square feet can be treated with a material having sound-absorption coefficients of 0.20, 0.27, and 0.25 at 128, 512, and 2048 cycles, respectively. It may be impossible or impractical to select a material having these exact values, but certain acoustical plasters furred out from the walls and suspended from the ceiling (see Table A.2, Appendix 1) approximate the required coefficients. Deviations of 0.1 second from the specified optimum reverberation times are acceptable, and therefore considerable latitude is allowable in the selection of the absorptive materials. Another way to supply the required absorptive treatment is to treat 550 square feet of the upper walls (preferably the side and rear walls) with an acoustical tile having coefficients of about 0.40, 0.70, and 0.65 at 128, 512, and 2048 cycles, respectively, and to finish the entire ceiling and the remainder of the walls with ordinary plaster on suspended or furred-out metal lath, which has a coefficient of 0.06 at the three frequencies. The latter treatment has several advantages, acoustically and otherwise, over acoustical plaster: (1) It adds the absorptive material on the side and rear walls and thus provides a means of eliminating flutter echoes and (if applied in panels, strips, or patches) of aiding the diffusion of sound in the room. (2) It leaves the front wall and ceiling reflective—an aid in the reinforcement of the sound reaching the rear part of the room. (3) It allows the choice of acoustical materials that have absorption coefficients determined by factory-controlled processes, that can be decorated repeatedly with any kind of paint, and that can be cleaned without impairing the absorptivity or otherwise damaging the material.

The questions of cost, fire resistance, appearance, and all other aspects of acoustical materials, besides absorption, must be considered in choosing the most suitable materials for the treatment of classrooms.

If the entire walls and ceiling of the classroom described above had been finished with hard-wall or cement plaster applied directly to clay tile partitions and concrete slab ceiling, the reverberation time at 512 cycles would have been 2.3 seconds with 32 students in the room, and about 4.5 seconds in the empty room. Although this would provide tolerable hearing conditions, it is much less satisfactory than the speech articulation of about 90 per cent in the acoustically treated room. Furthermore, *the assumption of reasonably quiet conditions* usually is not realized in the reverberant room. Owing to outside noise which is transmitted into the room, and noises originating within the room, a deficiency of absorption results in an elevation of sound level in accordance with Eq. (8.6). In fact, the acoustical treatment of classrooms is justified at least as much by the benefits of noise reduction as it is by the improvement in speech articulation. Furthermore, a quiet environment probably instills quiet habits among the pupils in the room.

Lecture Rooms

The acoustical requirements for classrooms become increasingly important as the size of the room increases. In small classrooms, a rectangular shape with level floor and ceiling is taken for granted. In larger rooms, such as those used for lectures, forums, and debates, where 100 or more persons are to be seated, every possible means, including the proper design of shape, should be utilized to give to the room the optimum properties for the hearing of speech. The volume should be kept as small as feasible (about 125 cubic feet per seat is desirable), the auditors should be seated near the lecture platform, the seats should be elevated appropriately, and the walls and ceiling should be designed to give beneficial reflections of sound. The ratio of length to width, although not critical, should be about 1.2:1.0. A room with a length of 60 feet may be made somewhat wider than the indicated width of 50 feet, but there should not be any appreciable extension of length, especially if the room is to be used without a sound-amplification system. In general, long, narrow rooms are less satisfactory than short, wide ones.

There are many possible designs that will prove satisfactory. The plan and section shown in Fig. 16.2 exhibit the more salient acoustical features. This room has a length of 60 feet, a width of 50 feet, and an average height of 16 feet. The volume is about 45,000 cubic feet. It will accommodate as many as 360 seats. The seats are elevated at an angle of 20 degrees, which is sufficient to give everyone an unobstructed view of the top of the lecture table—a feature of distinct value for demonstrated lectures, and usually of sufficient value acoustically to justify the added cost. If the room is to be used only for lectures without demonstrations, a limitation that may be anticipated at the time of planning, but one that probably will be short-lived, the slope of the floor can be reduced somewhat and the front portion can be level, but the rise of the floor toward the rear seats should not be less than 8° (see section on p. 184). The slopes of the various splays or sections of the ceiling are designed to give the optimum reflection of useful sound, as shown by the ray diagrams in Fig. 16.2.

After the size and shape of the lecture room have been determined, attention should be given to the control of noise. The average noise level in a good lecture room should not exceed 40 db. The noise-insulation factor required to insure this degree

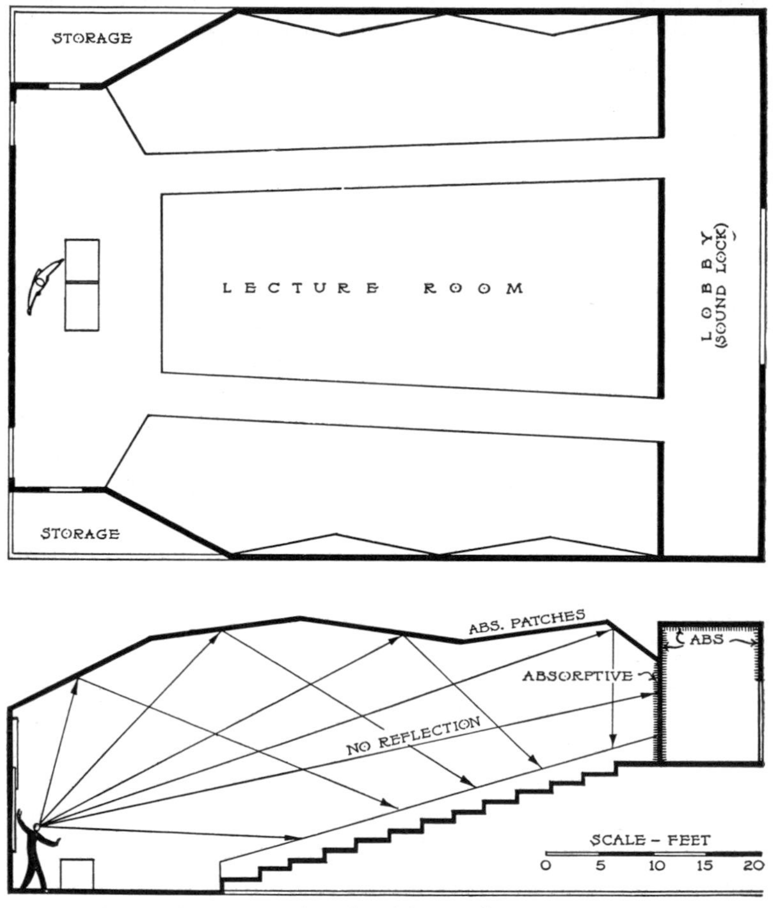

Fig. 16.2 Acoustical study of lecture-demonstration room.

of quietness will depend on the location of the room within the building and on the magnitude and nature of the noise in the immediate vicinity of the room, both within the building and outside. Suppose that the noise has a probable average level of 75 db (with a standard deviation of 5 db) for the location adjacent to the room. This corresponds to a moderately noisy location. The noise-insulation factor required to reduce the outside noise to the acceptable level in the room (40 db) (5 db being allowed for the standard deviation and another 5 db for uncertainties in construction and for a factor of safety) is $75 + 10 - 40 = 45$ db. The type of wall and ceiling construction required to give this amount of insulation can be calculated by the method presented in Chapter 10. From Eq. (11.3) the noise-insulation factor in decibels is given by

$$\text{Noise-insulation factor} = 10 \log_{10} \frac{a}{T} \quad \text{db} \qquad (16.1)$$

where T is the total transmittance, defined by Eq. (11.2), through the boundaries of the room, and a is the total amount of absorption in the room. The volume of the room shown in Fig. 16.2 is about 45,000 cubic feet, the interior surface 10,000 square feet, and the optimal reverberation time for frequencies of 512 to 2048 cycles is about 0.85 second. The total absorption in the room is therefore given by the reverberation equation, $t_{60} = 0.049V/[-2.3 \log_{10} (1 - \alpha)]$. Hence $\alpha = 0.233$, and $a = \alpha S = (0.233 \times 10{,}000) = 2330$ square-foot-units. Substituting in Eq. (16.1), $45 = 10 \log_{10} (2330/T)$ or $T = 0.074$ square foot. In order to reduce T to this value, it is necessary to dispense with windows or to use two panes of glass separated by an air space of at least 3 inches. Suppose that there are no windows and that there are two doors, each 5 feet by 7 feet and each having a transmission loss of 33 db (this requires solid wood doors, at least 1⅜ inches thick, carefully fitted to their frames so that all threshold cracks are closed). The transmittance of the doors will then be $(0.0005 \times 70) = 0.035$ square foot. The total transmittance of the walls and ceiling (if it is assumed that transmission through the floor will be negligible) must not exceed $(0.074 - 0.035)$, or 0.039 square foot. Since the area of the walls and ceiling (less the area of the doors) is 7120 square feet, the transmission coefficient τ of the walls and ceiling must not exceed $(0.039/7120) = 0.0000055$. Hence, by Eq. (11.1) the T.L. for the walls and ceiling must be as high as 52 db. Thus the walls and ceiling should have an insulation equivalent to that of a 10-inch brick wall. The use of sound locks for the doors will reduce the insulation requirements of the walls and ceiling by about 3 db. Consequently the T.L. of these partitions then could be reduced to 49 db, and this reduction would be provided by a 6-inch brick wall or its equivalent. The sound locks, in this case, provide about the same amount of noise insulation as would be furnished by increasing the thickness of brick walls from 6 to 10 inches. Sound locks also can be designed to serve as "light locks," which contribute greatly to the utility of the room for projecting either lantern slides or motion pictures.

The absence of windows implies that artificial ventilation will be provided for the room. It is necessary therefore to install an adequate amount of acoustical attenuation in the ventilating ducts to reduce the equipment noise level throughout the seating area in the room to not more than 35 db so that it will not contribute appreciably to the room noise. The heating and ventilating contractor should be familiar with the art of noise control and should provide such equipment and so install it that the fan and motor noise reaching the room from the duct openings, or from other sources, will not exceed this specified level. See Chapter 13 for further information.

The reverberation times in the lecture room shown in Fig. 16.2 should not exceed 0.85 second at 512 to 2048 cycles and 1.20 seconds at 128 cycles. If the room has a seating capacity of 360, it is advisable to provide the optimum reverberation times for two thirds of the capacity audience, that is, for 240 adult persons. The choice of absorptive material, the amount required, and its location in the room should be determined by the principles and calculations set forth in the section on p. 195. As a general rule, it is preferable to apply the absorptive material on the rear wall, and in panels, strips, or patches on the side walls, rather than on the ceiling, giving consideration to the need for preventing long-delayed reflections from the rear wall and flutter echoes between the side walls. The front wall, the front portions of the splayed side walls, and most of the ceiling should be finished with highly reflective materials in order to increase the amount of beneficially reflected sound. A reflective ceiling is advantageous for discussions from the floor; it facilitates the flow of sound from a speaker, located anywhere on the floor, to all auditors within the room. A wood floor and a 3/8-inch plywood wainscot (with an average height of about 6 feet and applied to furring strips), a suspended plaster ceiling, and furred-out plaster walls contribute to the required absorption at low frequencies. Applying approximately 1150 square feet of acoustical material, having coefficients of sound absorption of 0.35 to 0.40 at 128 cycles and of 0.65 to 0.75 at 512 to 2048 cycles, to the rear wall above the wainscot (the wainscot for the rear wall should be about 4 feet high) and to panels on the side walls will provide the optimum reverberation times for 240 persons in the room. Several materials that meet these requirements are listed in Appendix 1.

A lecture room, designed in accordance with the foregoing procedure, will be free from noise, excessive reverberation, and other acoustical defects; its shape and size have been adapted to its primary functions. The action of the splayed reflective surfaces at the front end of the room will enable the lecturer to be heard well, even when his head is turned away from the audience. Unamplified speech of the average speaker will be heard satisfactorily in all parts of the room under all possible conditions of use. In short, the room will have good acoustics.

Figure 16.3 is a plan and section of the Arnold Auditorium in the Bell Telephone Laboratories, Murray Hill, New Jersey (also see Figs. 6.8 and 15.4). The room, which was designed primarily for lectures and technical conventions and for testing the performance of sound-amplification systems, has a number of acoustical features which have been repeatedly advocated in this book. The shape and size of the room insure uniform and adequate sound levels throughout the seating area. Because the chairs are heavily upholstered the reverberation is almost independent of the size of the audience. The walls and ceiling are of furred-out 1/4-inch plywood panels. Certain panels, as indicated in Fig. 16.3, are perforated and backed with 2-inch mineral-wool blankets. These are designed in size, shape, number, and location to provide (1) optimum reverberation times for speech and (2) good diffusion. The auditorium serves unusually well the acoustical functions for which it was designed.

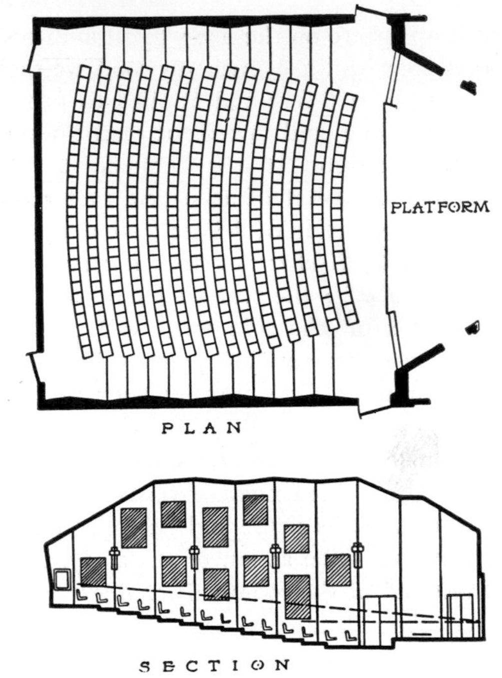

Fig. 16.3 Arnold Auditorium, Bell Telephone Laboratories, Murray Hill, New Jersey. (Voorhees, Walker, Foley and Smith, Architects.)

Music Rooms

The specifications for the shape and size of rooms to be used for band, chorus, or orchestra rehearsals and instruction will not differ greatly from those described in the preceding section for lecture rooms. However, there are some points of difference. The floor usually will be provided with a platform of two or three stepped-up levels to accommodate, in the conventional manner, the band, chorus, or orchestra. Diffusion is somewhat more desirable than it is for speech rooms; therefore it may be necessary to provide splayed walls or other irregularities in the wall and ceiling contour.

The optimum times of reverberation for music rooms depend on the kind of music to be performed. It usually is not practical to provide means for adjusting the reverberation characteristics of music rooms. In these cases, the reverberation time at 512 cycles should be midway between the range of values shown in Fig. 9.11;

the reverberation times at other frequencies can be calculated by means of Fig. 9.12. Where the best acoustical conditions are desired, means should be provided for changing the reverberation characteristics of the room in accordance with the number of performers, the size of the audience, and the type of music to be performed. The various constructions for accomplishing this have been described on p. 126.

The distribution of the absorptive material in a room requires considerable care. The walls and ceiling near the platform should be finished with hard, reflective materials, whereas the side and rear walls of the seated area should be fairly absorptive. Large areas of wood paneling and wood flooring are beneficial in music rooms. It is therefore desirable to use wood floors and wood wainscot, especially in the vicinity of the platform. These materials provide relatively high absorption for the low-frequency components of music. The use of velours or other hangings on the platform should be avoided. Such absorptive materials will impair the reverberant properties of the *generating* end of the room. As a result, the members of a chorus or other music ensemble may have difficulty in hearing each other sing or perform, and the sounds they generate will not be radiated effectively to the audience. Such hangings, unless lined and hung in very deep folds, absorb the high frequencies much more than they do the low ones, and thus distortion is introduced. In music rooms that are too dead at the *source* end of the room a portable reflective screen, of wood veneer or heavily painted canvas designed in the form of an "orchestra shell," often can be used to good advantage.

The insulation of sound deserves special attention in the design of music rooms. It is not only necessary to exclude extraneous noise, but also it is equally necessary to confine the music within the room so that other near-by rooms will not be disturbed. This can be accomplished satisfactorily by employing wall, floor, and ceiling constructions having a transmission loss of not less than 45 db, and doors and windows having a T.L. of not less than about 30 db. The windows, if any, should be double-glazed and should be kept closed. Therefore, artificial ventilation is necessary. It is very important that adequate measures be taken to suppress "cross-talk"—the transmission of sound from one room to another via common or connecting ventilating ducts. These measures have been discussed in Chapter 13. Where it is not practical to provide artificial ventilation, the room should be located in a separate wing and the windows should open upon a part of the campus which is well removed from assembly or classrooms.

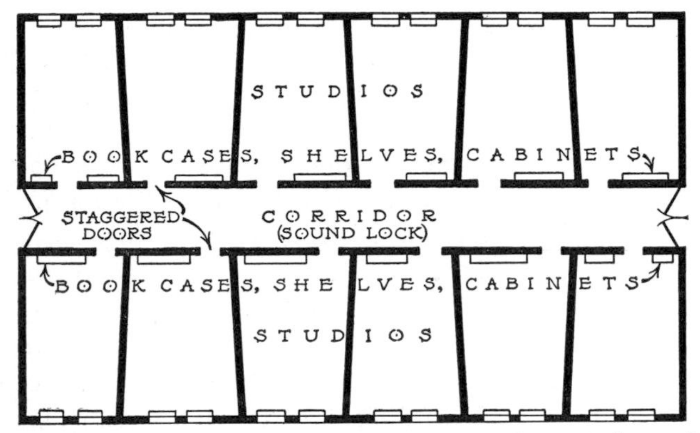

Fig. 16.4 A group of practice rooms for a music building. Ceilings of ½-inch fiberboard, slope 1 in 20. Book shelves on one end wall and non-parallel side walls prevent room flutter and aid diffusion.

The doors should be of solid panel construction and fit tightly in their frames. It is advisable to stagger the doors along a hallway so that no two are directly opposite each other. The entire corridor ceiling and upper 4 feet of its walls should be treated with a highly absorptive tile. Thus, the corridor will serve as a "sound lock" between adjacent rooms and also between these rooms and other parts of the building.

Figure 16.4 shows a plan of a group of adjacent practice rooms in a music building. There are no windows. The ceilings slope about 1 foot in 20 feet to avoid parallelism with the floor. The side walls similarly are kept non-parallel by partitions as shown in the figure. Flutter echoes between the end walls are eliminated by means of book shelves and doors. The floors are of wood; the ceiling is of fiberboard furred down from plaster; the walls are staggered studs and ordinary plaster. The fiberboard ceiling, the books on the shelves, and the furniture provide sufficient absorption to give approximately the optimum reverberation characteristics to each room.

Figure 16.5 shows a plan of a preliminary design for a music room to be used for choral and orchestral rehearsals in the

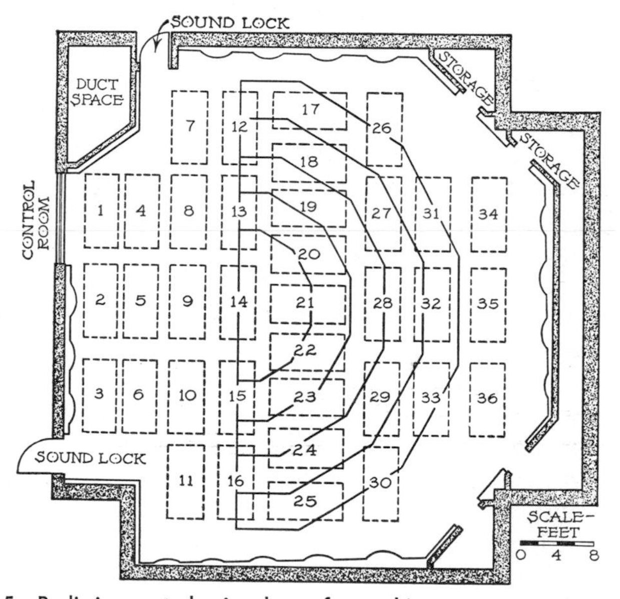

Fig. 16.5 Preliminary study, in plan, of a multi-purpose music room for the Music Building at the University of Washington. (Whitehouse and Price, Architects.)

Music Building of the University of Washington. The walls, no opposite pairs of which are strictly parallel, are of painted ⅜-inch plywood, and they incorporate several polycylindrical diffusers, some of which are separated by absorptive strips of panels of 2-inch Fiberglas covered with perforated plywood. The plywood panels have random spacing of the bracing and framing strips to "spread out" their resonant response. The floor is Oregon fir. The ceiling is ordinary plaster on suspended metal lath pierced with cylinders. The risers in the floor of the platform, the cylindrical diffusers, and the persons in the room provide adequate diffusion. The control of the reverberation characteristic is accomplished by means of rotatable cylinders in the ceiling. The convex surface of each cylinder is made of three different materials which differ greatly in their absorptive properties, each extending 120° around the cylinder as indicated in Fig. 7.7. These cylinders can be rotated by means of a motor drive so that any one of the three surfaces, or any desired combination of two of them, can be exposed. One of the three surfaces is 3/8-inch plywood providing a fairly reflective material having a uniform coefficient of absorption of about 0.10 at all frequencies; a second surface consists of a 2-inch blanket of Fiberglas covered with perforated plywood, and has an absorption coefficient that increases with frequency; the third surface is 1/8-inch plywood backed with a 2-inch layer of Fiberglas and has an absorption coefficient that decreases with frequency. The cylinders are divided into three banks; those numbered 1 to 16 are in the front section of the ceiling, 17 to 25 in the middle section, and 26 to 36 in the rear section. Since each bank has its separate motor control, it is possible to "set" the cylinders at those positions which will give each of the three sections of the ceiling almost any desired absorption characteristic.

Although such a system of rotatable cylinders is costly, it has many acoustical advantages. Thus, if the room is to be used at times as a lecture room, cylinders 1 to 16 should have the plywood portion exposed to the room, whereas cylinders 17 to 36 should have the Fiberglas portion exposed. For most musical purposes, cylinders 1 to 16 usually would have the Fiberglas exposed, and cylinders 17 to 36 would have the plywood exposed. When a small audience is in the room, most of the Fiberglas surfaces of all cylinders should be exposed. For chamber music, especially when such music is to be broadcast or recorded, as is planned for this room, much of the Fiberglas would be exposed; but, for other types of music, various combinations of the three surfaces would prove more satisfactory. Changes can be made, as required, for different numbers of the same program, or even for different parts of the same number. If such uses are contemplated, the cylinders and their motor control must operate quietly.

Gymnasium, Cafeteria, Library, Offices, Miscellaneous Rooms, and Corridors

The acoustical designing of the gymnasium, cafeteria, library, shops, study rooms, offices, lobbies, and miscellaneous rooms in school buildings does not present any special problems that have not already been considered. In general, however, all these rooms require careful consideration with respect to noise control, sound insulation, and reduction of reverberation.

GYMNASIUM. If the gymnasium is to be used as a dance hall, classroom or assembly room, even though only occasionally, the acoustical requirements include many of those we have considered in connection with classrooms and auditoriums. Rectangular shapes are almost always preferred for the primary functions of the gymnasium. Its uses as a dance hall, classroom, or for other purposes involving speech or music, will almost never justify major deviations from rectangular shapes. It is desirable, however, to avoid designs in which opposite pairs of smooth, reflective parallel walls will give rise to flutter echoes (see p. 170). Irregularities in the ceiling, such as beams, trusses, purlins, the proper placement of windows and doors on the side walls, and the distribution of absorptive material in strips, panels, or patches can be worked out to eliminate room flutter. The problem of noise insulation is important primarily in so far as the noise generated in the gymnasium may constitute a disturbance to near-by rooms. It is advisable to locate the gymnasium a considerable distance from classrooms, lecture halls, and other rooms that require a quiet environment. The reduction of the reverberation is the outstanding need for good acoustics in a gymnasium. The optimum reverberation times for speech (see Figs. 9.11 and 9.12) are recommended if the room is not to be used for music. If the room is to be used for dancing, the optimum times of reverberation will be the mean of the values for speech and for music. In making calculations of reverberation, it is satisfactory to assume that there will be one person in the room for each 20 square feet of floor area. Of course, there will be occasions (for example, at basketball games) when there will be many more persons in the room than this criterion predicts; but, unless the reverberation times are near the optimum values for a small number of persons in the gymnasium, the usual noise in the room will not be adequately reduced. The acoustical material selected for the control of reverberation must be very durable; for example, fiberboard applied directly to a hard backing, or mineral wool protected with a suitable perforated facing such as Transite. In general, parts of the upper walls, and most or

Fig. 16.6 Photograph of the natatorium at the Massachusetts Institute of Technology. The upper walls and entire ceiling have been treated with Stucoustic acoustical plaster, made with a portland cement binder and applied ½ inch thick over two portland cement base coats on metal lath. (L. B. Anderson and H. L. Beckwith, Architects.)

all of the ceiling, will require treatment with absorptive materials if the optimum reverberation times are to be attained. The problem of which materials, how much of them, and where and how to install them can be solved by the methods already described in this chapter.

Cafeteria. The acoustical problem in the cafeteria is simply one of the reduction of noise and reverberation. Usually, it will be solved satisfactorily by treating with highly absorptive acoustical tile the entire ceiling of the eating area, of the serving area, and of the adjacent areas or rooms in which the food is prepared. Often, the treatment of these adjacent areas or rooms is neglected because it is assumed that they are separated by a door, or doors, from the main eating room. Such doors are open much of the time and therefore the *separated* rooms actually are *coupled;* in order to control noise and reverberation in the eating room it is necessary to control them in all adjacent *coupled* rooms. The selection of the absorptive material should be made from those products that can be cleaned readily and painted repeatedly without impairing the acoustical or other physical or decorative properties of the material. In certain areas (for example, over the steam tables) absorptive materials that can withstand high humidity should be used. In general, if all ceiling surfaces are treated with acoustical material having coefficients of sound absorption of not less than 0.15 at 128 cycles and not less than 0.60 at 512 to 2048 cycles, satisfactory results will be obtained. If the ceiling is higher than about 12 feet, it may be necessary to use some additional absorptive material for upper walls of the cafeteria. It is especially important to choose for the treatment of the kitchen materials that can be washed. Since the noise level in an acoustically treated cafeteria, when in use, is about 60 db, the noise which comes into the room through windows, doors, ventilating grilles, etc., should not exceed about 55 db. Therefore, unless the room is located in noisy surroundings, no special sound-insulation measures are necessary. The floor should be covered with a soft, resilient material like battleship linoleum, asphalt tile, or rubber tile.

LIBRARY. The problem in the library usually involves only the proper control of noise and reverberation. It is necessary, however, that the noise level in the room be kept low. *Silence* is indeed golden in the library. If the noise level in the unoccupied room is reduced to 40 db, satisfactory acoustical conditions normally will prevail when the room is in use. Absorptive materials should be employed extensively in libraries. If the ceiling is not a high one and if the room has a simple, rectangular shape, it usually will suffice to treat the entire ceiling with a highly absorptive material. The walls often are lined with book shelves, which, when filled with books, will provide an absorptive surface (coefficient of about 0.25 at 512 cycles). The floor should be covered with cork tile or carpet or similar material that will minimize the noise of footfalls.

MISCELLANEOUS ROOMS. The ceilings of such rooms and areas as shops, study rooms, offices, corridors, stairways, lobbies, vestibules, and similar spaces where noise should be suppressed or prevented from being transmitted from one part of a building to another should be treated with absorptive material having sound-absorption coefficients of not less than 0.15 at 128 cycles and not less than 0.50 at 512 to 2048 cycles. If the offices have relatively high ceilings and highly reflective walls, it is desirable to add some absorptive material to the upper walls. Also, for corridors connecting noisy shops, band rooms, etc., to portions of the building where quiet conditions are required, the upper walls as well as the ceiling should be treated with highly absorptive material. Where practical, the floors of these rooms and areas, and especially those of offices, study rooms, corridors, vestibules, and lobbies, should be covered with soft, resilient material such as asphalt tile or rubber tile. Motors, fans, machinery, and other sources of solid-borne noise should be isolated from those portions of the building that require quiet surroundings. (See Chapter 12.)

17 · Commercial and Public Buildings

The major acoustical problems in commercial and public buildings are concerned with the exclusion of outside noise and the suppression of noise generated within the buildings. The general procedures for solving these problems already have been considered. In this chapter we shall consider specific types of commercial and pubic buildings in which acoustical design is a matter of importance, such as offices, stores, restaurants, industrial plants, libraries, club and recreational buildings, museums, hospitals, and government buildings. It has been indicated previously that, to attain an environment free from the annoyance of noise and excessive reverberation, the following steps should be taken:

(1) Plan the building with regard to noise insulation so that rooms which house sources of considerable noise are separated from rooms in which quiet is required. (See Chapter 10.)

(2) Provide adequate noise insulation for the rooms in which quiet is required (Chapters 11 and 12).

(3) Use enough sound absorption to reduce reverberation and noise generated within the room to an acceptable level.

All three steps should be taken. It is imprudent and uneconomical to neglect step 1 with the hope that noise-insulative measures or the application of acoustical materials or both will solve the resulting noise problems.

There is an increasing demand for the suppression of noise in all commercial and public buildings. It is well to recognize that uncontrolled noise in such structures will diminish their utility and contribute to their obsolescence.

Office, Bank, and Store Buildings

The deleterious effect of noise on workers engaged in various occupations and the benefits derived from the reduction of noise were discussed in Chapter 10. Employees in offices, banks, and stores generally prefer a quiet environment. Many find that they are less fatigued at the end of the day if they are not subject to the incessant bombardment of irritating noises. Even though there may be some uncertainty about the injurious effects of noise on speed and accuracy in the performance of many routine manual and mental operations, there is no question about the harmful effects of noise on the hearing of speech. Noise lowers the intelligibility of speech.

Reasonable precautions should be taken to insulate, against noise from adjacent rooms, machinery, ventilating ducts, and the outside, the rooms in which people work or converse. In very noisy locations it may be necessary to keep the windows permanently closed and install unit air conditioners in the rooms thus affected. The principal and most expedient means for providing good acoustics in such rooms, however, consists of the reduction of noise and reverberation by means of sound-absorptive materials. Such treatment usually is necessary and desirable even after all other reasonable sound-insulative measures have been taken.

The amount of sound-absorptive treatment required for noise-reduction purposes in stores, banks, and office buildings can be estimated by the following rule-of-thumb: The reverberation times for such rooms should not exceed the values given for speech rooms in Figs. 9.11 and 9.12. For example, an office with a volume of 100,000 cubic feet should not have a reverberation time greater than 1.4 seconds at a frequency of 128 cycles, and not greater than 0.9 second between 512 and 2048 cycles. From the nomogram of Fig. 8.11 the total absorption (including that from the furnishings and all wall surfaces) corresponding to these times is 3500 square-foot-units (sabins) at 128 cycles and 5400 sabins between 512 and 2048 cycles. The difference between this amount of absorption and that which exists in the untreated room must be supplied by the absorptive treatment to be added. If the total area available for treatment is known, the coefficients can be calculated. Thus, if a square-foot-units (sabins) of absorption are required, and S square feet of surface are to be treated, the absorption coefficient α of the material should be (a square-foot-units)/(S square feet). Where unusual noise conditions exist it is better to make calculations at octave intervals between 128 and 2048 cycles; but, for routine jobs of office-noise reduction, it is sufficient to make a single calculation at 512 cycles and select the absorptive material on the basis of its noise-reduction coefficient. In such procedures, however, materials that have a very low coefficient at certain frequencies should be avoided, especially if the noises to be absorbed have prominent components at these same frequencies.

Often the amount of absorptive material added to a store, bank, or office room in order to provide the optimum reverberation time is of the order of 6 times the amount present in the untreated room. Figure 11.12 shows that the absorptive treatment in this case results in an 8-db reduction in the noise level of the room, whether the noise is of inside or outside origin. As judged by the average person, this amounts to about a twofold reduction in the loudness of the noise. The addition of this amount of absorptive material will make a marked improvement in the acoustical properties of the room for the recognition of the sounds of speech, and this factor will contribute greatly to the individual's satisfaction with the room. Also, the effect of absorptive treatment in facilitating the localization of noises to the immediate vicinity of their origin is a factor which contributes to the personal comfort of the worker. He will notice that a noise which originates in a remote part of an acoustically treated room is not only reduced in level, but it also seems to originate at a relatively greater distance; it is "pushed back."

The choice of material and type of treatment is determined not only by the coefficients of absorption, but also by a number of factors such as combustibility, maintenance, appearance, ease of decoration, and cost. A difference of 10 per cent in the absorption coefficients of two materials will make a difference, at most, of only 1 db in the average noise level in the room. Therefore it would be a mistake to choose one particular material merely because it was 10 per cent more absorptive than another, if the choice of this more absorptive material entailed an appreciable sacrifice in structural quality, appearance, or cost.

In large, rectangular rooms, where the ceiling height is less

Fig. 17.1 Banco do Estado de São Paulo, São Paulo, Brazil. Acousti-Celotex has been applied to both wall and ceiling areas. (Camargo and Mesquita, Architects and Engineers.)

than about one-half the smaller horizontal dimension, the ceiling is the best surface in the room for the absorptive treatment, although it may be necessary to treat also portions of the upper walls in order to obtain the required amount of absorption. In a room with a relatively high ceiling and highly reflective walls, the treatment of the ceiling may lead only to disappointing results. Noise may excite normal modes of vibration in which the sound bounces back and forth between the walls of the room, giving it a resonant character that is annoying to the occupants. Therefore, in such rooms it is necessary to distribute the absorptive treatment over portions of the walls as well as the ceiling. An example in which acoustical treatment was applied to the upper portions of the walls and ceiling of a bank is shown in Fig. 17.1.

Restaurants

The acoustical treatment of the ceilings of restaurants, cafeterias, and dining rooms has become common practice, especially in modern construction. It has been demonstrated again and again that such treatment is justified by economic as well as other reasons. Many medical authorities advocate a period of rest and relaxation before eating as an aid to digestion and good health. No less important is an atmosphere of quiet during the meal. Many persons have observed that they leave a noisy and reverberant dining room with a feeling of exhaustion. On the other hand, these same persons have remarked, after acoustical treatment of the same dining room, that it has become a joy to eat in the room. One feels refreshed after eating in a quiet environment where the noise of clashing dishes is "muffled" and where conversation is almost effortless.

The acoustical treatment of dining rooms, especially large ones, should be based on the procedure already outlined for school cafeterias (see Chapter 16). In small restaurants, sufficient absorption may be provided by a completely carpeted floor, window hangings, wood-paneled walls, or other absorptive boundaries or furnishings. Carpeting of the floor is not, in general, the equivalent of treating the ceiling with highly absorptive material. Most carpets are not very absorptive at low frequencies, and much of the carpet may be under tables, where it loses much of its absorptive value at high frequencies. Where the ceiling height is more than 12 feet, absorptive material should be applied to the upper walls as well as to the ceiling. But it is not always a complete or even a satisfactory solution of the acoustical problem merely to add absorptive treatment to the ceiling or walls or both, of a restaurant.

In order to reduce the noise transmitted to the dining room from the outside, a "sound lock" may be needed between dining room and the outer door. A sound lock is almost always needed between the dining space and the kitchen. Noise from other adjacent rooms or from mechanical equipment in the building should be reduced to the acceptable noise level of about 50 to 55 db (see Chapters 11 and 12). Windows opening on a noisy street or court should be avoided.

Industrial Buildings

Many factories, laboratories, assembly plants, and other industrial establishments present a variety of acoustical problems, some of which require special attention. The problems are of two principal types: (1) the protection and well-being of workers within the building, and (2) the noise resulting from normal operations within the building that constitutes a nuisance in its vicinity.

Among buildings of the first type are factories for the manufacture of nearly all the materials required by an industrialized society for the fabrication and assembly of everything from toys to ocean liners. Many noise levels encountered in these factories are not merely annoying; they make speech almost impossible, and they may inflict permanent as well as temporary injury on the hearing of those who are long exposed to them. It is generally regarded that noises in excess of about 100 db should be avoided in order to eliminate the risk of impaired hearing. Yet there are many factories in which the noise level is 120 db or more, a level that not only inflicts permanent injury to the sense of hearing but also induces dizziness and other vestibular disturbances. Every feasible means should be utilized to reduce such hazardous noises to acceptable levels, that is, below 80 db. Segregation of the noisiest operations in sound-insulated rooms, in which both the inner walls and ceilings are treated with highly absorptive material, is advisable. A 2- or 3-inch pad or blanket of mineral wool protected by muslin and hardware cloth often can be used for this purpose. Industry should cooperate by reducing the noise at its source as much as possible. If, after such measures have been taken, the noise remains above the tolerable level, the workers should protect their hearing by means of suitable ear plugs. When properly fitted, good ear plugs provide a noise reduction as great as 30 db.

Among industrial buildings of the second type, namely, those that constitute a noise nuisance in their neighborhoods, are electric and gas generating and distributing stations, factories, mills, assembly plants, and many other offenders. It is necessary to protect those who work in the immediate vicinity of these hazardous noises by the means suggested in the preceding paragraph. It also is necessary to confine these noises within legitimate premises by means of appropriate sound-insulation and sound-absorption structures. Frequently the acoustical problems are of such complexity and importance that the services of an acoustical engineer are required for their satisfactory solution.

Libraries, Club and Recreational Buildings, and Museums

Many persons (including too many architects) are of the opinion that reference and reading rooms in libraries are not in need of acoustical treatment since they are not used for speaking purposes. However, in large untreated rooms, especially if the walls and ceilings are of hard plaster, the noise resulting from the closing of a door, the dropping of a book, coughing, talking, or other activities incidental to the conduct of routine business is so loud and reverberant as to constitute a real annoyance, one which greatly reduces the utility of such rooms. It is axiomatic that silence is advantageous for study or reading in the library or for the contemplation of the works of art on exhibit in museums. Every reasonable effort should be made to secure a quiet environment in these buildings. The reverberation times in reading, reference, museum and work rooms, and in all lobbies, corridors, and stairwells in these buildings, should be reduced to the times recommended for the absorption of noise in offices and workrooms (see p. 352).

Recreational and club buildings and hospitals and sanitariums often contain swimming pools. If the pools are enclosed, they always need acoustical treatment. Water is nearly as reflective as solid concrete; therefore, when the walls and ceiling of a swimming pool are finished with glazed tile or painted plaster, as is customary, the room will be excessively reverberant. The reverberation times are often as long as 6 to 8 seconds. It is advisable to avoid this condition by treating the ceiling and upper walls with an absorptive material that will not deteriorate or lose its absorptivity by the action of the water and high humidity in the room. Cork tile and special types of acoustical plaster have been developed that withstand the moisture and yet have high absorption.

Hospitals

The first order of the doctor for the patient confined in a hospital frequently is "the patient must have quiet." The hospital usually is located in a busy part of the city where the noise level is 70 db or more; the windows of the rooms may be open for ventilation, and there may not be more than 10 or 20 db of isolation. The "quiet" ordered for the patient is seldom realized (see Fig. 10.4). Instead, there is a din of about 50 db from traffic noise as well as noises originating inside the building, such as the crying of infants, the groaning or screaming of suffering adults, and the noises of attendants, utility wagons, and building equipment. Such is the situation in many hospitals where no real effort has been made to secure quiet. An array of acoustical problems is presented by modern monolithic structures with hard plaster walls and ceiling, long unbroken corridors, and patients' rooms containing no absorptive carpets or hangings and having open windows and poorly fitting doors. Practical methods for solving these problems are indicated here.

The first step in the acoustical design of a hospital is to determine the sound-insulation requirements. A noise survey should be made at various locations on the proposed site for the hospital at various times of the day and the night. It is then necessary to make calculations on sound insulation similar to those indicated in the typical problem worked out in detail in Chapter 11.

In laying out the general plan of the hospital, much can be done to minimize the sound-insulation problem. The building should not be located near busy traffic arteries, trolley lines, or other sources of noise and should be set back from the streets. Shielding from noisy streets by garden walls or by tall trees and dense planting is helpful. (See p. 223.) The patients' rooms should be far removed from the approaches for ambulances, doctors' automobiles, and delivery trucks. The heating and ventilating equipment room, the X-ray room, utility rooms, the kitchen, employees' dining room, the administration offices, and the elevators should be thoroughly insulated from the patients' rooms. Covering the inner walls of the elevator shafts with sound-absorptive material will reduce the noise from the elevators and will also suppress the transmission of noise through these shafts. Special sound insulation should be provided for the maternity and nursery rooms and for the rooms of acute sufferers who are likely to be noisy. It is often possible to segregate rooms for such patients in separate wings of the building. The cries of a patient may create sound levels of 85 db in his room. Therefore, an insulation of 50 to 55 db is required between such rooms and those which are to be shielded from these disturbances.

In general, this amount of insulation cannot be attained by a single wall, much less by a single door. It is necessary to use double walls or discontinuous constructions and at least two separate doors between units which require these large amounts of sound insulation. An inspection of the tables of transmission coefficients for different materials and types of wall structure (see Appendix 2) will enable one to select the required type of wall construction.

Although the amount of noise that hospital patients can tolerate without being disturbed varies greatly for different individuals, a level of about 35 to 40 db will be found to be acceptable by nearly all, and this is about as low a level as can be attained without going to extreme types of structures involving prohibitive costs. Furthermore, if the noise level in a room is reduced much below 30 db, the patient may hear his own heart beat or other internal noises which may be a source of annoyance or even worry. However, there is little probability of attaining this degree of quiet under the existing conditions associated with hospitals and the construction of buildings. In order to attain noise levels even as low as 35 to 40 db in the patients' rooms, it is necessary to choose a location in which the average outside level is not greater than about 50 db if open windows are depended on for the ventilation. If the average outside noise level is greater than 50 db, the building should be designed with permanently closed windows, in which case air conditioning must be provided. The use of closed windows for buildings located in noisy sites is the first and the most important expedient in providing quiet in hospitals; they will insure a reduction of at least 25 db of all outside noise.

If the cost of construction or other factors make it mandatory to depend on open windows for ventilation, the use of suitable window mufflers will provide an effective insulation of as much as 10 db. An effective window muffler can be made from a 3-foot length of lined ventilating duct having the width of the window, a height of about 6 inches, and containing a dust filter. The dust filter consists of a section of light-density glass wool which

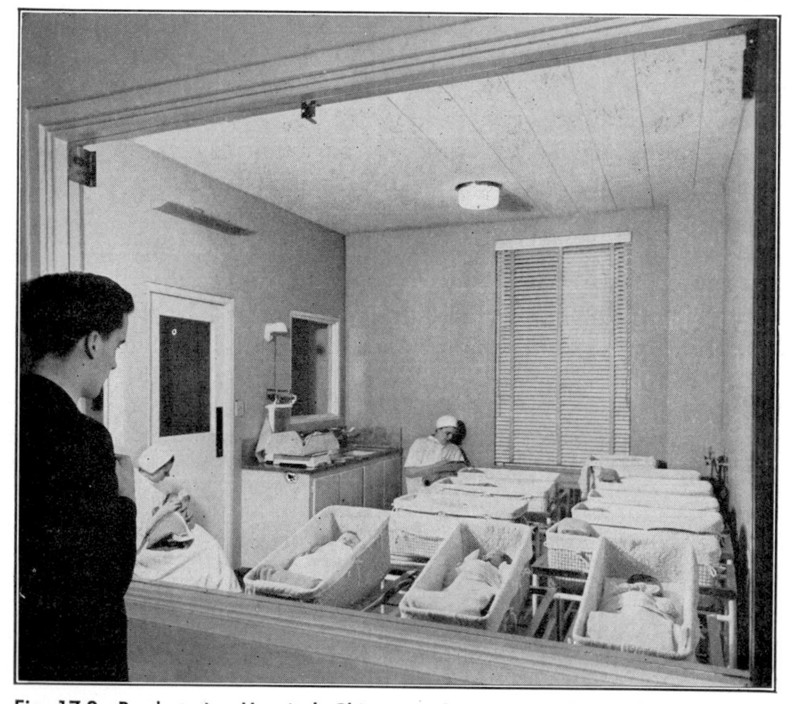

Fig. 17.2 Presbyterian Hospital, Chicago. Acoustone ceiling reduces the noise level produced by crying infants. (Courtesy U. S. Gypsum Co.)

is usually removable. If the muffler is equipped with an electric fan, it can provide a good supply of air for the room. Although a window muffler contributes materially to quiet conditions in the room, it should be borne in mind that its sound-insulation value is not equal to that of a closed window.

A hospital may have rooms which are used for instruction purposes, for clinics, for conferences, and for technical or administrative meetings. These rooms should be treated in such a manner as to provide the optimum acoustical condition for the hearing of speech. (See p. 194.) Wards, private patients' rooms, labor and delivery rooms, nurseries, operating rooms, clinic or classrooms, dining and utility rooms, corridors, and in fact, nearly all rooms in a hospital should be treated with sound-absorptive material to reduce noise. However, it should be clearly recog-

Fig. 17.3 Children's Hospital, Denver, Colorado, Exercising Pool. The upper side walls and the ceiling are treated with Corkoustic. (Courtesy Armstrong Cork Co.)

nized that this is a measure which is additional to, not a substitute for, adequate sound insulation. If the ceilings of such rooms are covered with an acoustical material having absorption coefficients of not less than 0.20 at 128 cycles and not less than 0.60 at 512 to 2048 cycles, the control of reverberation and of sound transmission along corridors will be satisfactory. If less absorptive material, such as acoustical plaster, is used, it should be applied to the upper half of the walls as well as the ceilings. The choice of acoustical materials should receive the most careful consideration. First of all, the material must not interfere with the sanitary requirements of the hospital. This requires a hard, clean surface which will withstand washing or cleaning every two or three months and redecorating every year or so without disintegrating or without losing an appreciable amount of its absorptive value. Mineral-wool blankets covered with a perforated membrane of metal, Transite, or similar material are especially suitable, for they can be cleaned with soap and water. Contrary to the opinion of many lay persons, these acoustical treatments are no more likely to be germ breeders or nests for the proliferation of bacteria than are wood, hard plaster, or other materials used for the walls and ceilings.[1] The floors of all corridors and utility rooms should be covered with a resilient floor covering, such as rubber tile or battleship linoleum. These coverings help to take up the noise of footfalls and utility wagons. Rubber heels on the shoes of attendants and soft tires on utility wagons are advocated as standard "equipment" for all hospitals.

Legislative, Administrative, and Judicial Buildings

Governmental activities have increased enormously in recent years. With this growth there has been a corresponding expansion of the number and size of government buildings and of the diverse functions they serve. The legislative, administrative, and judicial activities taking place in these structures involve hearings, conferences, discussions, court trials, parliamentary debates, and many other activities that involve speaking. Naturally, good acoustics is a prime requirement. There are three principal types of rooms that require acoustical consideration: (1) assembly rooms, including council chambers, legislative chambers, court rooms, committee rooms, and other rooms where meetings or conferences may be held; (2) work rooms, public and private offices, typing and mimeographing rooms; and (3) rooms for rest, relaxation, and refreshment. Everything possible should be done to provide the optimum conditions for the hearing of speech in the first type of rooms. The second type should be designed to facilitate the speed and ease of the work to be performed; the

[1] C. S. Neegaard, *J. Acoust. Soc. Am.*, **2**, 106 (1930).

rooms should be quiet and also free from reverberation. For the third type, quietness is the prime requirement.

Since parliamentary buildings are usually located in a noisy part of the city, it is necessary that they be designed to reduce outside noises to a level of not more than 45 db. Thus, a court room, a council chamber, or an assembly room in an urban locality cannot be maintained quiet enough for parliamentary purposes unless artificial ventilation is provided so that all windows can be kept closed. In such rooms, the windows should be of heavy glass; and in extreme cases they should be double-glazed and separated by an air space. Since assembly rooms are used exclusively for speech, their reverberation times should be in close agreement with the value for speech rooms given by Figs. 9.11 and 9.12. It is particularly important that good acoustical conditions be provided in council chambers and court rooms, where both spectators and participants should be able to hear easily and perfectly. To this end, high ceilings and unnecessarily large rooms should be avoided. A flat ceiling of hard plaster or other highly reflective material is desirable if both the walls and floor can be treated to give a suitably low time of reverberation. A ceiling height of 20 feet should not be exceeded.

A small council chamber, based on an acoustical study, is shown in Fig. 17.4. The dimensions of the plan are 29 by 38 feet, and the ceiling height is 14 feet. The volume of the room is only 15,000 cubic feet. The room should be designed so that its average noise level when unoccupied does not exceed 40 db. In general, this requires tightly closed windows (double-glazed if the room is in a very noisy location); as a result, air conditioning will be necessary. The lobby has its ceiling and the upper half of its walls treated with a highly absorptive tile, and its floor covered with cork linoleum, so that this space acts as a sound lock. The floor for the council table is elevated 1 foot above the main floor. The ceiling and the front wall are reflective (ordinary suspended or furred-out plaster or ¼-inch plywood), the rear wall above a 4-foot wainscot is absorptive, and the side walls have panels of absorptive material above the wainscot. The elevated portion of the floor, the aisles, and the floor between the audience area and council table are carpeted over ½-inch felt pad; the floor under the seats is covered with cork linoleum. The absorptive panels on the side wall facing the windows are staggered to suppress flutter and aid diffusion. The coefficients of absorption of these panels, and of the rear wall, are values that will give the optimum times of

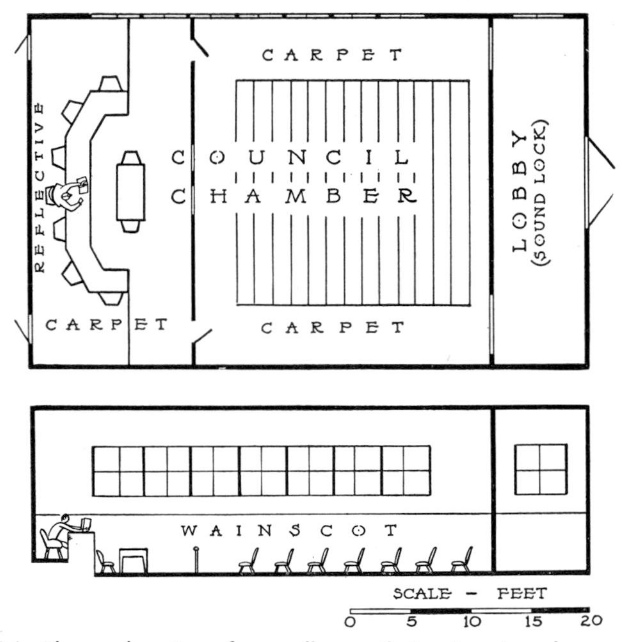

Fig. 17.4 Plan and section of a small council chamber, based on acoustical study. (See text.)

reverberation, namely, 1.0 second at 128 cycles and 0.75 second at 512 to 2048 cycles—the values indicated by Figs. 9.11 and 9.12 as optimum for a speech room with a volume of 15,000 cubic feet. For a room of this size, treated as recommended in this paragraph, a sound-amplification system will not be needed, and all auditors will be able to hear without effort.

Figure 17.4 is suggestive also of the acoustical features that should characterize larger parliamentary rooms (for assemblies, court chambers, etc.). The absorption required to provide the optimum reverberation characteristics for speech can be furnished by the carpet (over ½-inch felt pad) and by the use of absorptive tile or panels on the rear end wall and on the upper side walls, the amounts of absorptive material being calculated as in Table 16.1. When the volume of such a room is larger than about 50,000 cubic feet, it is necessary to use a sound-amplification system; if the background noise is more than 40 db, sound amplification may be necessary even in smaller rooms.

18 · Homes, Apartments, and Hotels

Most people desire quiet residences; those who do not generally want sounds of their own making or choosing. Insulation against unwanted sounds is an all-important problem in the acoustical design of homes, apartments, and hotels, but it is not an easy problem to solve. Although costs frequently preclude a complete solution, much can be done, within the budget prescribed, by good judgment or even by unadorned necessity. Intelligent planning includes good acoustics as one of many desirable features that a residence should have. When all these features are considered together, it often is found that the extra cost of providing a good acoustical environment is fully compensated for by other advantages. Thus, a brick wall may not only furnish the desired sound insulation, which a wood-stud and stucco wall does not furnish, but also its added value for heat insulation, freedom from plaster cracks, and permanence may more than justify its extra cost. The substitution of a sound-absorptive fiberboard in place of plaster on lath for portions of the interior finish of a room may not only reduce the noise level in the room and provide the optimum reverberation, but it may also improve the utility, heat insulation, and appearance of the room.

Homes

Most homes are designed and built without any thought of their acoustics. The outcome often is deplorable: the halls and stairways are "speaking tubes," the dining and living rooms are excessively reverberant, the bedroom windows face a noisy street, and the sound insulation between adjacent rooms is so poor that both quiet and privacy are non-existent.

The most important acoustical problems in the design of residences are the insulation against outside noise and the insulation of sound between rooms. Attention should be given to these problems in laying out plans for a house. Bedrooms and other rooms where quiet is required should be ventilated from windows that open upon the quiet side of the lot. Windows and doors, especially those facing the street or those near the neighbor's house, should be of heavy construction and should fit tightly in their frames. The insulation of bathrooms from the other rooms in the home is a difficult problem. Great care should be exercised in the selection of bathroom fixtures. The architect should obtain data and make careful observations on the amounts of noise produced by the different types of toilets and bathtubs. In order to prevent turbulent flow, water pipes should be large enough to allow relatively low speeds of the water. The more noisy pipes should be wrapped with a fibrous blanket covered on both sides with tough paper, and they should be insulated from the rigid frame of the building by means of flexible connectors. Wherever practical, the entire bathroom floor should be insulated from the rest of the building by means of flexible chairs or by means of felt, mineral wool, or cork board. The floor and ceiling section under the bathroom should have a sound-transmission loss of not less than 45 db.

The rooms in which telephones are located should be reasonably free from reverberation. A telephone located in a reverberant hall imposes difficulties on both hearing and talking. If the telephone is located in a hall or separate room, absorptive treatment of the ceiling and upper portion of the walls of the hall is desirable. Reverberant halls are not only noisy and annoying to persons walking through them, but they also act as effective conveyors of noise between different parts of the home. If the floors and stairways are tile or hard wood, the ceiling or the walls or both should be treated with an absorptive material having a coefficient of sound absorption at 512 cycles of not less than 0.25. If the hall floors and stairways are 75 per cent carpeted, it usually is not necessary to use additional absorption on the walls and ceiling.

Large areas of highly absorptive material in a dining room are very desirable. If the floor is completely carpeted over a ½-inch felt pad, and if the windows are draped with heavy hangings, absorptive material for the walls and ceiling may not be necessary. In one dining room designed for good acoustics, the entire ceiling is treated with an acoustical tile having an absorptivity of about 0.50, the entire floor is heavily carpeted, and doubly lined draperies are used for all windows and for one large door opening. The social conversation of as many as 12 persons sitting at dinner in this room does not produce the usual din of noise that would result if this same conversation took place in an untreated room. In fact, it is easy to converse at one end of the table while general conversation is in progress among the other guests in the room. In like manner, the playful chatter of children is reduced to a level that can be readily tolerated, whereas this same chatter in a reverberant dining room may be well-nigh intolerable.

The breakfast room should be treated with absorptive material, especially if the floor is finished with tile, linoleum, or other highly reflective material. An absorptive blanket with a perforated, washable covering, such as metal, is one of the most satisfactory methods of treatment. This same material can be applied also to the ceilings of the kitchen, especially in large homes. Most acoustical plasters available at the present time are not suitable for use in residences since they offer difficulties in cleaning, washing, and decorating.

Even the modest home which has no other music than that furnished by the radio should have its living room (or the room that houses the radio) designed for good acoustics. The reverberation time should not exceed 1.5 seconds at 128 cycles and should be about 1.0 second at 512 to 2048 cycles. The walls at the end of the room where the radio and the piano are to be located should be somewhat more reflective than the "listening" end. A corner of the room usually is the best location. It often is possible to provide the optimum reverberation by the suitable selection of rugs, hangings, and furniture, if the floor is wood and the walls and ceiling are wood paneling or ordinary plaster backed with air space. Figure 18.1 shows a suggested arrangement of an upright piano, the rug, and the furnishings in a living room 14 feet by 20 feet by 9 feet based on acoustical considerations. With furred-out plaster walls and ceiling and a hardwood floor, the room will have approximately the optimum reverberation if (1) three fourths of the floor is covered with or-

Fig. 18.1 Suggested arrangement of a piano, rug, and the furnishings in a living room 14 feet by 20 feet by 9 feet, based on acoustical considerations.

dinary carpet over a ¾-inch felt pad, (2) 100 square feet of heavy draperies are hung around the windows, and (3) two or three pieces of overstuffed furniture are in the room. Of course, many other arrangements are possible, but the carpet and absorptive furnishings should not be in the "live" end of the room reserved for the radio, television set, or piano. Book shelves, pictures, and other wall "irregularities" should be used to facilitate diffusion of sound and to suppress flutter echoes. Wood seats or other non-absorptive furniture on the uncarpeted portion of the floor will prevent room flutter between the wood floor and the reflective ceiling. In larger residences containing a separate music room, the acoustical design should be worked out along lines suggested on p. 342.

Apartment Houses

Building and health codes provide for many health, safety, and comfort measures in all types of residential buildings—heating, lighting, plumbing, structural strength, etc. Some of the codes in Sweden, England, and Hungary also specify the minimum amount of sound insulation that must be provided in the partitions that separate family units in apartment houses. In Great Britain, the Building Research Station and other agencies have issued several helpful bulletins[1] that provide the prospective builder with authentic information and detailed drawings for obtaining good sound insulation at reasonable costs. Holland has constructed a number of experimental or model apartment houses incorporating several practical methods of sound insulation, so that its builders and prospective home owners can know in advance of construction how much sound insulation will be furnished by each of several methods of construction. The United States has not kept pace with European countries in these matters. Education of the building trades and the public and revision of building codes are long overdue.

The transmission loss (T.L.) of partitions between separate units in an apartment house should not be less than 40 db in low-cost housing, and not less than 45 db in moderate and high-cost apartments. The over-all noise reduction between adjacent units usually will be much less than 40 db when the windows are open, especially when these windows open on a common court. But when the windows are tightly closed the over-all noise reduction will be determined largely by the T.L. of the separating partition. Therefore, this T.L. should not be less than the values recommended above. A number of satisfactory constructions are included in the tables of Appendix 2. Figure 18.2 is a photograph of the Chandler (Arizona) Housing Project. Thick adobe walls separate the family units and project out from the main building walls, providing privacy and more than usual sound insulation.

[1] See R. Fitzmaurice and W. Allen, "Sound Transmission in Buildings," His Majesty's Stationery Office, London (1939); R. C. Bevan and W. Allen, "Party Walls between Houses," Special Report No. 4, Department of Science and Industrial Research, His Majesty's Stationery Office, London (1948).

One of the most troublesome sources of noise in apartment houses is the transmission of impacts through floor and ceiling sections. The "floating" of the finished floor on a sound-insulation blanket or on resilient supports or chairs is an effective means for reducing such noises. A number of approved methods

Fig. 18.2 Photograph of the Chandler (Arizona) Housing Project. Thick adobe walls separate the family units and project out from the main building walls, providing privacy and more than usual sound insulation. (Burton D. Cairns and Vernon DeMars, Architects.)

of floating-floor construction are shown in Figs. 12.2, 12.4, and 12.5. If, in addition to supporting the floor on flexible chairs or blankets, it is completely carpeted over a felt pad at least ½ inch thick, the noise of footfalls and other impacts against the floor will be adequately reduced for even very high standards of noise control.

All plumbing fixtures, pipes, ducts, elevators, and other mechanical equipment should be selected and installed in accordance with the recommended procedures described in Chapters 12 and 13. Covering the floors and ceilings of hallways with sound-absorptive material greatly reduces noises originating in the hallways and often aids in preventing the transmission of noise from one apartment to another. Similarly, the transmission of noise from one part of the building to another often can be minimized by treating the inner walls of elevator shafts with sound-absorptive material.

Hotels

Two types of rooms in hotels require consideration with respect to acoustics: (1) community and social rooms, such as lobbies, corridors, dining rooms, ball rooms, recreation and game rooms, convention rooms; and (2) guest rooms. The principal requirements in the public and social rooms are the proper control of noise and reverberation and adequate insulation against noise from the outdoors and from one room to another. Both these requirements are flagrantly neglected in most hotels. Recently, for example, a scientific meeting and a dinner meeting of a men's club were scheduled for the same evening in a modern metropolitan hotel, the former in the ball room and the latter in a banquet room. The folding doors that "separated" the two rooms provided a transmission loss (T.L.) of not more than 20 db. The speakers in the ball room used a sound-amplification system which had to be operated at a level of about 80 db to emerge above the 60-db roar of the ventilation system. Even so, the raucous singing in the banquet room often drowned out the distorted shouting of the distinguished guest speakers in the ball room. The T.L. between such rooms in a hotel should not be less than 50 db. This and the other acoustical requirements for the community and social rooms in a hotel have been described in Chapter 17.

The acoustical problem in guest rooms is primarily one of sound insulation. The amount of insulation between adjacent rooms in a hotel usually is determined by the window route if open windows are depended on for ventilation. Under such circumstances, the T.L. would be less than 25 db; little would be gained by providing walls with a high T.L. However, when the windows are tightly closed, as they would be if the rooms were air-conditioned, the T.L. via the window route between adjacent rooms may be as high as 55 db. For this reason, the partitions between guest rooms in air-conditioned hotels should be designed with a higher T.L. than in those where the ventilation is furnished by open windows. A T.L. of not less than 40 db is recommended for low-cost hotels, and not less than 45 db for first-class ones. The noise from the air-conditioning ducts and other equipment should not exceed a level of about 30 db. If separate guest rooms are connected by doors, there should be two doors in each frame. Both doors should be of solid panel type; they should fit tightly in their frames and be separated by an air space of not less than 3 inches.

The exterior walls of hotels in noisy locations should be designed to provide a T.L. of not less than 40 db. In very noisy locations, the T.L. should be not less than 45 db, in which case double-sashed windows separated by an air space of at least 4 inches are required to match the insulation of the walls. This calls for cased windows and therefore air conditioning. A number of wall and ceiling-floor partitions providing the required T.L. will be found in the tables on sound insulation in Appendix 2. The floors of all corridors should be carpeted over a ½-inch carpet pad.

All pipes and ducts should be isolated from partitions and the solid structure of the building by hair felt or other flexible materials. Plumbing fixtures that operate quietly should be selected. Tank-type toilets are usually less noisy than the valve type. The wrapping of pipes with paper-covered mineral-wool blankets helps to minimize the noise from near-by toilets, showers, and other parts of the plumbing installation. Mechanical equipment, such as fans, motors, elevators, should be well isolated, so that both guest and social rooms will be quiet places for relaxation and rest.

19 · Church Buildings

The multifarious functions of the church and its rituals, traditions, and strivings for architectural beauty have affected profoundly the design of church buildings. Nearly all existing forms have evolved from the oblong, the circle, or the Greek or Latin cross. The more complicated forms consist of a number of spaces coupled together. The acoustical design of churches with these complex shapes involves consideration of the acoustical properties of each of the spaces separately as well as in combination. Thus, the organ chamber and choir loft require the best environment for the generation of music; the chancel should provide optimum conditions for the spoken service; the nave and transepts require the properties of a good listening environment for both speech and music; and all spaces within the church require quiet surroundings which are conducive to undisturbed meditation and prayer.

It is necessary to recognize the general nature of the acoustical requirements for churches of different denominations. The optimum reverberation time for churches of different faiths and denominations depends on their size. The values of reverberation time at 512 cycles given in Fig. 19.1 are for a two-thirds capacity audience, and they have proved satisfactory for purposes of design of a large number of churches. This chart is based upon the requirements for both speech and music and upon the nature of the services conducted. Observations and measurements in many churches which have excellent acoustics show that these churches have times of reverberation comparable with those shown in the chart. The lower part of the band should be used for Christian Science churches because of the predominant importance of the spoken service; the upper part of the band should be used for cathedrals and Roman Catholic churches;

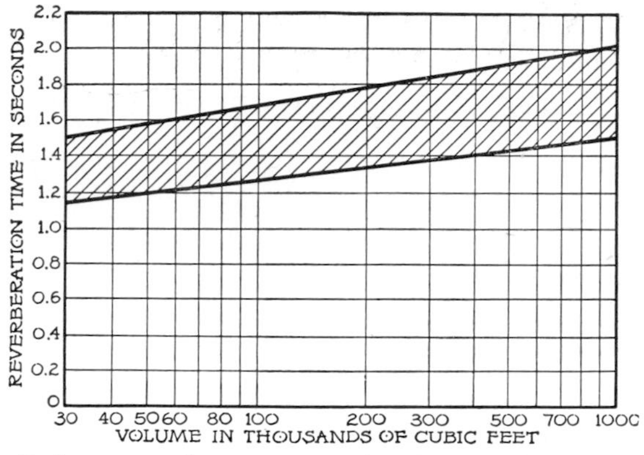

Fig. 19.1 Optimum reverberation times for church auditoriums. The lower part of the band should be used for churches in which speech is especially important, as in Christian Science churches; the medial part for churches in which speech and music are about equally important; and the upper part for cathedrals and churches in which music is a predominant part of the church service.

and the medial part of the band should be used for most Protestant churches and Jewish synagogues. In the Christian Science church it is necessary to provide good acoustical conditions for speaking, reading, singing, and organ music which originate on or near the platform, and, in addition, good conditions for speaking which may originate at any seat in the auditorium. Many synagogues are nearly octagonal in plan and have a high, domed ceiling. They therefore must receive special acoustical treatment in order to avoid echoes, sound foci, and long-delayed reflections.

The necessity for insulation against outside noises, in all churches, cannot be too strongly emphasized. The church should provide a refuge where one is not disturbed by the noise and turmoil of the outside world. Chapels and confessionals require even greater sound insulation. As an example of good acoustical design with respect to the insulation against outside noise, consider the precautions taken in the construction of one church located on a busy thoroughfare where the average noise level is 65 to 70 db. The walls are of heavy reinforced concrete; there are two sets of heavy and tightly fitting doors between the outside and the inner auditorium; there is an additional wall between the auditorium and two surrounding corridors; the corridor and the vestibule are treated with absorptive material; the windows are of heavy, leaded glass, are permanently closed, and the ratio of the window area to the total wall area is low, so that the amount of noise transmitted through the windows is relatively small. The ventilating equipment room is well removed and insulated from the auditorium so that the noise reaching the auditorium from this source is negligible. The noise-insulation factor provided by the combined effects of insulation and absorption in the auditorium amounts to slightly more than 40 db; therefore the residual noise from the outside does not exceed 30 db.

We shall begin with a consideration of the simplest form of church—a rectangular room with a gabled ceiling. Then we shall consider larger churches which usually have a more complex form. No attempt will be made to propose acoustical designs for cathedrals, since each cathedral presents highly specialized problems that merit the services of an acoustical expert. However, one example is considered in the case studies at the end of the chapter because its acoustical defects are typical of those found in many large churches.

Small Churches

The small church is usually a long, rectangular room with the chancel at one end and the nave at the other. Figure 19.2 shows an arrangement of organ, choir, pulpit, and lectern which conforms to good acoustical design. The pulpit and lectern are both near the audience. The organ speaks directly through the chancel into the nave and is behind the choir, where it can support rather than submerge the singers. This location of the organ is far better than the one so frequently assigned to it, namely, on either side (or both sides) of the chancel or at the rear of the church.

Reasonable precaution should be taken in insulating the church against outside noise, or against noise which may originate in adjacent rooms, in order to reduce it to an acceptable level of 35 to 40 db. If the building is located on a quiet country road it is probable that no special measures need be taken, but if it is located on a busy highway or if there are other sources

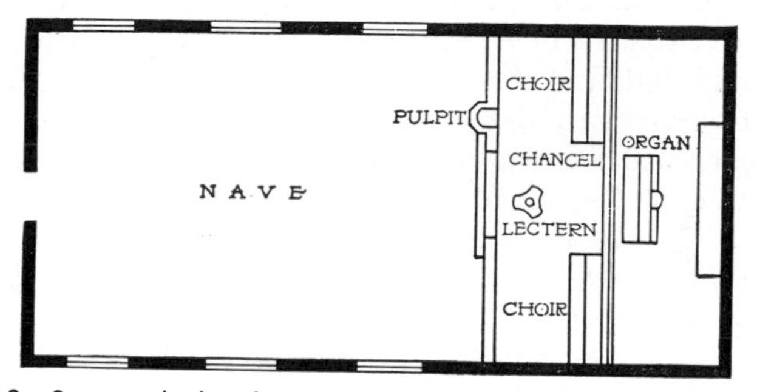

Fig. 19.2 Suggested plan for country church, based on acoustical requirements, showing location of organ, choir, pulpit, and lectern.

of noise in the vicinity of the church, the walls, ceiling, doors, and windows should be designed to provide adequate insulation against all existing or probable future noises. The manner of making the calculations and choosing the proper types of construction for the required amount of insulation is worked out in Chapter 11, and an example of such calculation is given in Chapter 16. If the walls are of stone or brick and if the ceiling has a comparable degree of insulation, it is likely that the problem of sound insulation will consist only of the addition of a narthex and the provision of heavy and tightly fitting doors and windows.

The prime consideration in the acoustical design of the small church is the proper control of reverberation. In general, the walls, floor, and ceiling of the chancel should be finished with acoustically reflective materials, such as wood sheathing, furred-out wood paneling, or ordinary plaster on lath. The walls and ceiling of the nave should be finished with such materials as will provide the optimum reverberation. The method for calculating the required amount of acoustically absorptive materials has been described previously. (For an example of similar calculations see Table 15.1, p. 311.) There are many acoustical materials listed in the tables of Appendix 1, any one of which, if applied in the required amount to suitable wall and ceiling surfaces of the nave, will provide the desired reverberation characteristics. A ½-inch fiberboard ceiling often will provide most of the required absorption. (See also Chapter 7 for special sound-absorptive constructions, which in some cases may yield a more practical solution of the reverberation problem than will standard acoustical materials.)

Suppose that the small church shown in plan in Fig. 19.2 has a length of 90 feet, a width of 40 feet, and an average height of 35 feet. Because the church accommodates 300 persons the optimum time of reverberation should be provided for 200 persons. Suppose that the entire floor is of pine and that the 1100 square feet of aisle and chancel floor are covered with carpet strips over ½-inch felt padding. With the carpet on the floor of the chancel, no other absorptive material need be used for the walls and ceiling of this space if its walls are furred-out plaster and its ceilings are wood.

The volume of the church is 126,000 cubic feet and the interior surface is 16,300 square feet. The optimum time of reverberation is 1.4 seconds at 512 cycles. In order to determine which materials, when applied to the walls and ceiling of the nave, will provide this time of reverberation, it is necessary first to calculate the total amount of required absorption. This can be done readily by the use of the nomograph in Fig. 8.11. The absorption supplied by the audience, wood floor, carpet, pews, and ceiling should be listed in tabular form, as in Table 15.1, and the additional required sabins of absorption determined as on p. 313. For the small church here considered, this additional absorption can be supplied by an acoustical plaster listed in Table A.2, Appendix 1. If the plaster is applied to the walls of the nave above the wainscot, it is possible to obtain the optimum reverberation time *vs.* frequency characteristic. Deviations of less than 0.15 second will not be recognized by the average listener, and they are permissible.

If the acoustical design of a small church conforms to the principles and recommendations set forth in this section, speech will be heard satisfactorily in all parts of the nave and chancel, and music will have the required reverberation to balance the separate tonal components and blend the harmony.

Large Churches

The acoustical problems of the large church differ from those of the small church principally because (1) the volume is larger, (2) the shape is more complex, and (3) it usually is in a noisier location. The large church occasionally is of simple rectangular plan with a relatively low ceiling. More often it is either (*a*) of cruciform plan with a high, vaulted nave, narrow aisles, fluted columns, long and narrow chancel and adjacent transepts, or (*b*) of octagonal plan with a high, domed ceiling, with or without a balcony. The acoustical problems become more important and more difficult as the size of the auditorium increases and as its shape becomes more complex. Also, since large churches are likely to be located near sources of traffic noise or near industrial plants, careful consideration must be given to the shutting out of these disturbances.

In plan, a large church differs from the small church shown in Fig. 19.2 in two essentials: (1) both the length and the width will be larger; and (2) it is likely that transepts and a narthex will be added to the nave. The addition of transepts to the nave will alter the distribution of sound in the auditorium. Since this type of church is nearly always of large dimensions, the problem of supplying an adequate amount of sound to all auditors is likely to present a real difficulty. The pulpit and lectern should be well elevated and near the audience, and, if possible, there should be large and near-by reflecting surfaces either above or behind both the pulpit and the lectern. The shape of each church design should be studied with regard to the proper distribution of sound to all auditors and to the elimination of echoes, interfering reflections, and sound foci. This usually can be done by a study of sound rays on pencil sketches of the proposed design. (For example see Fig. 9.9.) Three-dimensional models, described in Chapter 9, can be used as a check, and they are especially useful for very large and complicated structures.

The addition of transepts to the nave should not affect the arrangement of the chancel, choir, or organ chamber; the arrangement of these spaces shown in Fig. 19.2 for small churches should be followed as closely as possible in the larger church. A close liaison between the architect and organ builder should be maintained in planning a church. Frequently, the space allotted for the organ is inadequate. Poor location of the organ with respect to the choir is a characteristic of most churches. Usually, the organ is above both the choir and the organ console and in such a position that too little organ tone reaches the choir and organist. The singers complain that they cannot hear the organ; and, the organist, in the effort to hear his own playing, registers too heavily for the choir or soloist. The softer stops used for accompanying the singers should speak into the choir loft directly; the heavier diapasons and reeds may be farther away.

In general, it is necessary to use a certain amount of absorptive material on the walls and ceilings of the nave and transepts to obtain the optimum condition of reverberation. The absorptive material should be distributed (preferably in panels, strips, or patches) throughout the nave and transepts so that there will be good diffusion and approximately the same rate of growth and decay of sound in all parts of the auditorium. As in the small church, the chancel should not be *overtreated* with absorptive material. If as much as 50 per cent of the floor area of the chancel is covered with carpet, the walls and ceiling should be finished with materials having an absorptivity comparable with that of wood or ordinary furred-out plaster. If no carpet for this area is contemplated, a portion of the upper walls, or the ceiling, should be treated with a suitable absorptive material, but the chancel should not contain large amounts of absorptive material.

The calculations for the proper control of reverberation will be similar to those outlined for the small church; the procedure to be followed is given in Table 15.1. These computations will indicate the amount of absorption required at different frequencies to provide the optimum reverberation time *vs.* frequency characteristic. The materials in the tables of Appendix 1 will provide a choice wide enough to make possible the selection of materials that will supply the necessary absorption and will also be in keeping with the requirements of permanence, maintenance, and the entire plan of the architectural treatment of the interior. It generally will be preferable to apply absorptive material to the walls rather than the ceiling, as the treatment of the walls with such material will prevent multiple reflections between parallel surfaces, and will help to provide a more uniform rate of growth and decay of sound in the entire enclosure.

Acoustical tiles of a masonry-like type are especially suitable for the control of reverberation in large city churches. Type I prefabricated units (see Table A.1, Appendix 1) are recommended for wall ashlar or vaulted surfaces. They form a structural part of the wall and are entirely fireproof. Often acoustical plaster or a sprayed-on plastic absorbent can provide a better treatment for vaulted and other curved surfaces than other materials. When acoustical plaster is used, thorough supervision of its application is required to make sure that its absorptivity is not sacrificed by the inexperience or carelessness of the plasterers. The special sound-absorptive constructions described in Chapter 7 may provide the means of controlling the reverberation without resorting to acoustical tiles or plasters. For example, selected wall areas comprising a lattice of bricks, wood, or wrought iron, backed with mineral wool, for selected wall areas may furnish both the required absorption and the type of interior treatment the architect wants to use.

The problem of sound insulation can be worked out along the lines already discussed in connection with the small church. The addition of the narthex to the nave provides a means of securing two sets of doors for better sound insulation between the nave and the outside. Provision should be made to keep both sets of doors—those between the nave and the narthex and those between the narthex and the outside—closed during services. The ceiling of the narthex should be treated with absorptive material, and in noisy locations both the walls and the ceiling should be treated. The use of carpets on the aisles not only contributes to the amount of absorption required for the optimum condition of reverberation, but also reduces the noise of footfalls of those who leave or enter the church during services. With heavy, rigid walls and ceiling, with heavy and tightly fitting doors and windows, with the proper amount and distribution of absorptive material within the nave, transepts, and chancel, and with a good sound-amplification system, the large church of conventional size and shape can have good acoustics.

The modern church in the city is much more than the traditional place of worship in which the formal services of the church are conducted. It usually is a religious and social center that provides a Sunday School assembly room, a social hall, club rooms, dining room, classrooms, and rooms for many other activities. Good acoustical design in the modern city church requires that careful consideration be given to the treatment of all these rooms. Most of them require acoustical treatment not essentially different from those of school buildings, and their acoustical design should be worked out along the lines described in Chapter 16. In general, where the rooms are of conventional rectangular form with ceiling heights less than 30 feet, it is sufficient to treat the walls or ceilings, or both walls and ceilings, of the several rooms with materials of suitable absorptivity and to provide an adequate amount of sound insulation to prevent noise interference between adjacent or near-by rooms which are to be used at the same time.

The privacy desired for confessionals often calls for special sound-insulation and sound-absorption measures, especially where there are a number of confessionals grouped together. The Burgess Acousti-Confessional Panels have been designed especially for the purpose. They are rigidly constructed, with a solid outside panel and a perforated inside panel backed with a sound-absorptive blanket. They are 1¾ inches thick and come in a variety of sizes from 36 inches by 42 inches to 42 inches by 84 inches, and they can be incorporated easily into the walls of the confessional by means of nails, screws, or bolts.

A large church, especially one in which the spoken service is important, requires a sound-amplification system. It is highly desirable that this be designed so that only one loudspeaker, or a single group of loudspeakers, need be used. The loudspeakers should be located as near the sound source (pulpit or lectern) as possible, so that the sound the audience hears appears to come from its true source and not from some far-removed loudspeaker. The sound-amplification system should be so designed, installed, and operated that the audience will not be aware of its use.

This requires a restricted type of amplification—an increase in sound level for those portions of the church that otherwise would have an inadequate level. This often can be done by the judicious use of directional high-frequency loudspeakers. When sound-amplification equipment is used, it usually is necessary to treat the rear wall of the nave, and possibly other surfaces, with absorptive material in order to prevent echoes and long-delayed reflections from these surfaces.

Acoustical Studies of Four Churches

The nature of the acoustical problems that often arise in churches and practical methods which have proved satisfactory for solving these problems will be further considered by outlining the results of surveys in four churches.

I. An acoustical study was made of a Protestant church located on one of the busiest traffic arteries in Los Angeles. It has a volume of 350,000 cubic feet. The nave is 49 feet wide, 119 feet long, and has an average height of 45 feet. The narthex, the ceiling of which is covered with absorptive material, opens onto the noisy street. Absorptive material in the church consists of a ½-inch stenciled fiberboard in the gabled ceilings of the nave, transepts, and chancel, and of acoustical plaster on the walls of the aisles, transepts, and chancel. The other wall surfaces are hard plaster, masonry, or glass. The aisles of the nave are carpeted, and there are velours-covered cushions in the seats of all pews.

Measurements made in the empty church indicated fairly uniform reverberatory conditions in all parts of the seating area. The reverberation times were 2.2 seconds at 150 cycles, 2.7 seconds at 300 cycles, 2.5 seconds at 500 cycles, 2.3 seconds at 1000 cycles, 2.2 seconds at 2000 cycles, and 1.7 seconds at 4000 cycles. The optimum reverberation time for this church is about 1.45 seconds at 512 to 4096 cycles and about 2.0 seconds at 128 cycles. With the average audience of 600 present, the reverberation time will be somewhat greater than optimum. In order to reduce it to the optimum value, absorptive materials should be applied to the side walls of the nave and to the side and rear walls above the balcony; about 1850 square-foot-units (sabins) at 512 cycles and about half this amount at 128 cycles and at 2048 cycles should be added.

Noise measurements were also conducted in the church at a time when outside traffic was quite heavy. The doors between the narthex and the nave were closed, but one of the three doors between the narthex and the outside was open, simulating conditions of use during Sunday morning services. Noise from the ventilation system was not objectionable. The noise level at the rear of the nave (measured with a standard sound-level meter with the 40-db weighting network) often reached values as high as 55 db when busses or speeding automobiles were nearest to the church—about 40 feet from the outside doors. The level was as low as 35 db during intervals of minimum traffic, but above 40 db at least 25 per cent of the time and above 45 db about 10 per cent of the time. The noise level in the central part of the nave is about 5 db less than it is in the rear. Thus, the noise is greatest in the seats that are farthest removed from the pulpit and lectern, an undesirable situation that often exists in churches located on busy streets. The rear seats, therefore, are the poorest ones for hearing not only because the ratio of direct to reverberant sound is low, but also because the noise level is high.

The survey indicated that there were three acoustical defects in the church auditorium: (1) poor sound insulation, (2) slightly excessive reverberation, and (3) an inadequate sound level for the hearing of speech. A study of the data also revealed that, for a limited expenditure of funds, more improvement for the hearing of speech would be derived from the installation of a good sound-amplification system than from spending the same funds for improving either (1) or (2). Accordingly, a modern sound-amplification system was installed. This increased the average syllable articulation of speech from about 70 per cent to about 85 per cent. Because of the high noise level in the church it is necessary to operate the sound system at a rather high level. Since the correction of the inadequate sound insulation would be prohibitively costly, the correction of the excessive reverberation has been recommended as the second step in providing this church with good acoustics.

II. An acoustical study was made in a church in which it was required that speakers located at any position in the seating area be heard well in all other parts of the auditorium. The plan of this auditorium is shown in Fig. 19.3. It is a Christian Science

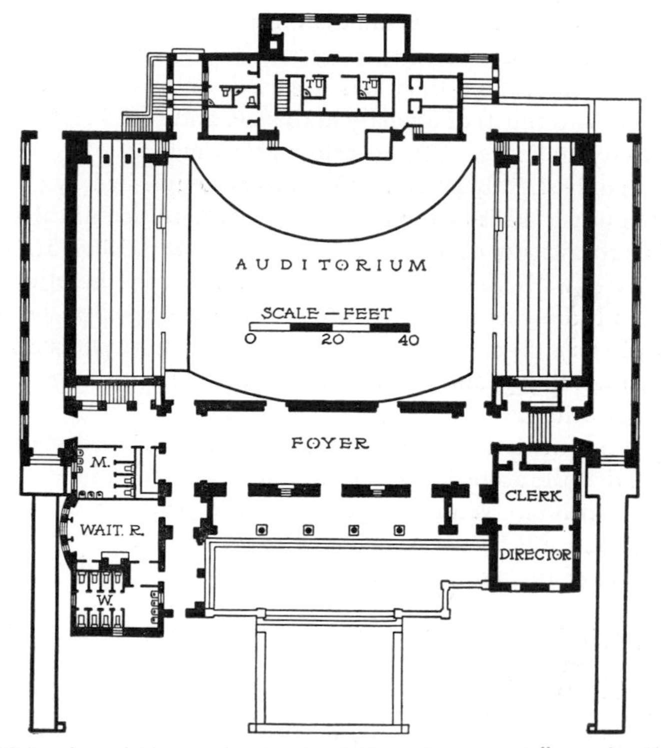

Fig. 19.3 Plan of Thirteenth Church of Christ Scientist, Hollywood. (Allison and Allison, Architects.)

church of moderate size (about 300,000 cubic feet), seating 1250 persons. The wall behind the readers' platform gives a good reinforcement to the readers' voices, and the relatively low ceiling provides a beneficial reflection to speech originating at any seat in the auditorium. The platform is well elevated, and it has good audition lines as well as good sight lines.

All the usual precautions were taken to provide a quiet auditorium. The church is located on a rather quiet side street where the noise level is rarely above 50 db. However, the insulation was made good enough to provide acceptable conditions inside the auditorium even with an outside noise of 60 db. All doors and windows are of heavy, rugged construction and are fitted carefully into their frames so that all threshold cracks are closed. The entire floor of the foyer is carpeted over a felt carpet lining.

The reverberation time was reduced to 1.5 seconds at 512 cycles for two thirds of capacity audience, and to 1.3 seconds with a capacity audience. The reverberation times at lower frequencies are only slightly longer. This was accomplished by (1) treating the upper walls of the auditorium with an acoustical plaster having coefficients of 0.16 at 128 cycles and of 0.18 at 512 cycles (the lower walls were left in hard plaster to give helpful reflections and to prevent possible rubbing or dusting off of the acoustical plaster), (2) covering the entire floor with carpet strips over ½-inch felt lining, and (3) installing upholstered seats throughout the auditorium.

Speech-articulation tests conducted in the finished church revealed that the acoustical properties were very good in all parts of the auditorium. With the speaker at the platform, the syllable articulation was 85 per cent in the front and 82 per cent in the rear part of the auditorium.

III. Churches of some faiths customarily have a high, domed ceiling. Special acoustical problems arise for shapes of this type. The nature of these problems and their solution is indicated by an acoustical study made in the design and construction of a synagogue in Los Angeles. Figure 19.4 shows an outline of the plan of the main auditorium. It is 100 feet wide, 100 feet deep, has a ceiling height to the soffit of the dome of slightly over 100 feet and a volume of about 800,000 cubic feet. All seats are located as near as possible to the speaker's platform, and they receive advantageous reflections from wood paneling directly behind the platform and from the diverging side walls (also of wood paneling up to a height of about 16 feet from the floor). The platform is elevated 4 feet above the main floor, and the rear seats are elevated sufficiently to allow all auditors on the main floor an abundant supply of direct and once-reflected sound. The choir loft is elevated 15 feet above the speakers' platform and is backed by wood paneling. The organ is located just above and behind the choir loft, and it speaks directly into the main part of the auditorium.

During the early stages of the design, acoustical studies were made by means of spark photography in small sectional models of the auditorium. These studies indicated the presence of two troublesome reflections: (1) from the ceiling surface above the choir loft, and (2) from the rear half of the dome. The first reflection was converted into a beneficial one by elevating the choir loft. The second reflection, which would have produced a distinct echo in the front central part of the main floor, was overcome by penetrating the soffit of the dome with deep coffers, 12 to 16 inches deep, and treating the panels of the coffers with highly absorptive tile.

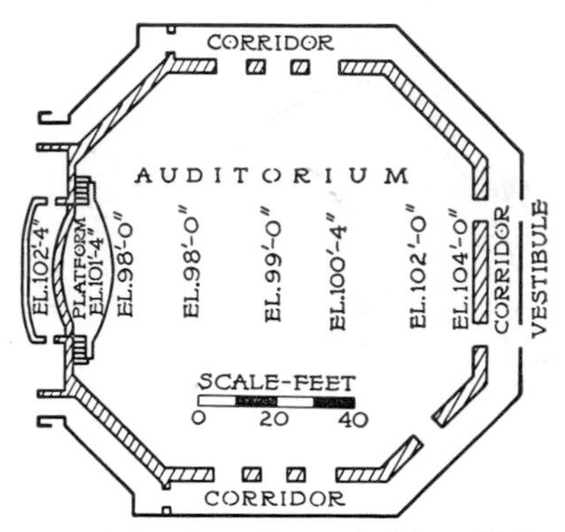

Fig. 19.4 Plan of Temple B'nai B'rith, Los Angeles. (A. M. Edelman, Architect; Allison and Allison, Consulting Architects.)

In order to secure optimum times of reverberation it was necessary to use large areas of highly absorptive materials for nearly the entire inner boundaries of the auditorium because the volume is so large. In fact, all surfaces except the wood wainscot and the frescoed frieze above the wainscot are acoustical materials: the panels in the coffers of the dome are a highly absorptive acoustical tile (0.30 at 128 cycles and 0.75 at 512 cycles); the ribs of the coffered dome are of cast acoustical plaster; the walls above the frescoed frieze are of acoustical plaster (0.12 at 128 cycles and 0.19 at 512 cycles); the walls of the aisle are of acoustical tile; the main and balcony floors are completely carpeted; and all pews are made up of heavily upholstered chairs with both backs and seats padded and covered with porous fabric.

Unusual care was exercised in the selection of the acoustical plaster. Nine small Sunday School rooms were plastered, each with a different type of acoustical plaster, and absorptive tests were conducted in all rooms to determine the absorptivity of the different plasters at frequencies of 128, 512, and 2048 cycles. The final selection of the acoustical plaster was based upon the results of these tests and upon the other physical properties of the plasters, such as tensile strength, texture, color, and ease of maintenance. After the selection was made, the plastering contractor was required to duplicate the approved plaster with respect to thickness, texture, color, and porosity. Porosity tests were made with apparatus similar to that illustrated in Fig. 6.12 during the application of the plaster.

After the auditorium was plastered, but before it was furnished with carpet and upholstered seats, absorptive tests were conducted to determine the grade and amount of carpet required to give the optimum condition of reverberation. The times of reverberation in the unfurnished auditorium were:

128 cycles	4.1 sec
512 cycles	3.4 sec
2048 cycles	2.6 sec

These tests indicated that all aisles on the main floor and in the balcony should be covered with a heavy grade of carpet strips over ¾-inch Ozite; the remainder of the floor space, both on the main floor and in the balcony, should be carpeted with carpet strips over ½-inch Ozite; and heavily upholstered chairs should be used throughout the auditorium. The resulting times of reverberation in the completely furnished auditorium were:

	128 Cycles	*512 Cycles*	*2048 Cycles*
No audience present	3.2 sec	2.0 sec	1.8 sec
600 persons present	3.0 sec	1.9 sec	1.6 sec
1200 persons present	2.9 sec	1.8 sec	1.5 sec
1800 persons present	2.8 sec	1.7 sec	1.4 sec

These times of reverberation are highly satisfactory for either small or capacity audiences and for either the spoken or musical service. Speech-articulation tests in the finished auditorium gave a syllable articulation varying from 88 per cent in the front part of the main floor to 77 per cent in the rear of the balcony.

IV. The acoustics of a large cathedral in New York has been investigated by a group of Bell Telephone Laboratories engineers headed by C. F. Eyring. The cathedral is a large Gothic structure with a volume of 2,230,000 cubic feet and a seating capacity of 2240. The room is approximately 300 feet long and 100 feet wide. The ceilings are arched, and in the nave and transepts the average ceiling height is 106 feet. A rough plan of the auditorium is shown in Fig. 19.5. The interior surfaces are of stone, plaster on wood lath, marble, and wood.

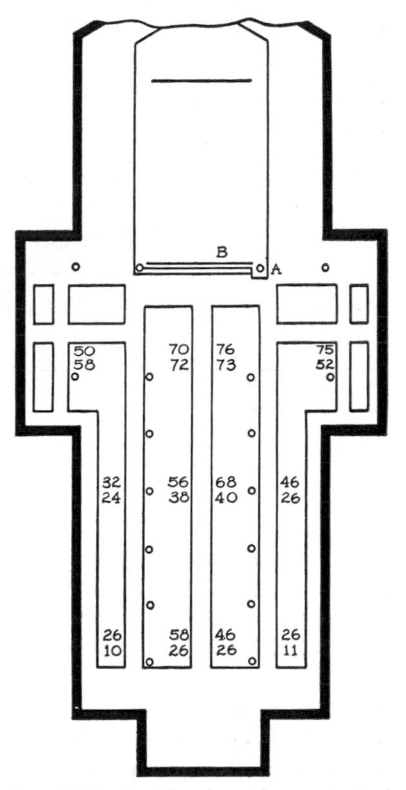

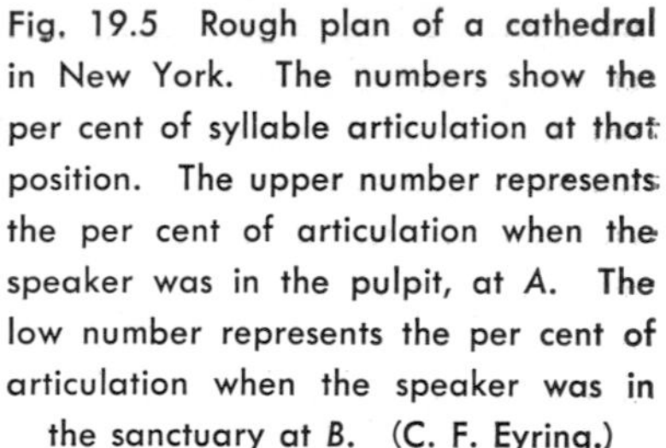
Fig. 19.5 Rough plan of a cathedral in New York. The numbers show the per cent of syllable articulation at that position. The upper number represents the per cent of articulation when the speaker was in the pulpit, at *A*. The low number represents the per cent of articulation when the speaker was in the sanctuary at *B*. (C. F. Eyring.)

Noise measurements taken in the cathedral during the early part of the day, with the doors open as is customary, revealed that the average level is 47 db, with peak levels often rising to 55 db. Figure 19.6 shows the reverberation time *vs.* frequency in the empty cathedral. The decline in the reverberation time for frequencies above 300 cycles is due to absorption in the air which had a relative humidity of about 60 per cent. Measurements of the reverberation time at 4000 and 6000 cycles, for different relative humidities, are in good agreement with calculated values based on the curves of Fig. 8.10, if it is assumed that the average absorption coefficients of the interior surface materials are the same (0.07) at 4000 and

6000 cycles as at 500 cycles. As an indication of influence of the humidity on the reverberation time at 6000 cycles, the measured reverberation time was 2.5 seconds for relative humidity of 63 per cent and only 1.9 seconds for a relative humidity of 50 per cent.

When this survey was conducted, there was a large parabolic fan-shaped "sounding board" above and behind the pulpit; it

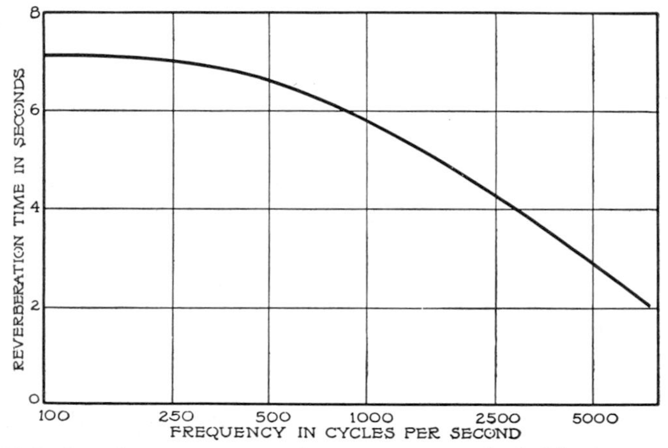

Fig. 19.6 Reverberation time vs. frequency in a cathedral having a volume of 2,230,000 cubic feet, the plan of which is shown in Fig. 19.5. (C. F. Eyring.)

had a width of about 14 feet and a focal length of about 3½ feet. Syllable-articulation tests conducted in the cathedral are summarized in Fig. 19.5. The top number in each pair of figures is the percentage syllable articulation at that listening location for the speaker at the pulpit (*A*) where he is "aided" by the sounding board; the second number gives the articulation for the speaker in the position (*B*) away from the sounding board. The articulation varies greatly at the twelve different locations throughout the nave, transepts, and aisles. The reflector is beneficial in the central and rear portions of the nave (along the beam) if the speaker is at the focus of the reflector. Thus, the average articulation in the nave, when the speaker uses the reflector, is 63 per cent, but when he does not use it (that is, when he is at position *B*) the average articulation in the nave is only

46 per cent. Although a directional reflector provides improved hearing conditions along the axis of its beam, the conditions off the axis are usually worse. Although such a reflector is often helpful, especially in noisy or reverberant rooms, it should not be regarded as a substitute for good acoustics.

The acoustics of the cathedral could be improved by the following measures:

(1) The installation of sufficient absorptive material (carpets, curtains, tapestries between the columns, and acoustical plaster or tile on certain wall and ceiling surfaces) to reduce the reverberation time to 2.2 seconds at 512 cycles and to correspondingly appropriate times at other frequencies. A reduction of the reverberation time below 2.2 seconds would, of course, greatly improve the speech intelligibility; however, such a reduction would not only be very costly, but it also would impair the acoustics for organ and choral music and for ritualistic chanting.

(2) The removal of the directional reflector and the installation of a sound-amplification system capable of providing an average level of undistorted speech of 70 db throughout the auditorium. If the reverberation time is not reduced, the amplification should be greatest at high frequencies. Such a selective increase of level at the high frequencies will greatly improve the recognizability of the consonants. Furthermore, since loudspeakers are directional at high frequencies the amplified sound can be directed toward the rear of the church, where the need for an increase in the sound level is greatest.

20 · Radiobroadcasting, Television, and Sound-Recording Studios

Acoustical problems of a special nature arise in the design of rooms used primarily for microphone pick-up. These rooms include radiobroadcasting, television, and sound-recording studios. Since the acoustical requirements are essentially the same for these, the word *studio,* in this chapter, will refer to any room in which the sound is usually picked up by a microphone. The special nature of the problems is primarily a consequence of the differences between monaural and binaural hearing.

When we listen to sound with both ears, as we usually do, the vibrations reaching one ear differ in loudness, quality, and time from those reaching the other ear. These differences enable us to determine with fair accuracy the location of a source of sound and to focus our attention upon this source, excluding to a large extent sounds coming from other locations. Thus, when we listen to an orchestra, we are aware of the locations of the various instruments that make up the ensemble. However if we listen with only one ear to the orchestra, our ability to localize the different instruments is diminished, *the room seems to be much more reverberant, and adventitious noises seem to have been greatly augmented.* When we listen to music that has been picked up by a microphone, it sounds very much as though we were in the room, listening with only one ear. Consequently, microphone pick-up is subject to about the same peculiarities as is monaural hearing, and therefore a studio should have about the same acoustical properties as would be required for the most favorable conditions for listening with one ear.

The four most important acoustical requirements that should be considered in the design of studios are:

(1) The optimum reverberation time over a wide range of frequencies.

(2) An unusually high degree of insulation against extraneous noise and vibration.

(3) The optimum diffusion of sound.

(4) Freedom from objectionable room resonance. The "transmission-frequency" characteristic of the room (for example, see Fig. 8.1) should be as uniform as possible.

The frequency range for which both sound insulation and reverberation time are considered in the design of studios is from about 50 to 10,000 cycles, which is greater than it is for most other types of rooms. For the routine calculations of reverberation and sound insulation in studios, the methods of "geometrical acoustics" are useful. However, consideration should be given the principles of "physical acoustics" (see Chapter 8) in determining the most favorable location of absorptive material in the studios and in determining the dimensions of the studio.

In view of the foregoing requirements, affected as they are by the peculiarities of monaural hearing, it is not surprising that a room which is held in high repute as a concert hall may not be ideal for the broadcasting or recording of music.

Optimum Reverberation Times for Studios

It is almost universally recognized that the optimum reverberation times for broadcasting and recording studios are shorter than the corresponding times for rooms in which speech and music are listened to directly with both ears. One reason for this results from the differences between monaural and binaural hearing. Another reason concerns the fact that sound that is picked up in a studio is usually reproduced in another ("listening") room. The apparent reverberation time in the listening room is greater than the actual reverberation time of the room. It depends to some extent on the reverberation time of the studio. A room that has the optimum reverberation characteristics for speech or music, as determined by Figs. 9.11 and 9.12, seems excessively reverberant when judged by the quality of reproduced sound which has been picked up in that room.

This apparent exaggeration of reverberation, which characterizes sound picked up by a microphone, is dependent (1) on the distance separating the microphone and the source of sound, and

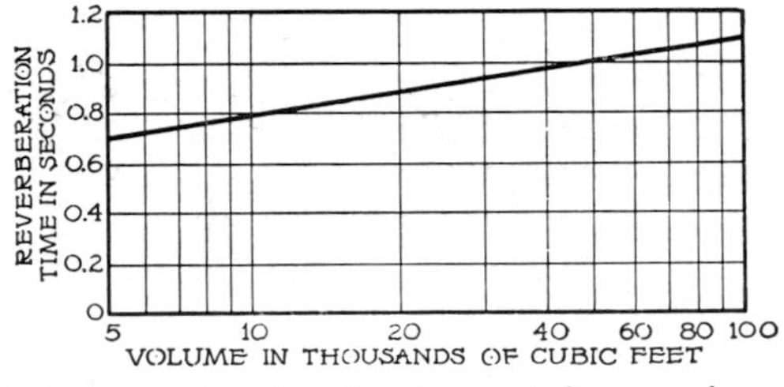

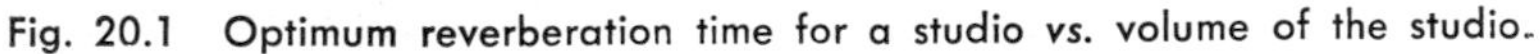
Fig. 20.1 Optimum reverberation time for a studio vs. volume of the studio.

(2) on the directional characteristics of the microphone. When the microphone is very near the source the greatest percentage of the sound it picks up comes directly from the source. Then, the reverberation in the room exerts relatively little influence on the sound picked up by the microphone. As the distance between the microphone and source increases, the direct sound diminishes about 6 db for each doubling of the distance, whereas the average level of the reverberant sound remains practically the same. When the reverberant sound is more prominent than the direct sound, amplified speech tends to become unintelligible, and amplified music tends to become a confusion of sounds.

Opinions differ somewhat among acoustical experts regarding the optimum reverberation time *vs.* frequency characteristic for studios. The trend of both performer and listener preference is toward a nearly flat characteristic for the entire audible-frequency range. Figure 20.1 gives the average reverberation times as a function of studio size. These times are based on the recommendations of leading studio engineers in the United States, Great Britain, and Europe, weighted and averaged by the authors.

To some extent the optimum time for a given room depends on the type of microphone technique employed. Hence, considerable deviation from the average values given in Fig. 20.1 are possible. These data give values that are satisfactory for musical programs, dramatic productions, and other spoken pro-

Fig. 20.2 Variable absorptive treatment in Radio Station KSL, Salt Lake City, Utah. One side of the rotatable panel is flat and is covered with absorptive tile; the other side is convex and is treated hardboard.

grams. It should be borne in mind that the optimum reverberation for one type of music may be far from the optimum for another (see p. 192). For example, music composed for performance in a relatively reverberant church will not sound like church music when it is produced in, and broadcast from, a small "dead" studio. Since studios are used for diverse musical programs, many are equipped with means for varying their reverberation characteristics, either electrically or by varying their total absorption. In Chapter 7, several variable absorptive treatments were discussed that are especially useful in studios, such as hinged or movable panels that can present either reflective or absorptive surfaces to the room, draperies that can be spread out over the walls or drawn into small alcoves, resonator absorbers that can be varied in size, and rotatable cylinders that can present different types of absorptive surfaces to the room. Some of these treatments are illustrated in Figs. 20.2 and 20.3.

Fig. 20.3 Side wall of NBC Studio 3A, Radio City, New York, showing heavily lined draperies and hinged panels.

Sound Insulation for Studios

Extraneous noises in a room that are unnoticed, when one is listening directly to speech or music in that room, may be prominent and annoying when the same speech or music is picked up by means of a microphone and then reproduced. This apparent exaggeration of noise, referred to earlier in this chapter, is a consequence of the fact that microphone pick-up is subject to many of the same peculiarities as monaural hearing. Hence, the noise level in studios without an audience should be exceptionally low—no greater than 25 to 30 db.[1] Since studios usually are located in noisy urban or metropolitan areas, their noise-insulation requirements are high if outside noise is to be excluded. A large transmission loss also is required to confine within a studio the high-level sounds that are often produced within that studio.

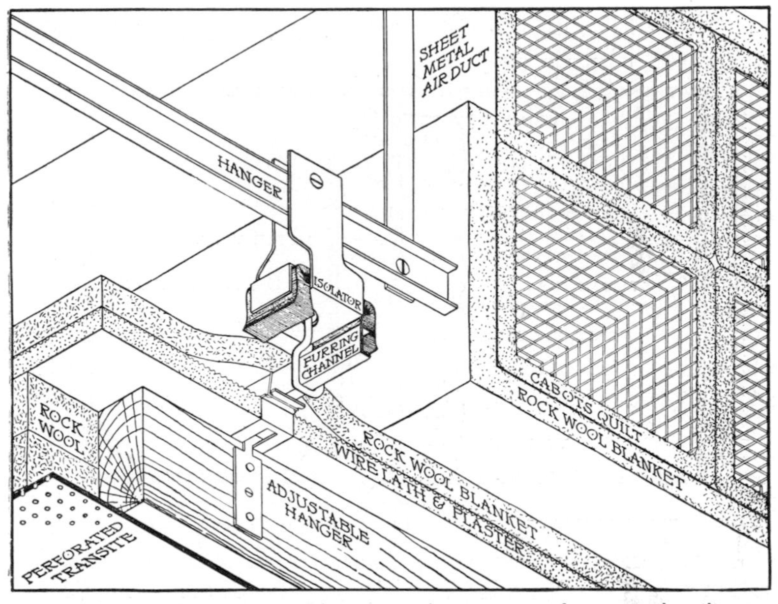

Fig. 20.4 Sound isolation and acoustical treatment of a typical ceiling at NBC studios in New York. (Courtesy National Broadcasting Co.)

Consequently, it is necessary to resort to special sound-insulation measures.

It is possible to calculate the over-all noise reduction that must be furnished by the combined effects of attenuation through the walls and sound absorption in the studio, in adjoining corridors, and in other intervening rooms, by the method worked out in Chapter 11 (see also p. 238). Chapter 12 describes appropriate measures for vibration control. The suppression of noise from ventilation systems, which is especially important in studios, has

[1] Measured with a sound-level meter incorporating a 40-db frequency-weighting network.

been discussed in Chapter 13. In general, the spectrum levels of the noise at low frequencies will determine the insulation requirements. The tables in Appendix 2 give the transmission losses of some compound partitions, among which will be found constructions that will provide the required attenuation for

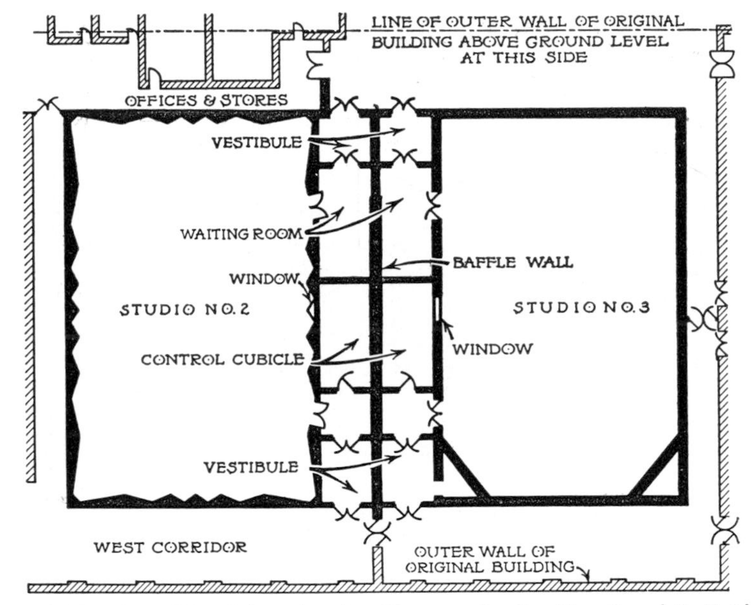

Fig. 20.5 Plan of British Broadcasting Company Studios Nos. 2 and 3, Maida Vale, London.

studios. Figure 20.4 shows sound isolation and acoustical treatment of a typical ceiling at the NBC studios in New York.

The control room should always be well insulated from the studio. This requires double or even triple plate glass windows between the two rooms. A tight window seal is essential, and it must be maintained. By using glass panes of two different thicknesses, by inclining one of the panes slightly, and by adding absorptive material around the peripheral surfaces separating the panes, the danger of transmitting noise at one or more resonant frequencies of the window is greatly reduced.

Figure 20.5, a plan of the British Broadcasting Company's Studios Nos. 2 and 3 at Maida Vale, London, is an example of good design with respect to planning for sound insulation. Each studio is a separate brick structure within an outer building, and the two studios are well insulated from each other by means of vestibules, control rooms, and waiting rooms.

Diffusion

Proper diffusion of sound is one of the prime acoustical requirements for studios designed especially for musical programs. Diffusion of sound in a room affects the rate of growth and decay of sound in the room. It also increases the uniformity of the distribution of sound throughout the room (see p. 139). Hence, microphone placement is not so critical in a studio having proper diffusion as it is in a studio having too little diffusion. On the other hand, if there is too much diffusion, apparent auditory perspective is diminished and critical listeners may object because they lose awareness of the size of the studio.

There are two principal means for providing diffusion in a studio:

(1) By a non-symmetrical distribution of absorptive material; for example, by patches of acoustical material scattered over the wall surfaces. One common method of obtaining an irregular distribution of absorptive material is to use wall surfaces of perforated Transite, selected areas of which are backed by an absorptive blanket.

(2) By irregularities on the wall surfaces, such as splays, convex "bumps," or other protuberances.

Convex surfaces made of thin plywood, plaster board, or other materials have been favored by a number of acoustical engineers. Figure 20.6 shows, by means of polar diagrams, the diffusive reflection of sound having a frequency of 1000 cycles from a convex panel compared with that from a flat panel. The convex panel consisted of a 4-foot by 8-foot piece of plywood attached to a cylindrical frame, bowed in such a way that the 4-foot arc had a chord length of 44½ and a maximum bulge of 7¾ inches. This convex surface is much more effective in "spreading out" or diffusing reflected sound than is the flat one. Similarly, other convex surfaces and prismoidal surfaces are better adapted for diffusing sound than are large flat ones, and they are less likely to give rise to high-frequency echoes or other prominent reflections. This is confirmed by measurements in studios treated with polycylindrical diffusers of the type shown in Fig. 20.6.[2,3] Such diffusive surfaces are usually made of plywood or plaster board about 1/4 inch to 3/8 inch thick, screwed or glued to a wood frame. The absorptive characteristics of the panels depend on

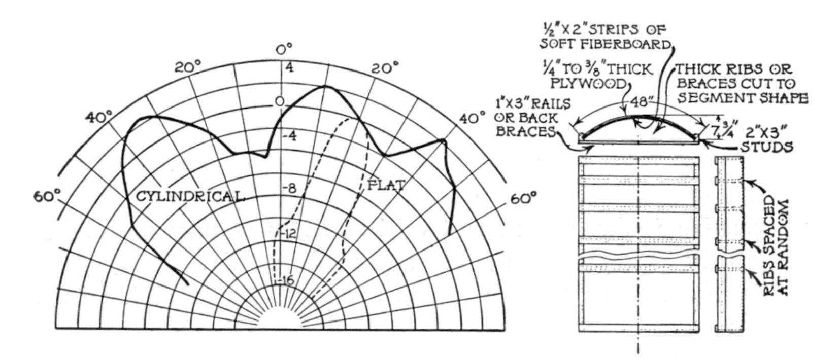

Fig. 20.6 Comparison of the polar distribution characteristics of reflected waves from the convex panel shown to the right and a flat panel of equal surface area. (J. E. Volkman.)

their resonant frequencies, and hence on their effective size. For this reason the braces are spaced at random to provide panels of various sizes and, hence, a relatively uniform absorption *vs.* frequency characteristic.

Polygonal shapes, such as those shown in Fig. 20.7, have most of the acoustical advantages of the cylindrical forms and require less skill to construct. A variety of sizes should be used in one studio, as is the practice with cylindrical diffusers, to provide better diffusion and more uniform absorption *vs.* frequency characteristics; the furring strips to which the plaster board or plywood is fastened should be randomly spaced. The ratio of the length of the base to that of the "bulge" should be about 4, although the dimensions of diffusers are not critical. However, they should be comparable with the wavelengths of the sound

[2] J. E. Volkmann, *J. Acoust. Soc. Am.,* **13,** 234 (1942).

[3] C. P. Boner, *J. Acoust. Soc. Am.,* **13,** 244 (1942).

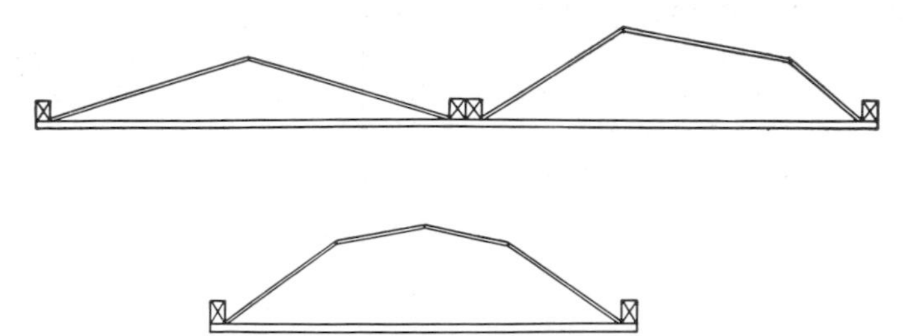

Fig. 20.7 Polygonal diffusing surfaces approximating the size and shape of cylindrical ones such as the one shown in Fig. 20.6.

Fig. 20.8 View looking from the control room into a studio of the Columbia Broadcasting System in New York. On the far walls are rotatable panels whose positions can be changed by push buttons located in the control booth.

they are to diffuse. Base lengths ranging from about 2 to 6 feet, with corresponding bulges of 6 to 18 inches, will meet most practical needs. Figure 20.8 is a photograph of the interior of a studio at the Columbia Broadcasting System in New York. On the far walls of the studio are rotatable panels whose position can be changed by push buttons in the control booth. When closed, these panels present a serrated surface which provides diffusion of sound in the studio.

Size and Shape of Studios

The size of a studio is determined principally by the number of performers and the size of audience it is to accommodate. A volume of 4000 cubic feet usually suffices for four or fewer artists; a volume of not less than 1000 cubic feet per artist is recommended for larger groups. Thus, a volume of about 50,000 to 100,000 cubic feet is regarded as the minimum size of studio for musical ensembles numbering up to about 50 performers. These recommended sizes are for studios without audiences; if an audience is to be accommodated, the size usually will be somewhat larger.

The room proportions can be determined either by rule-of-thumb or by considerations from physical acoustics. *The ratio of any two dimensions of the studio should not be a whole number or very close to a whole number.* A ratio of length to width to height of 1.6:1.25:1.0 is sometimes used in the design of small studios. These proportions would give very high ceiling height if applied to large studios. Among the ratios that have been recommended for *large* studios are 2.4:1.5:1.0 and 3.2:1.3:1.0. The auditorium studio at CBS in Hollywood, which has dimensions of 108 feet by 48 feet by 33 feet, corresponding to a ratio of 3.27:1.45:1.00, has been found to give highly satisfactory results. The above ratios are not the only feasible ones; many others are suitable and may be more desirable to conform with non-acoustical considerations, such as available or required floor space, costs, and the use of structural modules. Other favorable ratios can be determined, and poor ratios can be avoided by considerations of room resonance discussed in Chapter 8. It was noted that, if a sound source is enclosed in a room, the sound at certain frequencies is enhanced. For example, Fig. 8.1 shows the transmission-frequency characteristic for a small, rectangular room containing little absorptive material. The peaks in this curve correspond to the resonant frequencies. Equation (8.1) shows that these frequencies depend on the room dimensions.

It is desirable to have in a studio a relatively smooth transmission-frequency curve. Since the height and breadth of a peak are partially determined by the total amount of absorption at the frequency of the peak, proper selection of absorptive materials is an important factor in obtaining a smooth transmission-response curve. It is especially important to avoid materials or constructions that have excessive or deficient absorption in one or more narrow frequency bands. The other important factor in obtaining a smooth transmission-frequency curve is the uniformity of distribution of the resonant frequencies. If they cluster together at a few low frequencies, there will be peaks in the transmission-response curve at these frequencies; if they are well distributed, the response curve will be more uniform. A study of the distribution of the resonant frequency of rooms, including the effects of absorption, indicates the inadvisability of using a room dimension which is a small, whole-number multiple of another dimension, such as 2:1. A room which is cubic in shape would be particularly unsatisfactory. For example, consider a room 22½ feet by 22½ feet by 22½ feet. Its lowest 10 resonant frequencies are, from Eq. (8.1),

Mode	(1, 0, 0)	(0, 1, 0)	(0, 0, 1)	(1, 1, 0)	(1, 0, 1)	(0, 1, 1)	(1, 1, 1)	(2, 0, 0)	(0, 2, 0)	(0, 0, 2)
Resonant frequency, in cycles	25	25	25	35	35	35	43	50	50	50

The first 22 resonant frequencies for this room are plotted in the upper diagram in Fig. 20.9. Note that a number of them fall on top of each other. Now suppose that we keep the volume of the room approximately the same but alter the dimensions to 20 feet by 20 feet by 25 feet so that only two of the dimensions are alike. The new distribution of resonant frequencies, indicated in the center plot, is more uniform than that in the upper. If the dimensions are again changed, so that no two are the same, the distribution can be further improved. For example, if the dimensions are 20 feet by 22½ feet by 25 feet, the positions of the resonant frequencies are indicated in the lower portion of Fig. 20.9.

By this process, room dimensions that are desirable from other than acoustical considerations can be checked. If it is found that the resonant frequencies cluster together, the dimensions should

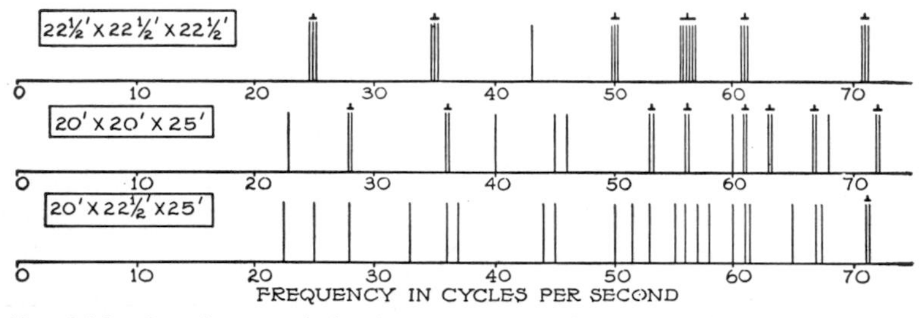

Fig. 20.9 Distribution of the lower resonant frequencies for rooms having different proportions. Each vertical line denotes a resonant frequency. A bar is shown above two or more modes when they have the same resonant frequency. Note that the distribution is least favorable for the cubic room 22½ feet on each side; it is best for the room 20 feet by 22½ feet by 25 feet, which has no two dimensions alike.

be altered slightly, as in the above example. By repeating the calculation for several sets of dimensions, it may be possible to find ratios that are much more practical than those obtained from a "rule-of-thumb" ratio.

Thus physical acoustics is seen to provide a logical and practical guide to the problem of studio dimensions, a problem that heretofore has been in the realm of empiricism. It is almost certain that in the years to come physical acoustics will make further important contributions to practical designing in architecture.

APPENDIX 1

Coefficients of Sound Absorption

See p. 104 for further information regarding these tables.

TESTING AUTHORITIES

1. Bureau of Standards
2. Acoustical Materials Association
3. National Physical Laboratory
4. Vern O. Knudsen
5. Average
6. P. E. Sabine
7. W. C. Sabine
8. Building Research Station
9. F. R. Watson
10. Wente and Bedell
11. Lydteknisk Laboratorium, Copenhagen
12. Swedish Broadcasting Company
13. R. W. Leonard
14. Knudsen and Harris (estimates)

Table A.1 Prefabricated

TYPES AND CLASSES OF MATERIAL

Type I Cast units having pitted or granular surface.
  Type I-A Mineral granules, Portland cement binder.
  Type I-B Mineral granules, lime or gypsum binder.
  Type I-C Mineral or vegetable granules, incombustible binder.
Type II Mechanically perforated surface.
  Type II-A Strong, durable facing backed with sound-absorptive material.
  Type II-B Circular perforations extending into sound-absorptive material.
  Type II-C Slots or grooves extending into sound-absorptive material.
Type III Fissured surface, usually mineral wool, cork, or vermiculite.
Type IV Units having felted fiber surface.

Material	Manufacturer	Type and Class	Composition	Thickness, inches	Unit Size, inches
Absorb-A-Sound	Luse Stevenson Co.	IV-A		2	36 x 36
				2	18 x 36
Absorb-A-Tone	Luse Stevenson Co.	IV-A	Compressed wood fibers with mineral wool binder	1	18 x 18
3″ rock wool batts behind tile				1	18 x 18
Acousteel Pad	Celotex Corp.	II-A	Metal pan backed with mineral wool sound-absorptive pad	1¼	12 x 24
Acoustex, Type 40R	National Gypsum Co.	IV-A	Shredded wood, cement binder	¾	12 x 12
Acousti-Celotex	Celotex Corp.	II-B	Compressed cane fiber		
Type C-2				⅝	12 x 12
				⅝	12 x 12
Type CS-1				½	12 x 12
				½	12 x 12
Type C-4				1¼	12 x 12
				1¼	12 x 12
				1¼	12 x 24
Type C-6				1¼	12 x 12
				1¼	12 x 12

* The coefficients are for the materials, mounted as specified.

† Unless otherwise specified, the light reflection coefficient is for factory-decorated material—usually painted an off-white color.

Note 1. Perforated, 441 holes per sq ft, 3/16″ diam., 17/32″ o.c.

Units—Tiles, Fiberboards, etc.

TYPES AND CLASSES OF MATERIAL (*Continued*)

Type IV-A	Long wood fibers, usually shavings or excelsior.
Type IV-B	Felted vegetable fibers or wood pulp.
Type IV-C	Felted mineral fibers.

TYPES OF MOUNTING

1. Cemented to hard, rigid backing, as concrete, brick, tile, etc.
2. Nailed or screwed to 1" x 2" wood furring strip, 12" o.c.
3. Attached to metal supports on 1" wood or metal furring strips.
4. Laid on laboratory floor.

Nature of Surface	Absorption Coefficients *						Mounting	Authority	Wt., lb per sq ft	Light Refl. Coeff.†
	128 cycles	256 cycles	512 cycles	1024 cycles	2048 cycles	4096 cycles				
Unpainted	.12	.30	.75	.70	.66	.78	4	1	4.81	...
Unpainted	.04	.24	.54	.88	.53	.70	4	1	4.90	...
Unpainted	.08	.24	.55	.78	.78	.74	4	1	2.25	.60
Unpainted	.58	.98	.90	.77	.79	.87	3	1	2.25 tile; 1.01 batts	.60
Perforated, enameled metal	.25	.52	.99	.99	.81	.60	3	2	Pad 1.08	...
Unpainted	.06	.17	.37	.68	.82	.74	1	1	1.75	...
	.15	.22	.61	.93	.79	.69	2	1	1.54	...
Perforated, painted	.09	.26	.69	.86	.67	.62	1	2	.83	.78
Perforated, painted	.12	.51	.65	.73	.66	.58	2	2	.83	
Perforated;[1] painted[2]	.09	.15	.61	.77	.70	.64	1	2	.66	.78
Perforated;[1] painted[2]	.14	.46	.52	.71	.72	.64	2	2	.66	.78
Perforated;[1] painted[2]	.14	.42	.99	.74	.60	.50	1	2	1.34	.78
Perforated;[1] painted[2]	.25	.58	.99	.75	.58	.50	2	2	1.34	.78
Perforated;[1] painted[3]	.53	.58	.93	.70	.55	.48	Note 6	2	1.58	.81
Perforated;[4] painted[3]	.15	.34	.99	.94	.61	.61	1	2	1.51	.76
Perforated, painted	.27	.57	.91	.91	.67	.58	2	2	1.51	

Note 2. Face painted before perforating.
Note 3. Painted after perforating.
Note 4. Perforated, 441 holes per sq ft, 1/4" diam., 17/32" o.c.

Table A.1

TILES, FIBERBOARDS

Material	Manufacturer	Type and Class	Composition	Thickness, inches	Unit Size, inches
Acousti-Celotex	Celotex Corp.	II-B			
Type C-7				1	12 x 24
				1	24 x 24
				1	24 x 24
Type C-9				¾	12 x 12
				¾	12 x 12
Type M-1				⅝	12 x 12
				⅝	12 x 12
Type M-2				1	12 x 12
				1	12 x 24
Acoustic Panels	Cincinnati Mfg. Co.	II-A	Perforated metal on each face, filled with 4-lb density Fiberglas	3½, bulged at center to 4″	24 x 60
			Similar to above except filled with 6-lb density Fiberglas	3½	24 x 60
Acoustic Panels	Industrial Sound Control	II-A	Perforated sheet-metal facing	3½	36 x 48
			Similar to above, filled with 5.4-lb density mineral wool		36 x 48
Acoustical Panel Assembly	Celotex Corp.	II-A			
½″ Q-T element plus perforated asbestos board facing				1 5/16	24 x 24
1″ Q-T element plus perforated asbestos board facing				1 7/16	24 x 24
Acoustifibre	National Gypsum Co.	II-B	Compressed wood fibers	⅝	12 x 12
Acoustilite	Insulite Co.	IV-B	Compressed wood pulp	¾	12 x 12
			Compressed wood fibers	½	12 x 12

Note 5. Perforated, 676 holes per sq ft, 5/32″ diam., 7/16″ o.c.
Note 6. Attached to special metal supports mounted on 2″ x 2″ wood furring.

(Continued)

(Continued)

Nature of Surface	Absorption Coefficients						Mount-ing	Author-ity	Wt., lb per sq ft	Light Refl. Coeff.
	128 cycles	256 cycles	512 cycles	1024 cycles	2048 cycles	4096 cycles				
Perforated; [1] painted [2]	.41	.48	.68	.79	.75	.55	Note 6	2	1.40	.78
Perforated,painted	.18	.35	.86	.87	.63	.56	1	2	1.31	
Perforated,painted	.25	.49	.69	.78	.61	.48	2	2	1.54	.81
Perforated; [1] painted [2]	.12	.45	.79	.89	.61	.60	2	2	.96	.78
Perforated,painted	.11	.23	.80	.93	.58	.50	1	2	.96	.78
Perforated; [5] painted [3]	.12	.48	.50	.79	.93	.82	2	2	1.31	.80
Perforated,painted	.07	.21	.64	.86	.93	.83	1	2	1.31	.80
Perforated; [5] painted [3]	.08	.27	.92	.95	.80	.71	1	2	1.81	.80
Perforated; [5] painted [3]	.40	.44	.79	.99	.77	.71	Note 6	2	2.23	.80
4608 holes per sq ft, 0.075″ diam.	.50	.98	.99	.92	.82	.80	4	1		
Same as above	.51	.99	.99	.96	.85	.92	Note 7	1		
4608 holes per sq ft, 0.075″ diam.	.38	.87	.93	.86	.84	.85	4	1		
Same as above	.60	.97	.97	.93	.91	.78	4	1	4.65	
Perforated, unpainted	.21	.51	.73	.83	.91	.76		2	.7 1.3	
Perforated, unpainted	.45	.49	.76	.89	.89	.71		2	1.3 1.3	
Perforated; [8] painted	.10	.16	.62	.97	.81	.73	1	2	.56	
Same as above	.13	.38	.72	.89	.82	.66	2	2	.56	
Mill-painted	.16	.34	.79	.72	.69	.64	1	1	.59	
Mill-painted	.07	.20	.53	.77	.74	.74	3	1	....	...
Mill-painted	.12	.49	.80	.85	.80	.83	2	1	.57	
Spray painted, 2 coats	.11	.53	.82	.82	.72	.68	2	1		
Spray-painted, 4 coats	.19	.73	.67	.55	.39	.32	2	1		

Note 7. Nailed on 2 x 8's, 12″ o.c., unless otherwise indicated.

Note 8. Acoustifibre is perforated with holes 3⁄16″ diam., 441 per sq ft.

Table A.1

TILES, FIBERBOARDS

Material	Manufacturer	Type and Class	Composition	Thickness, inches	Unit Size, inches
Acoustimetal	National Gypsum Co.	II-A	Type P pad, spacer, and facing	2½	12 x 24
Acoustone	U. S. Gypsum Co.	III	Artificial stone		
F				1 1/16	12 x 12
				1 3/16	12 x 12
				1 3/16	12 x 24
				1 3/16	12 x 24
				⅞	12 x 24
Airacoustic (for lining ducts)	Johns-Manville Corp.	IV-C	Rock wool plus binder	½	24 x 36
				1	24 x 36
Akoustolith Tile	R. Guastavino Co.	I-A	Artificial stone		
Grade B-2				1	6 x 12
				1½	6 x 12
				2	6 x 12
Grade C				5	12 x 12
Arphon Type PP3	Ab. Arki, Stockholm, Sweden	II-A	Mineral wool covered by hard, perforated fiberboard	1	18 x 18
Arrestone	Armstrong Cork Co.	II-A	Absorptive pads plus metal facing plus furring	2½	12 x 24
Auditone	U. S. Gypsum Co.	II-C	Wood fiber		
B				1	12 x 12
				1	12 x 24
C				¾	12 x 12
				¾	12 x 24
Celocrete Blocks	National Brick Corp.	I-A		5¾	7⅝ x 15⅝
Corkoustic	Armstrong Cork Corp.	III		1½	12 x 12
Cushiontone	Armstrong Cork Corp.	II-B			
A				½	12 x 12
				½	12 x 12
				¾	12 x 12
				¾	12 x 12
				⅞	12 x 12
				⅞	12 x 12

Note 9. Acoustimetal is a perforated, enameled pan backed with sound-absorptive mineral wool pad. Perforations are .093″ diam., 2016 per sq ft.

Note 10. Laid on 24 gauge sheet iron, nailed to 1″ x 2″ wood furring 24″ o.c.

(Continued)

(Continued)

Nature of Surface	Absorption Coefficients						Mounting	Authority	Wt., lb per sq ft	Light Refl. Coeff.
	128 cycles	256 cycles	512 cycles	1024 cycles	2048 cycles	4096 cycles				
Perforated, enameled metal [9]	.26	.47	.97	.99	.88	.88	3	2	Pad .91	.72
Fissured, painted	.08	.25	.76	.84	.78	.73	1	2	1.35	.80
Fissured, painted	.12	.31	.85	.88	.75	.75	1	2	1.53	.80
Fissured, painted	.23	.50	.72	.77	.74	.69	2	2	1.54	.80
Fissured, painted	.38	.64	.62	.68	.76	.75	Note 6	2	1.54	.80
Fissured, painted	.38	.60	.64	.74	.78	.74	7	2	1.68	.80
Unpainted	.13	.41	.40	.72	.78	.72	Note 10	2	.80	...
Unpainted	.29	.51	.70	.82	.79	.80	Note 10	2	1.50	...
Unpainted	.09	.17	.46	.77	.77	.56	4	1	4.6	...
Unpainted	.14	.30	.67	.87	.82	.57	4	2	6.1	...
Unpainted	.21	.50	.85	.81	.70	.70	4	2	8.5	...
Unpainted	.43	.92	.91	.88	.86	.74	4	2	24.4	
Perforated fiberboard	.13	.50	.85	.70	.57	.50	2	Swedish	.....	...
Perforated, enameled metal [11]	.25	.56	.99	.99	.77	.60	3	2	1.20	.76
Slotted, painted	.13	.33	.79	.85	.85	.72	1	2	1.15	.74
Slotted, painted	.28	.48	.64	.86	.77	.75	2	2	1.23	.74
Slotted, painted	.09	.28	.68	.80	.79	.81	1	2	.91	.72
Slotted, painted	.13	.49	.60	.72	.80	.81	2	2	.91	.72
Unpainted, joints not sealed	.63	.84	.46	.37	.59	.49	4	1	23.4	...
Fissured, painted	.04	.13	.44	.76	.43	.55	1	2	.84	.80
	.14	.25	.61	.43	.52	.54	2	2	.84	.80
Perforated; painted [12]	.05	.18	.56	.76	.77	.73	1	2	.79	.73
Perforated; painted [12]	.07	.47	.55	.70	.77	.74	2	2	.79	.73
Perforated; painted [12]	.10	.28	.66	.91	.82	.69	1	2	1.05	.73
Perforated; painted [12]	.14	.51	.59	.87	.81	.70	2	2	1.05	.73
Perforated; painted [12]	.09	.28	.74	.98	.78	.70	1	2	1.17	.73
Perforated; painted [12]	.17	.51	.73	.95	.75	.72	2	2	1.17	.73

Note 11. Arrestone is a perforated enameled pan, backed with mineral wool sound-absorptive pad. Perforations are .093″ diam., 1105 holes per sq ft.

Note 12. Perforated, 484 holes per sq ft, 3⁄16″ diam. Painted by manufacturer two coats, face and bevels.

Table A.1

TILES, FIBERBOARDS

Material	Manufacturer	Type and Class	Composition	Thickness, inches	Unit Size, inches
Cushiontone	Armstrong Cork Co.	II-B			
A				⅝	12 x 12
				⅝	12 x 12
F				¾	12 x 12
Danish Acoustical Materials Snedkermestrenes Trae-og Finerskaereri					
STF 0			4″-mm plywood		
			1″ wood board attached to plywood, 20-mm rock wool		
STF 1			4-mm plywood, 20-mm rock wool		
STF 2			½″ perforated soft fiberboard [14]		
STF 4			4″-mm perforated plywood, 2 x 60 mm, 9.4-mm spacing, 20-mm rock wool		
STF 5			1″ perforated soft fiberboard [14]		
STF 6			4-mm perforated plywood [15]		
STF 7			Wood panel with studs, 44-mm width, 6-mm slots between studs		
Econacoustic	National Gypsum Co.		Compressed wood fibers	½	12 x 12
		IV-B		½	12 x 12
				1	12 x 12
Fiberglas	Owens-Corning Fiberglas Corp.	IV-C			
NC-9 board			Felted glass fibers	1	24 x 36
PFD			Felted glass fibers	1	24 x 36
PFL			Felted glass fibers	1	24 x 36
			Felted glass fibers	2	24 x 36
PFT			Felted glass fibers	1	24 x 36
Acoustical Board			Felted glass fibers	1	24 x 48
Acoustical Tile, Type A			Felted mineral fibers	¾	12 x 12
			Felted mineral fibers	¾	12 x 12
			Felted mineral fibers	¾	24 x 24
			Felted mineral fibers	1	12 x 12

Note 13. Perforated, 484 holes per sq ft, 5⁄32″ diam.
Note 14. Perforated, 4-mm holes, 13-mm spacing.
Note 15. Perforated, 4-mm holes, 10.5-mm spacing, 20-mm rock wool.
Note 16. Some units are cross-grooved to give appearance of tile 12″ x 12″ or 24″ x 24″.

(Continued)

(Continued)

Nature of Surface	Absorption Coefficients						Mounting	Authority	Wt., lb per sq ft	Light Refl. Coeff.
	128 cycles	256 cycles	512 cycles	1024 cycles	2048 cycles	4096 cycles				
Perforated, painted	.09	.25	.60	.80	.78	.73	1	2	.92	.73
Perforated, painted	.13	.45	.55	.73	.79	.75	2	2	.92	.73
Perforated; painted [13]	.05	.14	.55	.90	.78	.78	1	2	1.13	.73
Perforated	.80	.32	.15	.08	.07	.09		11		
Perforated	.13	.30	.15	.11	.09	.08		11		
	.12	.25	.45	.41	.43	.46		11		
	.08	.16	.55	.60	.63	.50		11		
	.07	.21	.47	.52	.67	.51		11		
	.07	.17	.71	.65	.60	.30		11		
Perforated	.21	.52	.50	.51	.51	.35		11		
Painted	.05	.17	.62	.83	.77	.74	1	2	.40	...
Painted	.09	.32	.75	.78	.74	.78	2	2	.40	...
Painted	.13	.43	.78	.81	.75	.81	1	2	.62	...
Painted	.10	.30	.35	.37	.24	...	1	2	.76	...
Unpainted	.24	.32	.65	.77	.73	.81	1	2	.23	...
Unpainted	.20	.41	.75	.87	.86	.82	1	2	.34	...
Unpainted	.41	.60	.99	.99	.84	.81	1	2	.47	...
Unpainted	.25	.41	.86	.94	.84	.81	1	2	.47	...
Smooth; [16] painted [17]	.67	.67	.82	.89	.91	.98	Note 18	2	.90	.75
Factory-painted fiber surface	.12	.32	.75	.83	.70	.63	1	2	.67	...
Factory-painted fiber surface	.08	.44	.79	.83	.74	.65	2	2	.67	...
Factory-painted fiber surface	.47	.65	.75	.84	.83	.81	Note 18	2	.....	...
Factory-painted fiber surface	.11	.31	.79	.94	.84	.79	1	2	.94	...

Note 17. Spray-gun application of water-base paint mixed slightly thinner than brushing consistency.

Note 18. Mechanically mounted to metal furring strips 24″ o.c., 10½″ air space.

Table A.1

TILES, FIBERBOARDS

Material	Manufacturer	Type and Class	Composition	Thickness, inches	Unit Size, inches
Fiberglas	Owens-Corning Fiberglas Corp.	IV-C			
Acoustical Tile, Type A			Felted mineral fibers	1	12 x 12
			Felted mineral fibers	1	24 x 24
Plain type				¾	12 x 12
				¾	12 x 24
				1	12 x 12
				1	12 x 24
Perforated type		II-B		¾	12 x 12
				¾	12 x 24
Fibracoustic	Johns-Manville Corp.	IV-B	Felted mineral fibers	1	12 x 12
Fibretone	Johns-Manville Corp.	II-B	Perforated felted fibers	½	12 x 12
				13/16	12 x 12
				13/16	12 x 12
Fir-tex Acoustical Tile	Fir-tex Ins. Co.	IV-C	Fir wood and bark felted fibers	½	48″ widths
				1	48″ widths
Fissuretone	Celotex Corp.		Mineral fiber	¾	12 x 12
Flax-ni-lum			Felted flax fibers	¾	Varied
Flintkote Acoustical Tile	Pioneer Division, Flintkote Co.	II-B			
				½	12 x 12
				½	12 x 12
				¾	12 x 12
				¾	12 x 12
				1	12 x 12
				1	12 x 12
				1¼	12 x 12
				1¼	12 x 12
Heerwagen Acoustic Diaphragm Tile	Heerwagen Acoustic Decoration Co.				12 x 12
Insulite	Insulite Co.	IV-C	Compressed wood fibers	½	48″ widths
Lloyd Board	Lloyd Board, Ltd., London	IV-B	Compressed wood fibers	½	48″ widths
"L. W." Insulation Board	Ljrisne-Woxna, Sweden	IV-A	Compressed wood fibers	½	120 x 120

(Continued)

(Continued)

Nature of Surface	Absorption Coefficients						Mount-ing	Author-ity	Wt., lb per sq ft	Light Refl. Coeff.
	128 cycles	256 cycles	512 cycles	1024 cycles	2048 cycles	4096 cycles				
Factory-painted fiber surface	.17	.40	.83	.96	.85	.83	2	2	.90	...
Factory-painted fiber surface	.51	.52	.75	.87	.89	.91	Note 18	2	.94	...
Painted	.04	.20	.63	.91	.82	.82	1	2	.69	
Painted	.21	.50	.91	.99	.86	.80	7	2	.62	
Painted	.11	.33	.80	.96	.82	.76	1	2	.86	
Painted	.23	.57	.90	.98	.88	.84	7	2	.87	
Perforated, painted	.02	.16	.76	.99	.63	.44	1	2	.71	
Perforated, painted	.16	.47	.98	.83	.64	.46	7	2	.68	
	.18	.42	.81	.75	.71	.72	1	2	.54	.58
	.25	.62	.72	.72	.71	.76	2	2	.54	.58
Perforated; painted [19]	.08	.28	.58	.71	.68	.65	1	2	.75	.71
Perforated; painted [19]	.14	.37	.69	.80	.76	.73	1	2	1.17	.71
Perforated; painted [19]	.18	.54	.72	.74	.71	.72	2	2	1.17	.71
Spray-painted	.11	.36	.63	.64	.67	.68	2	5	Light	...
Spray-painted	.24	.51	.80	.83	.87	.89	2	5	Light	...
Fissured; painted	.16	.33	.68	.75	.80	.75	1	2	1.52	...
	.16	.24	.36	.46	.48	.46	2	4	.....	...
Perforated, painted	.08	.20	.63	.69	.72	.75	1	2	.54	
Perforated, painted	.12	.48	.56	.62	.70	.75	2	2	.54	
Perforated, painted	.06	.28	.76	.83	.81	.69	1	2	.83	
Perforated, painted	.12	.53	.65	.83	.80	.73	2	2	.83	
Perforated, painted	.09	.29	.85	.95	.80	.75	1	2	1.07	
Perforated, painted	.14	.57	.78	.92	.77	.74	2	2	1.07	
Perforated, painted	.21	.40	.99	.95	.67	.67	1	2	1.35	
Perforated, painted	.18	.57	.92	.93	.69	.70	2	2	1.35	
	.62	.75	.63	.76	.47			6	3/16	.72
	.22	.26	.29	.33	.37	.38	2	5	.....	...
	...	.40	.35	.35	.40	...	Note 26	3	0.77	...
Felted or dimpled	.20	.45	.35	.30	.35	.35	Note 26	3	0.75	...

Note 19. Holes 3/16″ diam., 484 per sq ft.

TABLE A.1

TILES, FIBERBOARDS

Material	Manufacturer	Type and Class	Composition	Thickness, inches	Unit Size, inches
Masonite	Masonite Co.	IV-C	Compressed wood fibers	7/16	48" widths
Maycoustic Perforated Tile	May Acoustics, Ltd., London, England	II-A	Asbestos	1 3/16	16 x 16
Muffletone	Celotex Corp.		Cast gypsum		
Standard		IV-C		1	12 x 12
Fissured		III		1	12 x 12
Nuwood Bevel Lap Tile	Wood Conversion Co.		Compressed wood fibers	1	
Paxfelt	Newalls Insulation Co., London	IV-C	Asbestos material	1	Varied
				2	
Paxtiles	Newalls Insulation Co., London	II-B	Asbestos fibers	1	18 x 18
Perfatone, Pad	U. S. Gypsum Co.	II-A	Pad, plus metal facing and pad supports, plus furring	1¼	12 x 24
Perforated Asbestos Panels	National Gypsum Co.	II-A		Pad 1; panel 1 3/16; furring 3¾	24 x 24
Porex	Porete Mfg. Co.	IV-A	Mineralized wood fibers	1½	20 x 40
Q-T Ductliner	Celotex Corp.	IV-B	Felted fiber or wood pulp	½	24 x 36
				1	22 x 32
Sanacoustic	Johns-Manville Corp.	II-A	Type KK, pad	2½	12 x 24
Type MA				Pad 1⅛; facing 1 9/16; furring 2½	12 x 24
Simpson Acoustical Tile	Simpson Industries	II-B	Compressed wood fibers		
Type S-1				½	12 x 12
				½	12 x 12
Type S-2				⅝	12 x 12
				⅝	12 x 12
Type S-5				1	12 x 12
				1	12 x 12
Softone	American Acoustics, Inc.	I-C	Cork granules and mineral binder	1	12 x 12

Note 20. Perforated non-absorbent face 3/16" thick, pierced with holes 3/16" diam., 0.44 o.c. May be painted or enameled.

Note 21. Nailed to 2" x 2" wood furring 18" or 20" o.c. 2" mineral wool between furring.

Note 22. Perforated, enameled metal pan backed with mineral wool around absorptive pad. Perforations are .068" in diam., 4608 holes per sq ft.

Note 23. Nailed to 1" x 3" wood furring 24" o.c. and filled in between furring with 1" mineral wool.

(Continued)

(Continued)

Nature of Surface	Absorption Coefficients						Mounting	Authority	Wt., lb per sq ft	Light Refl. Coeff.
	128 cycles	256 cycles	512 cycles	1024 cycles	2048 cycles	4096 cycles				
	.10	.21	.29	.30	.29	...	1	4	.....	...
	.18	.25	.32	.35	.33	.31	2	5	.....	...
Perforated; painted [20]	...	.70	.65	.65	.75	.70	...	3	.....	...
Integrally colored	.12	.30	.74	.76	.71	.67	1	2	1.80	...
Integrally colored	.09	.29	.83	.97	.77	.71	1	2	1.92	...
	...	...	.37	...	...	...	4	1	1.41	...
Firm and porous; unpainted	...	.50	.55	.65	.70	.75	2	3	0.75	
Same as above	...	.55	.65	.75	.80	.80	2	3	.....	...
Perforated; [27] painted	...	.55	.75	.85	.80	...	2	3	1.5	...
Perforated; enameled metal [25]	.23	.59	.98	.99	.87	.68	3	1	Pad .88	
Perforated panel	.23	.51	.95	.94	.76	.59		2	Pad .74	
Unpainted	.10	.19	.40	.79	.50	.77	4	1	3.81	...
Unpainted	.14	.38	.43	.76	.75	.75	Note 21	2	.57	...
Unpainted	.29	.41	.78	.89	.88	.78	Note 21	2	1.24	...
Perforated; enameled metal [22]	.25	.58	.96	.97	.85	.72	3	2	Pad 1.28	.76
50/50 perforated, unperforated	.14	.70	.61	.70	.56	.54	3	2	Pad 1.27	Perforated .76; unperforated .85
	.09	.16	.67	.84	.76	.72	1	2	.60	.77
	.07	.37	.71	.75	.77	.69	2	2	.60	.77
Perforated, painted	.06	.18	.74	.90	.78	.70	1	2	.80	.77
Perforated, painted	.14	.50	.70	.83	.78	.71	2	2	.80	.77
	.12	.31	.98	.94	.70	.64	1	2	1.10	.77
	.22	.51	.89	.98	.71	.66	2	2	1.10	.77
Factory-painted	.10	.26	.72	.90	.75	.65	2	1	.....	.85
	.11	.26	.66	.90	.74	.79	1	1	.....	.85

Note 24. Nailed to 2″ x 2″ wood furring 24″ o.c. 2″ mineral wool between furring.

Note 25. 4608 holes per sq ft, .073″ diam.

Note 26. Nailed to 1⅞″ x ⅞″ vertical battens on 16″ centers, and 1⅞″ x ⅞″ horizontal battens on 10′ centers.

Note 27. Paxtiles are made of pure asbestos fiber covered with a layer of asbestos paper and then perforated with holes 3⁄16″ diam., ½″ o.c.

TABLE A.1

TILES, FIBERBOARDS

Material	Manufacturer	Type and Class	Composition	Thickness, inches	Unit Size, inches
Stenit, Type P2	Ab. Arki, Stockholm, Sweden	II-B	Mineral wool	1	18 x 18
Thermacoust Slabs, Standard	The Cementation Co., Ltd., London	IV-A	Wood wool	1½ 2 2	84 x 24 84 x 24 84 x 24
Transite Acoustical Panels	Johns-Manville Corp.	II-A	Asbestos facing, Rock Wool Blanket	1³⁄₁₆	Varied
Transite Acoustical Unit	Johns-Manville Corp.	II-A		Pad 1; pad and facing 1⅛	12 x 12
Travertone	Armstrong Cork Corp.	III	Cork	¾	12 x 12
Treetex Acoustical Tile Type B1	Treetex, Stockholm, Sweden	II-C	Compressed wood fibers	1	23⅝ x 23⅝

(Continued)

(Continued)

Nature of Surface	Absorption Coefficients						Mounting	Authority	Wt., lb per sq ft	Light Refl. Coeff.
	128 cycles	256 cycles	512 cycles	1024 cycles	2048 cycles	4096 cycles				
Perforated	.12	.30	.63	.75	.77	.78	2	Swedish	.....	...
	.15	.30	.75	.55	.65	.70	..	3	.....	...
	.20	.30	.80	.75	.75	.75	1	3	.....	...
	.20	.45	.85	.55	.70	.70	2	3	.....	...
Perforated; unpainted	.17	.49	.94	.90	.70	.43	Note 23	2	1.20	...
Perforated; painted	.29	.57	.94	.93	.70	.48	Note 24	2	2.00	...
Perforated, painted	.32	.58	.72	.85	.76	.67	2	2	2.60	
Fissured; painted	.12	.27	.72	.79	.76	.77	1	2	1.13	...
Slotted; painted	...	.35	.50	.65	.60	.55	1	12	.....	...
	...	.40	.50	.70	.70	.65	2	12	.....	...

Table A.2 Acoustical Plaster and Other Materials for Plastic Application

TYPES OF MATERIAL

Type I Gypsum, cement, or lime with or without aggregate.   Type II Other plastic materials applied with trowel.   Type III Fibrous materials with binder applied with air gun or blower.

Material	Manufacturer	Type	Application	Thickness, inches	Base Coat	Surface Treatment	Absorption Coefficients						Authority
							128 cycles	256 cycles	512 cycles	1024 cycles	2048 cycles	4096 cycles	
Acoustic Plaster	Hollywood Stucco Products, Inc.	I	1st coat applied to dry base coat; 2nd coat applied 24 hr after 1st coat	½	¾″ gypsum plaster on metal lath	Finished with cork float	.10	.22	.42	.78	.78	.70	1
Acoustipulp	Val-Porter Co., Los Angeles	II	Applied to ½″ scratch coat on ½″ wood panels on 1″ x 4″ wood strips, 16″ o.c.	½	Porous plaster	Applied with trowel	.28	.43	.47	.50	.46	.42	4
				¾	Porous plaster	Same as above	.38	.51	.52	.52	.46	.42	4
Asbestos Spray	Newalls Insulation Co., England	III	Sprayed*	½			...	.30	.35	.50	.60	...	3
			Same as above	¾			...	.55	.60	.50	.60	...	3
			Same as above	1			...	.60	.65	.60	.60	...	3
			Same as above	2¾			...	.85	.95	.90	.80	...	3
"Fibrespray" Asbestos	Acoustics, Inc.	III	Sprayed on metal lath; 3¾″ air space back of lath	½		Unpainted	.47	.92	.82	.83	.90	.90	1
			Sprayed on ⅜″ plaster board; air space same as above	¾		Unpainted	.36	.32	.84	.91	.91	.90	1
			Same as above	¾		Spray-painted 2 coats emulsion paint at NBS	.45	.31	.87	.89	.87	.87	1

Hushcote Acoustic Plaster	Cleveland Gypsum Supply Co.	II	1st coat applied to dry base coat; 2nd coat applied 24 hr after 1st coat	½	¾″ gypsum plaster on metal lath	Finished with steel trowel	.13	.24	.45	.71	.56	.49	1
			Same as above	⅝	Same as above	Same as above	.16	.34	.50	.53	.43	.37	1
Kalite	Mission Lime Products Corp.	I	Over hard scratch and brown coats on metal lath	½			.18	...	.30	...	.34	...	4
			On metal lath	¾			.41	...	.51	...	.59	...	4
Grade "A" Acoustical Plaster			Over ¾″ Kalite Insulating plaster on metal lath	½	Mineral color pigment fixed integrally		.30	...	.60	...	.70	...	..
Kilnoise "A" Acoustical Plaster	Kelly Island Lime & Transport Co.		Applied to dry scratch coat	½	Scratch coat on metal lath	Brush-textured	.22	.37	.59	.62	.48	.38	1
Limpet (sprayed asbestos fibers)	Keasbey and Mattison Co.	III	Sprayed on metal lath; 5⅞″ air space back of lath	½		Finished with roller; unpainted	.25	.78	.97	.81	.82	.85	1
			Same as above	⅜		Same as above	.36	.92	.85	.81	.87	.91	1
			Same as above	⅜		Same as above, except spray-painted 5 coats	.43	.91	.82	.67	.62	.61	1
			Same as above	¾		Same as above	.41	.88	.90	.88	.91	.81	1
			Sprayed on gypsum wall board	¾		Same as above	.08	.19	.70	.89	.95	.85	1
Plastacoustic	R. Guastavino Co.	I	1st coat applied to dry base coat; 2nd coat applied 24 hr after 1st coat	½	¾″ gypsum plaster	Finished with steel trowel	.17	.22	.44	.81	.72	.72	1
Sabinite "M"	U. S. Gypsum Co.	I	1st coat applied on dry base coat; 2nd coat applied 24 hr after 1st coat	½	¾″ total scratch and brown coats on metal lath	Steel-troweled	.18	.27	.49	.79	.87	.85	1
			Same as above	½	Same as above	As above, except spray-painted 5 coats resin emulsion paint	.29	.36	.52	.65	.55	.49	1

* Asbestos fiber sprayed from one nozzle, adhesive from another; streams intermingle before striking treated surface.

Table A.2 (*Continued*)

Material	Manufacturer	Type	Application	Thickness, inches	Base Coat	Surface Treatment	Absorption Coefficients						Authority
							128 cycles	256 cycles	512 cycles	1024 cycles	2048 cycles	4096 cycles	
Spray-Acoustic	Spray-O-Flake Co.	III	Sprayed on metal lath; 3⅝″ air space back of lath	½		Unpainted	.47	.88	.87	.95	.95	.92	1
Type X			Sprayed on metal lath; 3⅝″ air space back of material	⅝		Finished with roller; surface sprayed with coat of binder	.59	.87	.85	.88	.94	.83	1
			Same as above	⅝		Same as above, except brush-painted	.65	.79	.80	.70	.83	.60	1
			Sprayed on gypsum wall board	1⅛		Finished with roller; surface sprayed with coat of binder	.18	.52	.95	.93	.91	.87	1
			Same as above	1⅛		Same as above, except painted 3 coats	.15	.47	.88	.92	.87	.88	1
Vermiculite Acoustic Plastic, Type B	Vermiculite Research Institute	II	1st coat applied to dry base coat; 2nd coat 24 hr of the 1st coat	½	¾″ total scratch and brown coat on metal lath	Sanded with smooth-sanded wooden darby	.31	.34	.52	.78	.83	.95	1
			Same as above	½	Same as above	Same as above, but spray-painted 2 coats casein paint	.31	.32	.52	.81	.88	.84	1
			Same as above	½	Same as above	Same as above, but spray-painted 4 coats casein paint	.30	.37	.59	.84	.74	.65	1

Table A.3 Acoustical Blankets, Felts, etc.

Material	Manufacturer	Thickness, inches	Type	Mounting	Absorption Coefficients 128 cycles	256 cycles	512 cycles	1024 cycles	2048 cycles	4096 cycles	Authority	Wt., lb per sq ft
Balsam Wool		1		1	.12	.25	.52	.81	.67	.53	5	.26
Felt, Mdse. 3567	American Felt Co.	½		4	.05	.07	.29	.63	.83	.87	1	.19
		1		4	.06	.31	.80	.88	.87	.87	1	.38
Fiberglas	Owens-Corning Fiberglas Corp.											
PF Insulation		1	Note 1	2	.24	.32	.65	.77	.73	.81	2	.21
		1	Note 1	2	.25	.41	.86	.94	.84	.81	2	.50
TW-F Wool		1	Note 2	1	.27	.30	.57	.69	.70	...	2	.25
		1	Note 2	1	.33	.40	.76	.91	.77	.73	2	.33
		2	Note 2	1	.44	.61	.96	.93	.77	.86	2	.50
		2	Note 2	1	.55	.79	.99	.99	.91	...	2	1.00
		3	Note 2	1	.55	.68	.95	.90	.79	.80	2	0.50
		3	Note 2	1	.69	.91	.99	.99	.91	.82	2	1.00
Hairfelt		1	Bare	1	.12	.32	.51	.62	.60	.56	5	....
Paxfelt	Newalls Insulation Co., England	⅜	Asbestos	Note 4	...	.20	.60	.65	.70	...	3	.28
Rock Wool		1	Bare	1	.26	.45	.61	.72	.75	...	4	.90
Sound Isolation Blanket, MK	Johns-Manville Corp.	1	Note 3	1	.15	.37	.89	.98	.89	.86	2	1.04
		2	Note 3	1	.43	.64	.97	.99	.87	.90	2	2.33
		1	Note 3	4	.22	.46	.86	.98	.88	.77	2	1.20
		2	Note 3	4	.39	.57	.91	.91	.80	.78	2	2.00

Note 1. No facing, Fiberglas blanket nailed to 1″ x 2″ furring, 12″ o.c.
Note 2. Perforated metal facing, 26 gauge, 0.076″ holes, 0.176″ o.c.
Note 3. Muslin-covered, blanket on floor.
Note 4. Sheets 10′ x 2′10″ x ⅜″ nailed on lattice composed of 2″ x ⅝″ battens.

Table A.4 Hangings, Floor Coverings, and Miscellaneous Materials

Material	Thickness, inches	Absorption Coefficients 128 cycles	256 cycles	512 cycles	1024 cycles	2048 cycles	4096 cycles	Authority
Carpets, lined		.10 *		.25		.40 *		7
Carpets, unlined		.08 *		.15		.25 *		7
Carpet, Amritza, on concrete	7/16	.09	.06	.24	.24	.24	.11	8
Carpet, Cardinal Batala, on concrete	7/16	.12	.10	.28	.42	.21	.33	8
Carpet, pile, on concrete	3/8	.09	.08	.21	.26	.27	.37	8
Carpet, pile, on 1/8″ felt	5/16	.11	.14	.37	.43	.27	.25	8
Carpet, rubber, on concrete	3/16	.04	.04	.08	.12	.03	.10	8
Cork flooring slabs, glued down	3/4	.08	.02	.08	.19	.21	.22	8
Cork flooring, as above, waxed and polished	3/4	.04	.03	.05	.11	.07	.02	8
Cotton fabric, 14 oz per sq yd, draped to half its area		.07	.31	.49	.81	.66	.54	6
Cotton fabric, as above, draped to 7/8 its area		.03	.12	.15	.27	.37	.42	6
Draperies, velours, 18 oz per sq yd		.05	.12	.35	.45	.38	.36	6
Draperies, as above, draped to half their area		.14	.35	.55	.72	.70	.65	4
Linoleum, on concrete floor		.02		.03		.04 *		4
Openings, balcony (see Table 9.1 and p. 198)				.25 to .80				14
Openings, stage, depending upon stage furnishings (see Table 15.2 and p. 313)				.25 to .80				14
Oregon pine flooring	3/4	.09		.08		.10		4
Ozite, 0.39 lb per sq ft	1/2	.06	.13	.20	.42	.47	.47	6
Rug, Axminster		.11	.14	.20	.33	.52	.82	10
Ventilators, 50% open		.30 *		.50		.50 *		9

* These coefficients are estimates made by the authors.

TABLE A.5 Hard Plasters, Masonry, Wood, and Other Standard Building Materials

Material	Thickness, inches	Absorption Coefficients						Authority
		128 cycles	256 cycles	512 cycles	1024 cycles	2048 cycles	4096 cycles	
Ashes damped, loose (2.5 lb water per cu ft)	11	.90	.90	.75	.80			3
Ashes damped, loose (2.5 lb water per cu ft)	3	.25	.55	.65	.80	.80		3
Brick wall, unpainted	18	.02	.02	.03	.04	.05	.07	7
Brick wall, painted	18	.01	.01	.02	.02	.02	.02	7
Concrete (see Poured concrete)								
Glass		.03 *		.03		.02 *		7
Gravel soil, loose and moist	4	.25	.60	.65	.70	.75	.80	3
Gravel soil, loose and moist	12	.50	.65	.65	.80	.80	.75	3
Interior stucco, smooth finish, on tile	½	.03 *		.04		.04 *		4
Marble		.01 *		.01		.01 *		9
Plaster, gypsum, on hollow tile		.01	.01	.02	.03	.04	.05	7
Plaster, gypsum, scratch and brown coats on metal lath, on wood studs		.02	.03	.04	.06	.06	.03	6
Plaster, lime, sand finish, on metal lath	¾	.04	.05	.06	.08	.04	.06	4
Poured concrete, unpainted		.01	.01	.02	.02	.02	.03	4
Poured concrete, painted and varnished		.01	.01	.01	.02	.02	.02	4
Sand (sharp), dry	4	.15	.35	.40	.50	.55	.80	3
Sand (sharp), dry	12	.20	.30	.40	.50	.60	.75	3
Sand (14 lb water per cu ft)	4	.05	.05	.05	.05	.05	.15	3
Water, as in swimming pool		.01	.01	.01	.01	.02	.02	4
Wood sheathing, pine	¾	.10	.11	.10	.08	.08	.11	7
Wood veneer, on 2″ x 3″ wood studs, 16″ o.c.	7/16	.11		.12		.10		4

* These coefficients are estimates made by the authors.

Table A.6 Audience, Individual Persons, Chairs, and Other Objects

Description	Total Sound Absorption, in square-foot units (sabins)						Authority
	128 cycles	256 cycles	512 cycles	1024 cycles	2048 cycles	4096 cycles	
Audience, mixed, seated in theater chairs, single padding on back		3.5	4.1	4.9	4.2		1
Audience, mixed, seated in church pews		2.7	3.3	3.8	3.6		1
Chair, American loge, fully upholstered in mohair			4.5				1
Chair, tabiet arm, upholstered with Durand plastic seat covering and mohair side vents, seats down	1.46	4.3	4.5	4.5	3.6	2.8	13
Chair, tablet arm, upholstered with Durand plastic seat covering and mohair side vents, seats up	2.0	4.1	4.2	3.8	3.6	3.7	13
Chair, box spring, pantasote seat and back, plywood on rear; seats up		1.4	1.6	1.3	0.71		9
Chair, plywood seat, plywood back; seats up		0.19	0.24	0.39	0.38		9
Chair, spring edge mohair seat and back, plywood panel on rear; seats down		3.1	3.0	3.3	3.5		9
Chair, as above, with thick, completely covered seat and back; seat up		3.3	3.5	3.7	3.8		9
Chair, theater, heavily upholstered		3.4	3.0	3.3	3.6		1
Person, adult	2.5	3.5	4.2	4.6	5.0	5.0	14
Person, adult, seated in American loge chair	3.0		4.5	5.0	5.2		14
Person, child, high school	2.2		3.8		4.5		14
Person, child, grammar school	1.8		2.8		3.5		14

APPENDIX 2

Tables of Sound-Insulation Data

(See p. 245)

(*Note:* The authority for a number of the following tables is The National Physical Laboratory. Their values of transmission loss are listed at 128, 256, 768, 1800, and 3500 cycles. These mean values are obtained by averaging the results at individual mean frequencies of 100 and 150, 200 and 300, 700 and 1000, 1600 and 2000, and 3000 and 4000 cycles, respectively.)

Table A.7 Wood-Stud and Steel-Stud Partitions

Description of Partition			Transmission Loss, in decibels						
No.	Construction	Weight, lb per sq ft	Average, 128 to 4096	128 cycles	256 cycles	512 cycles	1024 cycles	2048 cycles	4096 cycles
Wood Studs 2″ x 4″, 16″ o.c.			Authority: National Bureau of Standards						
162	Lime plaster ½″ thick on wood lath	15.6	**42**	27	36	41	50	55	60
201	Gypsum plaster ½″ thick on wood lath	17.1	**38**	35	24	34	37	45	61
164	Lime plaster ⅞″ thick on metal lath	19.8	**44**	26	41	44	52	56	58
165	Same as above	20.0	**39**	31	35	38	43	45	61
174	Gypsum plaster ¾″ thick on paper-backed wire-mesh lath	12.6	**35**	30	25	35	38	38	54
148	Gypsum plaster ½″ thick on ¾″ gypsum lath	15.2	**41**	33	31	39	46	49	66
177	Gypsum plaster ½″ thick on ⅜″ gypsum lath fastened to studs with solid-steel clips	14.4	**36**	19	29	35	42	42	60
202	Gypsum plaster ½″ thick on ⅜″ gypsum lath	15.0	**35**	33	25	28	36	42	59
203	Vermiculite plaster ½″ thick on ⅜″ gypsum lath	9.6	**33**	27	20	27	36	38	55
204	Vermiculite plaster ⅞″ thick on perforated gypsum lath	12.9	**37**	31	22	31	38	46	66
205	Gypsum plaster ½″ thick on ½″ fiberboard lath	12.6	**41**	28	31	41	46	47	66
179A	Plywood, ⅜″, with lightweight cotton fabric glued on one side and heavy cotton duck glued on other side	4.6	**31**	15	28	29	38	43	40
179B	Like No. 179A, with flame-proofed cotton bats 4″ thick between studs	4.8	**35**	14	33	34	42	46	44
206	Fiberboard, soft, wall finish, ½″ thick	3.8	**32**	16	22	28	38	50	52
213	Like No. 205, plus ½″ gypsum plaster on ½″ fiberboard fastened at top and bottom 2″ from one side of original wall by 2″ x 2″ wood strips	18.2	**51**	41	44	50	52	56	72

Table A.7 (*Continued*)

Description of Partition			Transmission Loss, in decibels						
No.	Construction	Weight, lb per sq ft	Average, 128 to 4096	128 cycles	256 cycles	512 cycles	1024 cycles	2048 cycles	4096 cycles
Wood-Stud Partitions			Authority: National Bureau of Standards						
175	Wood studs, 2″ x 4″, staggered; ⅞″ gypsum plaster on metal lath	19.8	**50**	44	47	47	50	52	63
211	Plywood facings ¼″ glued to 1″ x 3″ studs	2.5	**25**	16	18	26	29	37	33
212	Like No. 211 with ½″ plasterboard nailed to both plywood faces	6.6	**40**	27	33	39	46	50	51
214	Staggered studs 1″ x 3″, ¼″ plywood glued to both sides	2.6	**26**	14	20	28	33	40	30
215	Like No. 214 with ½″ plasterboard nailed to the existing plywood faces (joints must be tight)	7.0	**46**	40	39	48	51	54	55
216	Two sets of 2″ x 2″ studs, 2 sheets ½″ gypsum board inserted in 1″ space between studs, ¼″ plywood face glued to studs on each outer side, total panel thickness 4¾″	7.9	**35**	18	29	32	42	49	51
217	Two sets of 2″ x 2″ studs, ¼″ plywood sheet inserted in ¼″ space between studs, ¼″ plywood faces, slightly compressed paper-backed mineral wool inserted in both air spaces, total panel thickness 4¾″	5.1	**37**	20	31	37	41	49	50
218	Two sets of 2″ x 2″ studs, ½″ gypsum board nailed to inside face of each stud, leaving 1″ air space between gypsum boards, ¼″ plywood faces glued to outer face of stud	7.4	**39**	27	29	37	46	55	55
219	Two sets of 2″ x 2″ studs, ½″ fiberboard stood loose in 2″ air space between studs, ¾″ fiberboard faces; total panel thickness 7″	6.2	**43**	28	28	40	48	62	68
220	Similar to No. 219, with ¾″ fiberboard faces replaced by ½″ fiberboard and ½″ gypsum plaster for facings; total panel thickness 8″	44.3	**52**	42	48	49	55	54	73

Table A.7 (*Continued*)

Description of Partition				Transmission Loss, in decibels					
No.	Construction	Weight, lb per sq ft	Thickness, inches	Average, 200 to 2000	128 cycles	256 cycles	768 cycles	1800 cycles	3500 cycles
Wood Studs, 2″ x 4″, 16″ o.c.				Authority: National Physical Laboratory					
59	Plasterboard, ⅜″ on both sides of studs	5.1 *	4¾	**35**	13	27	35	44	42
60	Plasterboard, ⅜″ on both sides of studs; felt between board and studs; fixing nails passing through board and felt	5.1 *	5¼	**37**	11	27	36	49	49
61	Plasterboard, ⅜″ (baseboard) on both sides of studs; both faces plastered	12.5 *	5¾	**38**	24	28	41	42	53
62	Plasterboard, ⅜″ (baseboard) on both sides of studs; felt between board and studs, fixing nails passing through board and felt; both faces plastered	15 *	6½	**42**	24	30	46	50	57
63	Like No. 62, but with mineral-wool blanket 1″ thick hung between studs	16 *	6½	**42**	24	31	46	47	60
64	Plaster lath, ⅜″ on both sides of studs; nails fitted with expanded metal lath pads round their heads and driven into studs between edges of laths; both faces plastered	15 *	5¾	**42**	23	32	45	47	59
65	Like No. 64, but with felt between laths and studs; nails passing between edges of laths and through felt	15 *	6¼	**47**	23	34	49	59	60
67	Plaster, 3 coats, on metal laths on both sides of studs	12.5 *	5	**35**	29	29	38	38	51

* Including superficial weight of studding, ⅓ lb per sq ft, averaged over whole area. The variations in weight of partitions No. 61 to 67 are mainly due to unintentional variations in the thickness of the plastering.

Table A.7 (*Continued*)

Description of Partition			Transmission Loss, in decibels						
No.	Construction	Weight, lb per sq ft	Average, 128 to 4096	128 cycles	256 cycles	512 cycles	1024 cycles	2048 cycles	4096 cycles
Steel Studs			Authority: National Bureau of Standards						
143A	Steelex Channels, 1½″, 16″ o.c.; ⅞″ gypsum plaster on paper-backed wire-mesh lath	17.6	**30** †	18	21	27	43	39	58
143B	Like No. 143A, except that space between studs is packed with rock wool	....	**36** †	26	24	37	47	50	69
166A	Steel studs, 3″, 16″ o.c., ⅞″ gypsum plaster on expanded metal lath	19.6	**37**	30	28	35	40	43	53
166B	Like No. 166A, except that space between studs is packed with rock-wool bats, density 4.3 lb per cu ft	21.1	**38**	34	31	40	39	40	52
159	Metal studs of ¾″ channel-iron expanded-metal lath; gypsum plaster applied to only one side; panel *A* only (see drawing)	8.1	**33**	27	29	35	33	32	44
160A	Two panels similar to No. 159 placed back to back and resting on cork 1″ thick; distance from face to face of panels 10″	17.2	**55**	50	48	53	55	60	72
160B	Like No. 160A, except that distance from face to face is 8½″	....	..	..	..	..	..	..	..
160C	Like No. 160A, except that distance from face to face is 7″	17.2	**55**	49	46	53	54	58	72
160D	Like No. 160A, except that distance from face to face is 5½″	17.2	**54**	51	44	53	55	56	73
160E	Like No. 160A, except that distance from face to face is 4½″	17.2	**53**	43	45	52	51	61	73
160F	Like No. 160A, except that distance from face to face is 4⅜″, and braces at corners are in contact with each other	17.2	**53**	43	43	51	50	62	74

† Average value for "256 to 2048 cycles."

Table A.7 (*Continued*)

Description of Partition			Transmission Loss, in decibels						
No.	Construction	Weight, lb per sq ft	Average, 128 to 4096	128 cycles	256 cycles	512 cycles	1024 cycles	2048 cycles	4096 cycles
Steel Studs (*Continued*)			Authority: National Bureau of Standards						
160G	Like No. 160E, except that cork was removed and a 1″ board was placed under the panels to carry the load	17.2	**51**	45	44	46	50	56	70
160H	Like No. 160G, except that board was removed and concrete substituted for the board	17.2	**48**	46	44	48	46	49	60
160I	Like No. 160H, except that the two panels were tied together at two points with a shoe made of ¾″ channel iron, each point being approximately 18″ in the horizontal direction from the center of the panel	17.2	**46**	43	41	46	46	46	58

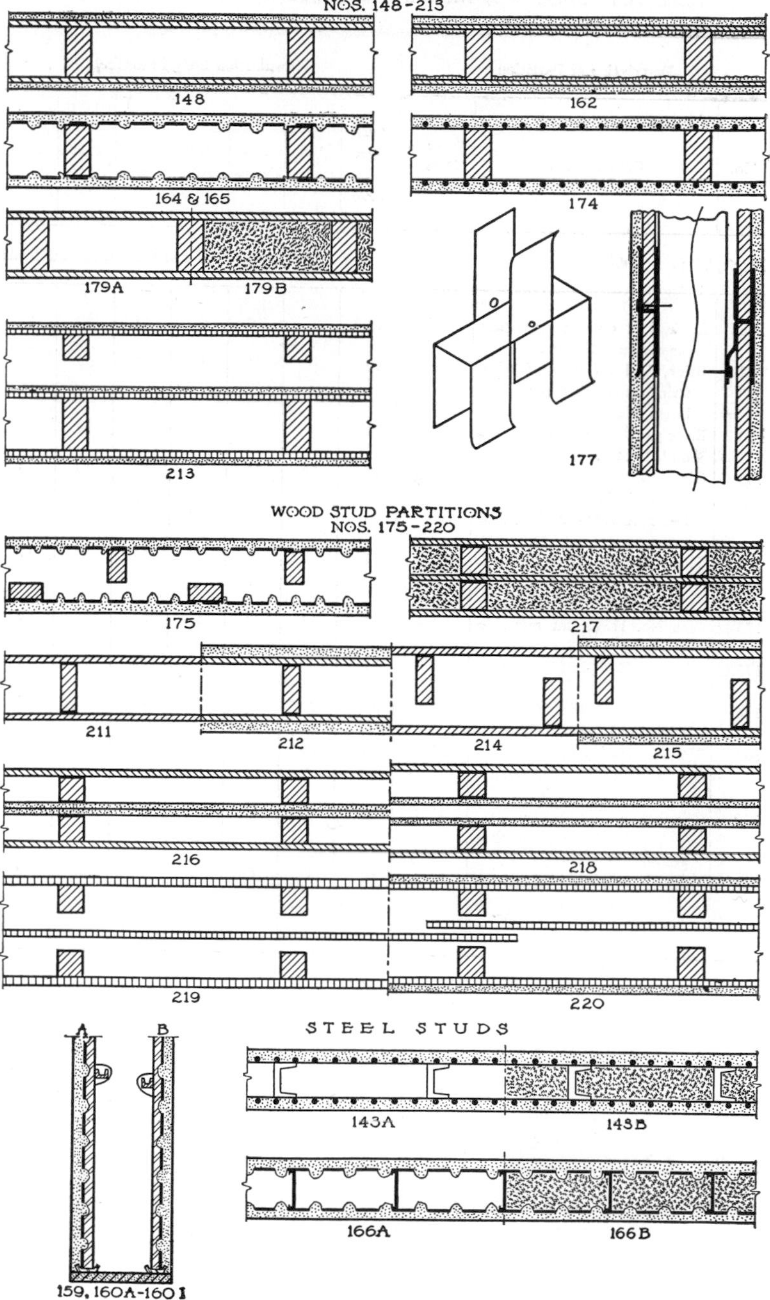
WOOD STUD PARTITIONS
NOS. 148-213
148
162
164 & 165
174
179A
179B
177
213
WOOD STUD PARTITIONS
NOS. 175-220
175
217
211
212
214
215
216
218
219
220
STEEL STUDS
A
B
159, 160A-160I
143A
143B
166A
166B

Table A.8 Brick, Tile, Masonry, and Poured-Concrete Partitions

Description of Partition			Transmission Loss, in decibels						
No.	Construction	Weight, lb per sq ft	Average, 128 to 4096	128 cycles	256 cycles	512 cycles	1024 cycles	2048 cycles	4096 cycles
Brick, Tile			Authority: National Bureau of Standards						
85	Brick, laid on edge; gypsum plaster on both sides	31.6	**42** *		40	37	49	59	
82	Brick, laid on edge; furring strips wired, gypsum plasterboard, plaster on both sides	36.5	**52** *		52	47	57	54	
80	Brick panel 8″, gypsum plaster on both sides	97.0	**51** *		48	49	57	59	
155	Glass brick partition 3¾″ thick, 3¾″ x 4⅞″ x 8″ bricks, mfg. Owens Ill. Glass Co.	....	**42** *	30	35	41	49	49	43
173C	Tile, hollow partition, 4″ thick, pumice-cement block, two cells 4″ x 8″ x 16″, no plaster	15.5	**11**	8	5	9	14	19	17
173B	Like No. 173C, but one side plastered	20.4	**35**	31	27	35	36	40	47
173A	Like No. 173C, but both sides plastered	25.3	**37**	32	34	36	39	42	52
161	Tile, gypsum, 3″ x 12″ x 30″, gypsum plaster on both sides	21.0	**38**	29	36	36	42	47	47
303	Tile, hollow-clay partition 4″ x 12″ x 12″, Vermiculite plaster on both sides	25.2	**38**	29	38	36	39	48	51
140	Tile, hollow-clay partition 4″ x 12″ x 12″, gypsum plaster on both sides	42.1	**38** *	31	31	36	47	50	58
145	Cinder block, hollow partition 3″ x 8″ x 16″, plaster on both sides	32.2	**45**	34	37	42	51	57	64

* Average value for "256 to 2048 cycles per second."

Table A.8 (*Continued*)

Description of Partition				Transmission Loss, in decibels					
No.	Construction	Weight, lb per sq ft	Thick-ness, inches	Aver-age, 200 to 2000	128 cycles	256 cycles	768 cycles	1800 cycles	3500 cycles
Concrete Slabs, Blocks				Authority: National Physical Laboratory					
28	Clinker concrete slabs, 3″ unplastered	17	3	**23**	17	18	22	30	40
33	Like No. 28, but with one face plastered	24	3½	**41**	27	32	40	52	58
34	Like No. 28, but with both faces plastered	30	4	**44**	26	33	44	56	57
29	Clinker concrete slabs, 2″ with 1″ ribs at 9″ centers on one face; other side plastered	19.5	2½	**41**	28	36	38	51	56
30	Like No. 29, but with combination of ½″ fiberboard and ⅜″ plasterboard nailed to ribs; skim coat of plaster	22.5	4½	**40**	31	32	37	53	60
31	Clinker concrete slabs, 2″ plastered on both faces	20	3	**38**	26 †	34	35	45	57 ‡
32	Like No. 31, but with ½″ cork insulation around edges (no plaster on cork)	20	3	**36**	37 †	33	35	43	55 ‡
35	Cellular concrete blocks, one face plastered	45	8½	**46**	24	35	47	56	55
38	Hollow foamed slag concrete blocks, filled with sand, unplastered	65	9	**28**	12	17	29	38	43
39	Like No. 38, but with skim coat of plaster on each face	68	9¼	**51**	31	40	52	59	66
Multiple-Block Partitions				Authority: National Physical Laboratory					
69	Two leaves, each of 3″ hollow blocks, separated by 2″ cavity and built on opposite sides of gap separating rooms; outer faces plastered (two partitions of nominally the same construction)	28	9	**54** **57**	38 39	47 50	49 57	69 65	77 76
70	Like No. 69, but with butterfly-shaped wire ties (0.14″ diam. wire) connecting leaves; ties staggered at 3′ horizontal and 18″ vertical intervals	28	9	**53**	32	41	54	64	78

† For 100 cps only.
‡ For 4000 cps only.

Table A.8 (*Continued*)

Description of Partition				Transmission Loss, in decibels					
No.	Construction	Weight, lb per sq ft	Thickness, inches	Average, 200 to 2000	128 cycles	256 cycles	768 cycles	1800 cycles	3500 cycles
Multiple-Block Partitions (*Continued*)				Authority: National Physical Laboratory					
71	Like No. 69, but with linked-wire ties (0.2″ diam. wire) connecting leaves; ties staggered at 3′ horizontal and 18″ vertical intervals	28	9	**45**	26	36	43	57	66
72	Like No. 69, but with galvanized iron bar ties, 1″ x 3⁄16″, staggered at 3′ horizontal and 18″ vertical intervals	28	9	**43**	26	35	42	52	63
73	Like No. 69, but with galvanized iron bar ties, ¾″ x ⅛″, tips covered with rubber 3⁄64″ thick for length of 3″; ties staggered at 3′ horizontal and 18″ vertical intervals	28	9	**55**	39	47	52	65	76
74	Two leaves, each of 2″ clinker concrete slabs plastered on outer faces, spacing between leaves 2″	33	7	**49**	39 †	45	46	56	65 ‡
75	Like No. 74, but with ½″ cork insulation around edges of each leaf (no plaster on cork)	33	7	**54**	40 †	48	52	64	76 ‡
76	Like No. 75, but with plaster over cork	33	7	**48**	44 †	46	44	58	75 ‡
77	Two leaves, each of 2″ clinker concrete slabs, separated by 14″ cavity and built on opposite sides of gap separating rooms; outer faces plastered	33	19	**71**	66 †	68	67	82	98 §
78	Two leaves, each of 3″ clinker concrete slabs, plastered on outer faces spacing between leaves 2″	40	9	**51**	40 †	44	49	60	69 ‡
79	Three leaves, each of 2″ clinker concrete slabs, plastered on outer faces and on one face of middle leaf; spacing between adjacent leaves 2″	50	11½	**49**	44 †	45	46	57	69 ‡

† For 100 cps only.
‡ For 4000 cps only.
§ For 4000 cps only, using subjective technique.

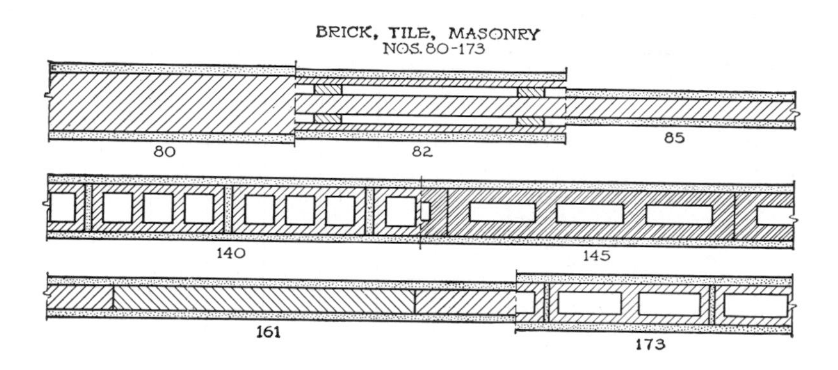
BRICK, TILE, MASONRY
NOS. 80-173
80
82
85
140
145
161
173

Table A.9 Clips and Special Nails

Description of Partition			Transmission Loss, in decibels						
No.	Construction	Weight, lb per sq ft	Average, 128 to 4096	128 cycles	256 cycles	512 cycles	1024 cycles	2048 cycles	4096 cycles
Wood Studs, 2″ x 4″, 16″ o.c., plaster facings of ½″ gypsum plaster, ⅜″ gypsum lath; lath held together by special nails with resilient heads, nails driven into the joints between the lath, or lath held by resilient spring clips			Authority: National Bureau of Standards						
401	Head of nail embedded in felt and covered with sheet iron; ¼″ felt pad between stud and gypsum lath	13.6	**41**	19	34	39	46	52	63
402	Nail similar to No. 401; no felt pad between stud and perforated gypsum lath	15.8	**42**	29	34	40	46	50	66
409	Nail similar to No. 401, gypsum lath snug against studs	15.2	**42**	31	35	39	47	50	64
406	Ordinary nail with head encased in metal-lath square; metal strap girdling the metal-lath square; gypsum lath snug against studs	14.8	**41**	31	31	39	45	48	62
407	Ordinary nail with head encased in corrugated cardboard and metal-lath square encompassing the cardboard but not touching nail; gypsum lath snug against studs	14.4	**41**	29	32	40	45	50	63
408	Ordinary nail with head encased in corrugated cardboard, metal strap girdling the cardboard square, but not in contact with nail; gypsum lath loose against studs, approximately 1/32″ play	14.8	**42**	34	32	40	45	51	64
410	Ordinary nail with head encased in thin cardboard, metal-lath square over cardboard which was highly compressed	13.6	**43**	31	33	42	48	48	65
411	Nail similar to No. 410, but head of nail was encased in felt and then covered by a metal-lath square; lath snug against studs	14.3	**43**	32	31	41	48	50	66
412	Same nail as No. 411; ¼″ felt pad between stud and gypsum lath	14.0	**47**	36	37	45	53	55	68
413	Clip as indicated in sketch	12.4	**42**	26	37	42	47	44	62
414	Same clip as No. 413, except that a resilient member was introduced in the clip	14.1	**46**	39	40	43	46	48	63

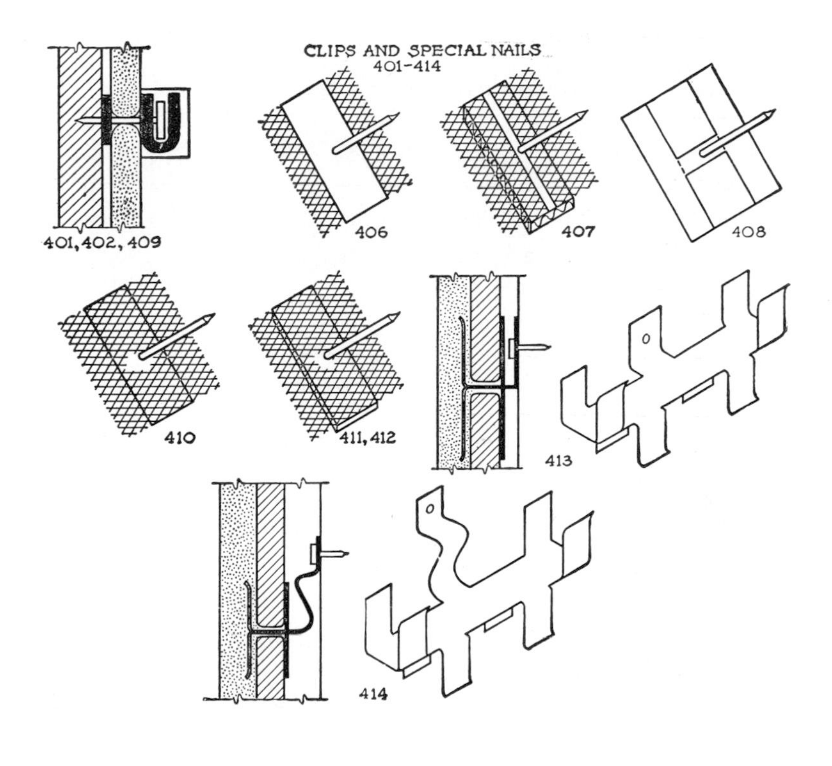
CLIPS AND SPECIAL NAILS
401-414
401, 402, 409
406
407
408
410
411, 412
413
414

Table A.10 Solid Plaster Partitions

No.	Construction	Weight, lb per sq ft	Average, 128 to 4096 cycles	128 cycles	256 cycles	512 cycles	1024 cycles	2048 cycles	4096 cycles
Solid Plaster with Studs, 2″ solid plaster partition, one-course lath, ¾″ steel channel studs			Authority: National Bureau of Standards						
170	Gypsum lath, perforated, ¾″; gypsum plaster, studs 12″ o.c.	19.4	**37**	30	33	31	38	48	53
171A	Metal lath, expanded; gypsum plaster; studs 12″ o.c.	16.4	**38**	36	30	34	39	47	54
501	Metal lath, expanded; Vermiculite plaster; studs 16″ o.c.	8.8	**34**	36	33	31	28	38	48
502	Like No. 501, but gypsum plaster	18.1	**38**	41	23	36	36	47	54
Studless Plaster Partitions			Authority: National Bureau of Standards						
503	Solid plaster, 2″ with expanded metal-lath core	18.4	**38**	37	29	36	38	48	55
504	Solid plaster, 2″ with ⅜″ gypsum-lath core	16.8	**37**	38	27	35	36	47	54
505	Double core 2″ partition; 2 sheets of gypsum lath spaced ¼″ apart with felt spacers, joints between lath covered with metal lath to prevent mortar from bonding two sides together, ½″ plaster faces	15.3	**38**	35	29	33	38	43	57
507	Double core of ½″ and ⅜″ plaster boards, held together at vertical joints partially by clip of No. 416 and by clip of sketch with ¼″ air space between plaster boards due to thickness of the clips; ⅝″ plaster facings applied to each plaster board; total panel thickness 2⅝″	12.9	**40**	31	32	38	40	50	62

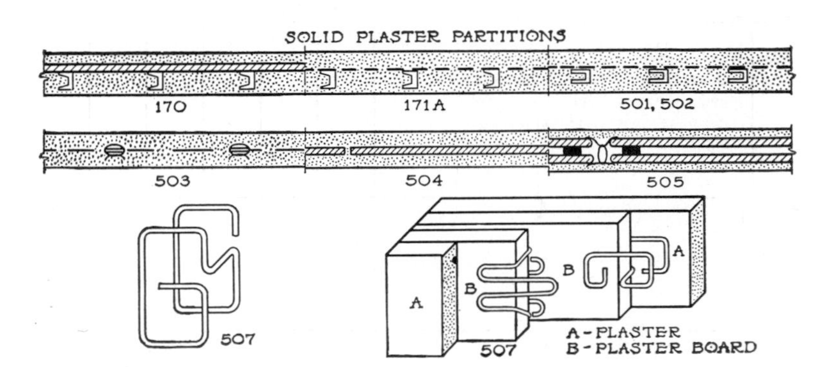
SOLID PLASTER PARTITIONS
170
171A
501, 502
503
504
505
507
A
B
B
A
507
A - PLASTER
B - PLASTER BOARD

Table A.11 Doors, Windows, and Miscellaneous Partitions

Description of Partition		Transmission Loss, in decibels						
No.	Construction	Average, 256 to 1024	128 cycles	256 cycles	512 cycles	1024 cycles	2048 cycles	4096 cycles
Doors (See p. 235.)						Authority: P. E. Sabine		
S1	Oak, solid 1¾″, with cracks as ordinarily hung	20	12	15	20	22	16	..
S2	Oak, like above, well-seasoned and airtight	..	15	18	21	26	25	..
				Authority: National Bureau of Standards				
182	Wood, heavy, approximately 2½″ thick, rubber gaskets around sides and top; special felt strip pushes down as door closes, eliminating any crack under door; 12.5 lb per sq ft	27	30	30	29	25	26	37

No.	Description of Structure	Transmission Loss, in decibels, Average, 125 to 2000 cps
Window Structures (See p. 235.) **Single**	Authority: P. H. Geiger	
E	DSA (double-strength, annealed) window glass	27
E2	⅛″ hammered glass	28
E3	⅛″ Louvrex	29
E1	¼″ plate glass	31
E5	¼″ safety plate glass, 0.030″ plastic	32
Thermopane	Authority: P. H. Geiger	
A	DSA: ¼″ air space: DSA	28
A5	DSA: ¼″ air space: ⅛″ hammered glass	30
A3	⅛″ hammered glass: ¼″ air space: ⅛″ hammered glass	31
A1	3⁄16″ sheet glass: ¼″ air space: 3⁄16″ sheet glass	32
A2	¼″ plate glass: ¼″ air space: ¼″ plate glass	33
B	¼″ plate glass: ½″ air space: ¼″ plate glass	36
Triple Thermopane	Authority: P. H. Geiger	
C	DSA: ¼″ air space: DSA: ¼″ air space: DSA	30
C3	3⁄16″ plate glass: ¼″ air space: ¼″ plate glass: ¼″ air space: ⅛″ plate glass	32
C1	¼″ plate glass: ¼″ air space: ¼″ plate glass: ¼″ air space: ¼″ plate glass	36
D	¼″ plate glass: ½″ air space: ¼″ plate glass: ¼″ air space: ¼″ plate glass	39

Table A.11 *(Continued)*

Description of Partition				Transmission Loss, in decibels					
No.	Construction	Weight, lb per sq ft	Thickness, inches	Average, 200 to 2000	128 cycles	256 cycles	768 cycles	1800 cycles	3500 cycles
Single Partitions				Authority: National Physical Laboratory					
1	Laminated building board, ⅜″, on wood frame	0.78 *	⅜	**22**	12	16	22	28	30
2	Fiberboard, ½″, on wood frame	0.78 *	½	**22**	12	15	22	28	29
3	Asbestos cement board, ¼″, in steel frame	1.7 *	¼	**26**	12	18	27	33	36
4	Plasterboard, ⅜″, on wood frame	1.9 *	⅜	**27**	16	20	27	34	31
5	Steel plates, 16 gauge, in steel frame	2.5 *	1/16	**29**	16	21	30	37	43
7	Boards formed of waste materials bonded with plastic	4.9 *	1¼	**35**	20	28	37	39	36
8	Plasterboard, 5 layers, each ⅜″ thick	8	1⅞	**32**	25	30	32	34	39
9	Plasterboard, ¾″, with ⅝″ plaster on both faces	13	2	**34**	29	32	31	42	50
Single Sheets on Studding				Authority: National Physical Laboratory					
11	Plasterboard ⅜″ on 4″ x 2″ studding at 16″ centers	1.9 †	⅜	**26**	14	20	26	32	30
12	Like No. 11, but with felt between plasterboard and studding	1.9 †	⅜	**27**	11	20	26	35	33
13	T and G boarding ⅞″ on 3″ x 2″ studding at 19″ centers (boards tightly cramped)	2.8 †	⅞	**21**	13	14	21	26	31
14	Like No. 13, but with junctions of boards and other cracks sealed	2.8 †	⅞	**24**	20	17	24	30	34
Single Panels, Composite Construction				Authority: National Physical Laboratory					
16	Two ⅛″ sheets of plastic material joined at 2″ intervals by ⅛″ webs	2.4	1	**22**	15	19	22	26	27

* Superficial weight of sheet (i.e., excluding frame).

† Superficial weight of sheet (i.e., excluding studding).

Table A.11 (*Continued*)

Description of Partition				Transmission Loss, in decibels					
No.	Construction	Weight, lb per sq ft	Thickness, inches	Average, 200 to 2000	128 cycles	256 cycles	768 cycles	1800 cycles	3500 cycles
Single Panels, Composite Construction				Authority: National Physical Laboratory					
19	Like No. 16, but with channels filled with foamed slag sand	5.2	1	**33**	21	25	35	38	39
18	Paper pulp between sheets of 1/16″ plywood	2.9	2	**25**	13	19	28	25	28
22	Wood-wool cement between 3/16″ asbestos cement boards (3 panels in wood frames)	7.5	1¾	**30**	21	26	30	33	37
24	Asbestos board between sheets of 18 gauge steel (8 panels in steel frame with cover strips)	7.6 ‡	½	**31**	22	27	30	36	45
Double Partitions of Lightweight Panels				Authority: National Physical Laboratory					
43	Two leaves, each of wood pulp between two sheets of 3/64″ plywood 1″ apart; spacing between leaves 2″	3.3	4¼	**34**	14	19	35	46	46
44	Two leaves, each of two sheets of ¼″ plywood, bonded together by plywood lattice; spacing between leaves 5½″	4.0	9	**26**	14	19	24	36	43
45	Two leaves, each of two ⅛″ sheets of plastic material joined at 2″ intervals by ⅛″ continuous webs; spacing between leaves 2½″	4.8	4½	**31**	16	20	31	40	45
46	Like No. 45, but with spacing between leaves 5″	4.8	7	**33**	19	24	33	40	45
47	Like No. 45, but with glass-silk blanket hung between leaves	5.2	4½	**38**	21	32	38	46	51
48	Like No. 45, but with spacing between leaves 5″ and with glass-silk blanket hung between leaves	5.2	7	**40**	25	33	41	44	53

Table A.11 (*Continued*)

Description of Partition				Transmission Loss, in decibels					
No.	Construction	Weight, lb per sq ft	Thickness, inches	Average, 200 to 2000	128 cycles	256 cycles	768 cycles	1800 cycles	3500 cycles
Double Partitions of Lightweight Panels (*Continued*) Authority: National Physical Laboratory									
54	Like No. 47, but with channels in both leaves filled with foamed slag sand	10.4	4½	**51**	31	43	51	60	66
55	Like No. 47, but with channels in both leaves filled with sawdust cement	12.2	4½	**47**	30	43	49	48	60
57	Like No. 47, but with channels of both leaves filled with sand	17.8	4½	**57**	36	48	58	65	70
49	Two leaves, each of two ¼″ asbestos boards spaced 1½″ apart by wood frames; spacing between leaves 8″	6.6	12	**46**	21	34	46	58	65
50	Like No. 49, but with spacing between leaves 2″	6.6	6	**44**	15	30	45	57	62
51	Like No. 49, but with strips of ¼″ fiberboard between asbestos board and frame on outer sides; fixing nails passing through fiberboard	7.4	12½	**49**	24	38	48	62	66
52	Like No. 51, but with glass-silk blanket between the asbestos boards in each leaf	8.0	12½	**52**	27	43	52	62	66
53	Like No. 52, but with spacing between leaves 2″	8.0	6½	**51**	21	41	51	61	65

Table A.12 Floor and Ceiling Partitions

Description of Partition			Transmission Loss, in decibels						
No.	Construction	Weight, lb per sq ft	Average, 128 to 4096	128 cycles	256 cycles	512 cycles	1024 cycles	2048 cycles	4096 cycles
Wood Joists			Authority: National Bureau of Standards						
180A	Joists, 2″ x 6″, 16″ o.c.; 1″ subfloor, 2″ x 2″ strips 16″ o.c.; 13⁄16″ finish floor; ¾″ gypsum plaster on metal lath for ceiling	16.3	**38**	35	24	34	42	50	62
180B	Same as No. 180A, but ½″ Balsam Wool laid on subfloor and strips attached with clips	16.6	**50**	33	38	48	55	65	76
180C	Same as No. 180B, but 1″ Balsam Wool used in place of the ½″ Wool	16.7	**50**	35	37	49	55	64	75
701	Joists, 2″ x 8″, ½″ fiberboard lath, and ½″ gypsum plaster ceiling; 1″ pine subflooring and 1″ pine finish flooring	14.3	**45**	23	34	47	55	54	69
702	Same joists and ceiling as No. 701; 1″ pine subfloor; ½″ fiberboard, 1″ x 3″ sleepers, and 1″ pine finish floor	16.2	**50**	30	37	50	57	65	79
703	Same as No. 701, except that a second ½″ fiberboard and ½″ plaster ceiling was added to the existing ceiling of No. 701 by means of 1″ x 3″ furring strips	19.0	**45**	31	32	45	48	54	79
704	Same joists and floor as No. 701, except that ceiling consisted of ½″ fiberboard, 1″ x 3″ furring strips, ½″ fiberboard lath, and ½″ plaster face	15.9	**47**	24	38	49	56	58	77
705	Joists 2″ x 8″, 16″ o.c., 1″ pine subfloor, 1″ pine finish floor; one ceiling consisting of ½″ plaster on ½″ fiberboard lath next to joists; an additional plaster and fiberboard lath ceiling on 2″ x 2″ joists was suspended by screw eyes and wire loops 36″ o.c., 4″ below upper ceiling, 5″ x 5″ x 2″ fiberboard block pads at fastenings	20.3	**56**	46	51	55	56	63	75

Table A.12 (*Continued*)

Description of Partition			Transmission Loss, in decibels						
No.	Construction	Weight, lb per sq ft	Average, 128 to 4096	128 cycles	256 cycles	512 cycles	1024 cycles	2048 cycles	4096 cycles
Wood Joists (*Continued*)			Authority: National Bureau of Standards						
706	2″ x 8″ floor joists, 2″ x 4″ ceiling joists, 2″ x 8″ spaced 4″ o.c. from 2″ x 4″; ½″ plaster on ½″ fiberboard lath ceiling; 1″ pine subfloor, ½″ fiberboard, 1″ x 3″ sleepers, 1″ pine finish floor	16.7	**54**	48	49	51	54	58	75
707	2″ x 8″ wood joists, ¾″ fiberboard ceiling, 1″ pine rough flooring, and 1″ pine finish flooring	9.6	**40**	22	31	40	44	55	62
708	Same as No. 707, except ceiling; ½″ fiberboard lath, ½″ plaster, and ¾″ fiberboard face	15.8	**42**	31	30	40	47	56	68
Steel Joists			Authority: National Bureau of Standards						
137B	Open-web 8″ steel joists 20″ o.c.; high rib metal lath attached to top of joists; 2½″ of concrete poured on top of lath; battleship linoleum cemented to concrete; ¾″ gypsum plaster on ribbed metal lath attached to under side of joist	....	**55**	40	48	54	66	63	72
129A	Combination floor panel constructed of 4″ x 12″ x 12″ 3-cell partition tile; the ceiling of this panel was finished with ½″ gypsum plaster; the floor surface consisted of 3⁄16″ oak flooring nailed to 2″ x 2″ nailing strips 16″ o.c., which were grouted into the concrete.		**45**	36	38	39	47	54	55
129B	Like panel No. 129A, except that U. S. Gypsum resilient-steel clips were inserted between the concrete and nailing strips		**61**	37	47	58	69	73	80

Table A.12 (*Continued*)

Description of Partition			Transmission Loss, in decibels						
No.	Construction	Weight, lb per sq ft	Average, 128 to 4096	128 cycles	256 cycles	512 cycles	1024 cycles	2048 cycles	4096 cycles
Steel Joists (*Continued*)			Authority: National Bureau of Standards						
129C	Like No. 129B, except that the oak flooring was removed and ½″ gypsum plasterboard was attached to the nailing strips and 1½″ Hydrocal was applied on top of the plasterboard		**64**	43	50	61	71	77	80
136A	A floor panel constructed by using steel floor section with flat top; top of this section was covered with 2″ of concrete and a suspended metal-lath and plaster ceiling attached to the bottom, leaving approximately 4″ between the metal and plaster		**53**	34	43	52	59	65	72
136B	A floor panel; like No. 136A, except that the 2″ concrete slab was removed and ½″ of emulsified asphalt applied directly to the top of the steel section; 2″ concrete slab was cast on top of this asphalt		**61**	42	53	60	67	77	83
156	A floor panel composed of a 4″ concrete slab; suspended ceiling of gypsum lath ¼″ gypsum plaster; finished with ½″ acoustic plaster, 3″ ground cork on top of gypsum lath; hangers were special coiled springs		**54**	39	44	51	60	68	70
158	Like No. 156, except that 4″ rock wool was used in place of the ground cork		**55**	37	47	51	60	69	77

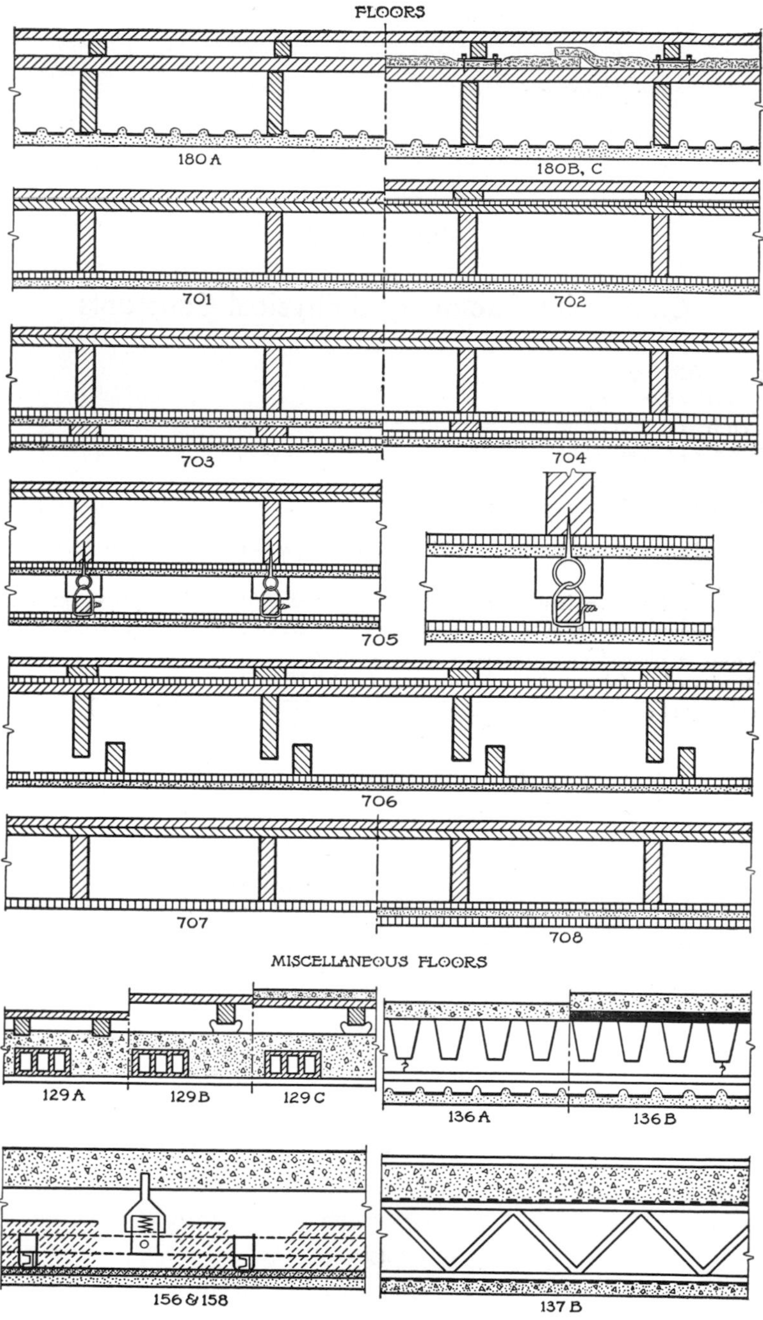
FLOORS
180 A
180B, C
701
702
703
704
705
706
707
708
MISCELLANEOUS FLOORS
129 A
129 B
129 C
136 A
136 B
156 & 158
137 B

APPENDIX 3

Conversion Factors and Physical Constants

1 ft = 0.305 m
1 sq ft = 0.0929 sq m
1 cu ft = 0.0283 cu m
1 lb per sq ft = 4.9 kg per sq m
1 lb per cu ft = 16.0 kg per cu m
1 ft per sec = 0.305 m per sec

Normal atmospheric pressure = 1.01×10^6 dynes per sq cm (at 32° F) = 14.7 lb per sq in.
0 db sound-pressure level corresponds to a pressure of 0.0002 dyne per sq cm
74 db sound-pressure level corresponds to a pressure of 1.00 dyne per sq cm
1 watt = 1,000,000 microwatts = 10^7 ergs per sec
Density of air (ρ) = 0.00120 g per cu cm (at 70° F)
Velocity of sound in air (c) = 1130 ft per sec = 345 m per sec (at 70° F)

Index